Palaeobiological Investigations

Research Design, Methods and Data Analysis

edited by

N. R. J. Fieller, D. D. Gilbertson and N. G. A. Ralph

Symposia of the
Association for Environmental Archaeology No. 5B

BAR International Series 266
1985

B.A.R.

5, Centremead, Osney Mead, Oxford OX2 0ES, England.

BAR-S266, 1985 :'Palaeobiological Investigations'.

Price £ 20.00 post free throughout the world. Payments made in dollars must be calculated at the current rate of exchange and $3.00 added to cover exchange charges. Cheques should be made payable to B.A.R. and sent to the above address.

ISBN 0 86054 339 0

For details of all B.A.R. publications in print please write to the above address. Information on new titles is sent regularly on request, with no obligation to purchase.

Volumes are distributed from the publisher. All B.A.R. prices are inclusive of postage by surface mail anywhere in the world.

Printed in Great Britain

Palaeobiological Investigations:

Research Design, Methods and Interpretation.

edited by

N.R.J. Fieller, D.D. Gilbertson and N.G.A. Ralph

(University of Sheffield, Sheffield, S10 2TN, U.K.)

Association for Environmental Archaeology

Symposium 5(ii)

held at Stephenson Hall, University of Sheffield

September 1983

CONTENTS

Page

N.R.J. Fieller, D.D. Gilbertson and N.G.A. Ralph
Palaeobiological Investigations: editorial notes 1

Vertebrates 3

P.M. Aird. On distinguishing butchery from other post-mortem destruction: a methodological experiment applied to a faunal sample from Roman Lincoln. 5

J.M. Maltby. Assessing variations in Iron Age and Roman butchery practices: the need for quantification. 19

S. Payne and P.J. Munson. Ruby and how many squirrels? The destruction of bones by dogs. 31

B. Levitan. A methodology for recording the pathology and other anomalies of ungulate mandibles from archaeological sites. 41

J. Coy. Assessing the role of pigs from faunal debris on archaeological settlements. 55

S.M. Stallibrass. Some effects of preservational biases on interpretations of animal bones. 65

R.D.S. Jenkinson. The environmental implications of changeover rates in vertebrate faunas. 73

R. Wilson. Degraded bones, feature type and spatial patterning on an Iron Age occupation site in Oxfordshire, England. 81

Insects

H.K. Kenward. Outdoors - Indoors? The outdoor component of archaeological insect assemblages. 97

A.K.G. Jones. Trichurid ova in archaeological deposits: their value as indicators of ancient faeces. 105

	Page
Molluscs	117
M.R. Deith. Seasonality from shells: an evaluation of two techniques for seasonal dating of marine molluscs.	119
K.D. Thomas. Land snail analysis in archaeology: theory and practice.	131
C. Ellis. Flandrian molluscan biostratigraphy and its application to dry valley infill deposits in east Sussex.	157
D.D. Gilbertson and A.B. Hawkins. Past and present distributions of _Hydrobia ulvae_ (Pennant), Gastropoda, and _Macoma balthica_ (L.), Bivalvia, along the southern shores of the Severn Estuary in relation to water pollution, field survey and laboratory methods.	167
Botanical approaches	179
C.O. Hunt. Recent advances in pollen extraction techniques: a brief review.	181
W.E. Boyd. The problem of the time span represented by pollen spectra in podzol turves, with examples from the Roman sites at Bar Hill and Mullins, central Scotland.	189
M. van der Veen. Evidence for crop plants from north-east England: an interim overview with discussion of new results.	197
R.G.W. Ward. Grazing damage to Scots Pine (_Pinus sylvestris_) by red deer (_Cervas elephas_) in highland Scotland during the mid-Flandrian: a short note.	221
Index 1: subjects	229
Index 2: biological taxa	247
Index 3: locations and people	251

PALAEOBIOLOGICAL INVESTIGATIONS: EDITORIAL NOTES

The 5th Annual Symposium of the Association for Environmental Archaeology was held at the University of Sheffield in September 1983. The conference was attended by well over 100 people. It had the inter-related themes of research design, research methods and interpretation. The topics discussed focussed on two subjects which we have termed "Palaeoenvironmental Investigations" and "Palaeobiological Investigations". These form the titles of the two symposia volumes 5(i) and 5(ii) which record the proceedings of this conference.

The emphasis on "investigation" is especially important in the development of the subject at this moment in time. In Karl Butzer's new book (1982) **Archaeology as Human Ecology**, Cambridge: Cambridge University Press, we have the first of a new generation of texts which emphasize the nature and development of conceptual models in environmental archaeology. The testing and elaboration of these models will only be achieved by investigation in which the skills of research design, methods and data analysis will be paramount.

In this day and age the academic and non-academic worlds are increasingly beset with questions of style, and guarantees of excellence. We wish to place on record the following information which concern the research papers in this volume. Every published paper has been subject to peer group review. Each paper has been scrutinised by two referees of international or national standing, and has also been read by an "informed reader" - usually a Sheffield post-graduate - as well as the editors.

These two symposium volumes have also presented major technical challenges to us. We have endeavoured to use the "new technologies" in their production, evidence of which is illustrated by the computer-generated set of indices. Several papers have been received in the form of magnetic tape or floppy disk. Unhappily, the other new computer technologies have failed to resolve problems of protocol and compatability. Consequently very difficult typing and re-typing tasks have fallen upon Miss Val Kinsler and Ms. Betty Gowland, who have coped admirably under difficult circumstances. Fortunately, the frustrations and production delays experienced have not affected the publication date of these volumes. This is because we did not receive the final corrected proofs from three contributors (who shall remain nameless) until mid-January 1985. Our best efforts were clearly of little avail.

The management and success of the Sheffield conference owed much to the courtesy of the staff of Stephenson Hall of the University of Sheffield and Ms. C.M. Griffin, Ms. M.V. Andrews, Mr Geraint Coles and Dr. C.O. Hunt, who assisted admirably with both administration and conference delegate management.

The field trip to the caves and magnificent Visitor Centre at Creswell Crags was a great success and we would like to record our gratitude to the Senior Ranger, Dr. R.D.S. Jenkinson, his staff and the County Councils of Nottinghamshire and Derbyshire.

Professor V.D. Barnett kindly made available word processing facilities and support in the Department of Probability and Statistics of the University of Sheffield.

N.R.J. Fieller
Department of Probability and Statistics
University of Sheffield
Sheffield S10 2TN

D.D. Gilbertson
and N.G.A. Ralph
Department of Archaeology and Prehistory
University of Sheffield
Sheffield S10 2TN

February 27th, 1985

Vertebrates

ON DISTINGUISHING BUTCHERY FROM OTHER POST-MORTEM DESTRUCTION: A METHODOLOGICAL EXPERIMENT APPLIED TO A FAUNAL SAMPLE FROM ROMAN LINCOLN.

P.M. Aird
Department of Prehistory and Archaeology
University of Sheffield
Sheffield S10 2TN

ABSTRACT

Actualistic studies have characterised the modifying effects of different taphonomic agents on bone. However faunal samples are usually the product of the interaction of several of these agents. The problem now facing archaeozoologists is the development of a methodology which can separate the effects of one agent from another in order to unravel the taphonomic history of the sample. This paper describes the method used to isolated the effect of butchery on an assemblage of animal bone from Roman Lincoln. Fragments of the major long bones were classified by shape into 'types' and the proportion of cut (as opposed to broken) edges of each estimated on a simple 1 - 4 scale. It was then possible to assess the degree of association of shape with butchery practice. The method could be developed with the aid of a computer.

INTRODUCTION

The study of butchery is now recognised as a valuable means of not only understanding economic strategy but also recognising differences in cultural and social organisation of societies in the past (Grant 1975, Noe-Nygaard 1977). However, the evidence is not always unequivocal. Most faunal samples are likely to be the cumulative product of a whole series of complex taphonomic processes and their interactions. Over the past few decades a number of researchers have turned to actualistic studies in order to characterise the effects of various taphonomic factors on bone (e.g. Muller 1976, Behrensmeyer 1978, Binford 1981, Brain 1981, Bahn 1983), but unfortunately it seems that splintering patterns, such as spiral fractures, depend more on the cross-sectional structure of the bone than the agent of fracture (Miller 1976, Gifford 1981). As a result it has not been possible to develop the same sort of criteria by which 'natural' and 'cultural' modifications are distinguished on stone.

Since regularity in fragmentation patterns within a bone assemblage can no longer intuitively be ascribed to deliberate breakage a methodology is now needed which will allow archaeozoologists to determine the differential effects of various taphonomic agents on any one faunal sample. This paper describes an attempt to fulfill such an objective using an assemblage of animal bone from Roman Lincoln.

THE SITE AND SAMPLE

Excavation at "The Park" in Lincoln began in 1968 in advance of development of the site for new municipal offices. The faunal remains were recovered during excavation of the rampart sequence of the lower colonia (Colyer 1975). Preliminary examination of the bones had shown that most were of cattle and only a very small proportion were unidentifiable. One context, "JO", had been deposited as a rubbish dump in the second half of the second or first half of the third century A.D. It contained mainly severely fragmented cattle long bones which showed evidence of gnawing but had also been cut or chopped. In addition, some breakage had occurred after recovery. The surface condition of the bones was fairly uniform, and generally good, so it seemed likely that they had been buried fairly quickly. Also, none of the fragments broken prior to recovery could be matched, so it was possible that fragmentation had occurred prior to burial, or even disposal, rather than in situ. The faunal sample therefore seemed to be ideal for developing a methodology to isolate the different agents which may have interacted to cause the breakage observed.

THE METHOD

Each long bone had broken in a fairly standard manner with great regularity, therefore fragments were grouped visually into types according to their shape. Conventionally, fragments are described according to an imposed formula by which the bone is divided into fractions along the medial/lateral, anterior/posterior and distal/proximal plans (e.g. Meadows 1978). However bones break naturally along lines of weakness or are cut (or broken) into portions which people find useful. The bone types 'identified' for each of the long bones are shown on Figures 1 - 27, and are described as follows:

Humerus (10 types)

1. medial distal articulation with part of shaft
2. distal articulation (usually trimmed on medial and lateral edges) with part of shaft
3. distal shaft excluding epiphysis
4. shaft only (more than 75%)
5. lateral edge of distal shaft excluding epiphysis
6. proximal shaft with up to 50% of distal shaft
7. proximal shaft
8. distal epiphysis
9. proximal articulation (fragmented) with part of shaft
10. deltoid ridge

Type 10 humerus was not always distinguishable from type 10 tibia.

Radius and ulna (8 types)

1. medial proximal articulation with part of shaft (radius only)
2. lateral proximal articulation with part of shaft (radius only)
3. medial distal articulation with part of shaft (radius only)
4. lateral distal articulation with part of shaft (radius only)
5. mid shaft (radius and ulna)
6. proximal articulation with part of shaft (radius and ulna)
7. medial shaft (excluding ulna)
8. lateral shaft (excluding ulna)

Patterning was not observed on ulnae separated from radii.

Femur (4 types)

The femur was originally divided into five types but subsequent analysis showed that two could be amalgamated.

1. shaft anterior including dorsal ridge (excluding epiphyses)
2. shaft including the supracondyloid fossa (excluding epiphyses)
3. shaft posterior including the trochanter minor (excluding epiphyses)

4/5. fragments from the distal articular surface

Fragment type 1 could not always be identified with confidence to species. In cases of uncertainty the presence of "possible cattle femur" was recorded but not included in further analysis of the data. Femur heads were not coded.

Tibia (5 types)

The tibia was initially divided into four types but later two were subdivided and one of the others was amalgamated with a sub-type.

1a. medial distal articulation including part of shaft
1b. lateral distal articulation including part of shaft
2. subsequently included with 1b.
3a. lateral side of proximal shaft posterior
3b. medial side of proximal shaft posterior
10. shaft including dorsal ridge. This type was coded 10 to make it easier to compare with humerus type 10 from which it was not always distinguishable.

Each fragment of one type was then plotted onto a single drawing of the bone from which it came in order to show the range of variation. A colour coded system was used to distinguish cut edges from those which were freshly broken, chewed, broken in antiquity, or for which the agent of fracture could not be determined. The criteria used for distinction were as follows:

cuts the edge was straight with a granular texture apart from the upper 0.5mm which was smooth and polished

fresh breaks easily distinguished by the colours

dog breakage denticulated or crenulated edges often associated with pitting, furrowing or scoring (Binford 1981)

Other broken edges were characterised by a conchoidal profile which was probably the result of the bone snapping under pressure.

Cut marks, puncture marks, flaking or any other surface scarring found on every fragment were described according to the terminology used by Sisson (1966) and Ryder (1969), and plotted onto a composite drawing for each bone type.

Wherever possible the direction from which the bone was cut has been noted. Horizontal cuts are thos made in the medial/lateral plane or anterior/posterior plane; vertical cuts are those made in the proximal/distal plane. For convenience, diagonal cuts have been coded according to the positions of the end points as follows:

M (medial)
L (lateral)
Pr (proximal)
D (distal)
Post (posterior)

As the main objective was to isolate butchery from the other taphonomic processes fragments were coded (adapting a system used in sedimentology to record angularity of pebbles) according to the percentage of deliberately cut edges.

1. well over 50%
2. approximately 50%
3. less than 50%
4. no cut edges

These figures may need to be adjusted if the same system is to be used on another sample. None of the bones from "The Park" had been cut on all edges. As breaks often continued along the same alignment as the cut edges it is possible that a practice of 'cut and break' was used for secondary butchery. Alternatively bones may have been fragmented after meat stripping along lines of weakness made by surface cuts, although elsewhere this has been discounted as a cause of breakage (Sadek-Kooros 1972).

From this experiment it was hoped that associations could be made between bone types and butchery practices.

Finally, wherever possible a description of each fragment was completed with details f epiphyseal fusion, and side of the body. From the latter it was thought that differences might be detectable

in the position of cutmarks on bones frm opposite sides of the animal which would perhaps show how the carcass was lying when it was butchered. Although there are difficulties with using epiphyseal fusion to assess age at death (Sue Stallibrass, personal communication) in this study the data have only been used in a relative sense so that the variable of age might be considered in terms of differential butchery.

RESULTS 1: FRAGMENTATION

A total of 299 fragments from "The Park" were examined to try and determine the causes of the observed breakage by the methodology described above.

Humerus

97 humerus fragments were recovered from "JO". The distribution of these fragments between bone types, with details of epiphyseal fusion and % cut edges, is shown in Table 1.

Type 1 (n = 12) (Fig. 1) All fragments were cut from the distal articular surface into the olecranon and coronoid fossas. Proximal to these parts, breakage type was varied. In some cases the cut edge continued on the anterior face as far as the teres tubercle but on the posterior face the cause of the break was usually indeterminate. The fragments from the left included proportionately less shaft than those from the right. The medial edge of the distal articulation was frequently cut off, particularly on the right-hand side. Evidence of canid gnawing was found on two examples of this type.

Type 2 (n = 11) (Fig. 2) Most fragments were almost completely cut to shape. The lateral edge of the distal articulation was frequently cut off. Fragments from the left-hand side tended to include less of the posterior shaft than those from the right. Evidence of canid gnawing was found on only one example of this type.

Type 3 (n = 7) (Fig. 3) In most cases the distal articulation seemed to have been broken rather than cut off. Otherwise the cut edges were concentrated at the proximal and distal ends of the breaks with the middle portion broken or indeterminate. Evidence of canid gnawing was found on two examples of this type.

Type 4 (n = 7) (Fig. 4) Breakage at the distal edges of fragments of this type was invariably the result of deliberate cuts, although at the proximal edges the cause was more uncertain, but associated in these examples with evidence of canid gnawing.

Type 5 (n = 6) (Fig. 5) Fragments of this type could either be confidently described as cut to shape or all breakage was of uncertain origin. There was no evidence of canid gnawing on any fragment.

Type 6 (n = 7) (Fig. 6) In most cases the cause of breakage for this type was indeterminate, although evidence of canid gnawing was found at the proximal edges of two examples.

Type 7 (n = 15) (Fig. 7) Examples of this type were usually cut along the proximal edges but the remaining breakage was either accidental or the cause was indeterminate. There was no evidence of canid gnawing on any fragment.

Type 8 (n = 19) (Fig. 8) Almost all examples of this type were cut to shape but the proportion of the distal articulation included varied considerably. Although some breakage was of indeterminate origin, particularly through the cancellous bone of the epiphyses the edges were remarkably straight, and therefore, probably cut. No traces of knife or saw were visible but this may be due to the spongy nature of the bone.

Type 9 (n = 3) (Fig. 9) So few examples of this type were found that little can be said about patterning or variability.

Type 10 (n = 10) (Fig. 10) Cut edges were concentrated around the deltoid tuberosity but otherwise most of the breakage on fragments of this type was of indeterminate origin. One example was gnawed along the medial edge of the break.

Summary

Types 1, 2, 4, 5 and 8 seem to display the best correlation between shape and cut edges and are most likely to be the result of deliberate butchering. Types 1 and 2 are complementary in shape and almost equal in terms of number of fragments present. Types 3, 4 and 5 complement type 8 and the sum total of fragments of the first three is approximately equal to the total number of the latter. It is perhaps then permissable to look at the evidence of the distal humerus in terms of two different kinds of butchery techniques: one being a longitudinal splitting of the bone with the epiphysis intact; the other the removal of the epiphysis in the medial/lateral - posterior/anterior plane. The difference in the amount of shaft present in right and left fragments of types 1 and 2 is more difficult to interpret without butchery experience or further work.

Only a third of the total number of humerus fragments were classified as types 6, 7, 9 and 10, which represent the proximal half. This may be due to differential destruction as this part of the bone is more fragile (Binford 1981) or the result of differential butchery and disposal practices. The evidence which has survived exhibits variable patterning although most of the cuts seem to occur below the head indicating that in butchering the carcass this articulation was deliberately removed from the rest of the bone.

Radius and ulna

54 radius and radius + ulna fragments were recovered from "JO". The distribution of these fragments between bone types, with details of epiphyseal fusion and % cut edges, is shown in Table 2.

Type 1 (n = 12) (Fig. 11) Most fragments were cut from the proximal articular surface as far as the fusion point, particularly on the posterior face, but few cuts continued into the shaft although in some cases breakage continued along the same line, terminating in a second cut edge. Fragments from the left-hand side usually included more of the shaft than those from the right.

Type 2 (n = 9) (Fig. 12) Fragments of this type which included very little of the shaft, tended to be cut to shape, whereas those with a large amount of shaft were usually cut only as far as the fusion point from which the line of the cut continued as a broken edge as far as the fusion point with the ulna. The latter group were confined to the right-hand side fragments. One example of this type showed evidence of canid gnawing.

Type 3 (n = 7) (Fig. 13) Most of the breaks on examples of this type were of indeterminate origin. Cut edges were more frequent between the distal articular surface and the fusion point, sometimes continuing as a break along the line of the cut which was occasionally interrupted by, or terminated in a second cut edge.

Type 4 (n = 6) (Fig. 14) The cutting pattern of this type was very similar to that of 3 (above) but cut edges more frequently continued into the shaft, in one example beyond the mid-point.

Type 5 (n = 1) (Fig. 15) Mostly broken apart from a cut edge on the ulna shaft

Type 6 (n = 7) (Fig. 16) Cut edges were concentrated in the area between the proximal articular surface and proximal fusion point, and on the posterior edge of the ulna opposite the articulation. The remaining breaks were of indeterminate origin. One example with evidence of canid gnawing.

Type 7 (n = 3) (Fig. 17) Mostly broken apart from a cut edge running parallel to the longitudinal axis of the bone. Two examples with evidence of canid gnawing.

Type 8 (n = 9) (Fig. 18) Apart from one example all edges were broken. The exception was a fragment with a cut running parallel to the longitudinal axis of the bone.

Summary

All the evidence suggests that deliberate breakage occurred only along the proximal/distal, anterior/posterior axes, or was concentrated at the articulations.

Femur

96 femur fragments were recovered from "JO". The distribution of these fragments between bone types, with details of epiphyseal fusion and % cut edges is shown in Table 3.

Type 1 (n =15) (Fig. 19) Cut edges only occur at the distal anterior and are concentrated around the dorsal ridge, orientated in a way that suggests that a blow from the proximal direction sliced down into the bone where it flares into the trochlea.

Type 2 (n =24) (Fig. 20) A number of examples of this type were gnawed at the distal fusion point and on some evidence of canid activity was also present on the medial edge. Much of the breakage was of indeterminate origin. However,cut edges were noted at medial and lateral edges of the fusion point, and these sometimes continued up the shaft beyond the proximal edge of the supracondyloid fossa.

Type 3 (n = 27) (Fig. 21) This type represents in most cases little more than the projecting trochanter minor and the medial part of the anterior proximal shaft. Cut edges are frequent on the posterior and anterior below the head (horizontal) and on the anterior shaft (vertical) and posterior shaft (diagonal - proximal/lateral to distal/medial). On some examples there is extensive evidence of canid gnawing.

Types 4 & 5 (n=26) (Fig. 22)lthough most examples of these types are only small fragments broken on all edges, or more than 75% of the edges, cuts do seem to have, on occasion, been directed diagonally through the distal articulation and horizontally across the bone at the distal fusion point, separating the epiphyses from the shaft.

Heads (n = 5) (not ill.) Only one head had definitely been cut from the rest of the femur. A further two were unfused epiphyses and the breakage on the remainder was of indeterminate origin.

Summary

Fragmentation of the femur seems to have been a complex process caused by a number of interacting agents. Cuts are centred on irregularities on the profile of the bone, which suggests that these were removed with a slicing action during some kind of butchery process. Breaks on some fragments indicate the removal of epiphyses in their entirety whilst on others the cut edges show that pieces were chipped from either edge of th distal articulation and the head was sometimes removed at the proximal articulation.

The double bevelled edge of many fragments of type 5 is similar to the sort of 'toggle' shape caused by trampling (Bahn 1983). In addition there is evidence to show that dogs have considerably modified some femur fragments.

Tibia

52 fragments of tibia were recovered from "JO". The distribution of these fragments between bone types, with details of epiphyseal fusion and % cut edges is shown in Table 3.

Type 1a (n = 4 (Fig. 23) Most fragments were cut from the distal articular surface as far as the fusion point. Some cut edges continued into the shaft.

Type 1b (n = 1 (Fig. 24)) Most fragments were cut from the distal articular surface as far as the fusion point and in some cases the breakage continued along the same line into the shaft terminating in a second cut edge.

Type 3a (n =11 (Fig. 25) Cuts were concentrated about the nutrient foramen and across the lateral edge of the distal fusion point. Other breakage was of indeterminate origin.

Type 3b (n =9) (Fig. 26) Most of the breakage was of indeterminate origin but a few cut edges were noted just above the medial edge of the distal fusion point.

Type 10 (n =16)(Fig. 27) Cuts were invariably concentrated on both sides of the dorsal ridge but otherwise breakage on fragments of this type was of an indeterminate nature.

Summary

Fragmentation of the tibia repeats the pattern noted from analysis of the other long bones: cut edges are concentrated around irregularities on the profile; the lateral or medial edges of articulations are frequently cut off, and sometimes splitting of the bone occurs along the proximal/distal and anterior/posterior axes; on occasion the epiphyses are cut off horizontally from the shaft.

RESULTS 2: SURFACE CUTS

Humerus

Cut marks were concentrated about the deltoid ridge; the deltoid tuberosity was often shaved off and chipping and facetting frequently continued as far as the coronoid fossa. Both horizontal and vertical cuts were clustered on the lateral side of the coronoid fossa but on the medial side diagonal cuts were more common. On both the anterior and posterior aspects just below the proximal articulation there was a series of diagonal cuts (proximal/lateral to distal/medial) particularly towards the medial edge. The teres tubercle was cut into heavily and occasionally completely shaved off. Often vertical cuts occurred in this area. Chipping and cutting was also common on the posterior face of the humerus where it flares into the olecranon fossa.

The pattern of cut marks described above would be consistent with a butchery practice in which the major deltoid and triceps muscles were detached at their ends by vertical, horizontal and diagonal cuts and then removed from the bone by a knife being run parallel to the shaft on the posterior and anterior sides from a proximal direction, chipping and slicing the bone as it came up against irregularities in the profile. Cuts on the distal articular surface may be related to longitudinal splitting of the bone (3: 1).

Radius and ulna

Surface cut marks were very rare and occurred only on the lip of the proximal articular surface of the radius and just below the ulna articulation, on the distal epiphysis and along the line of fusion between the ulna and radius shafts. All these cuts could result from detaching the lesser muscles of the legs at their points of attachment to the bone.

Femur

Surface cuts on the femur were concentrated around irregularities of the profile such as the supracondyloid fossa and trochanter minor or where the bone flared out into the distal articulation on the dorsal ridge and the ridge above the medial condyle. Frequently cuts in these areas were accompanied by shaving, chipping and facetting. Cuts were also common below the trochanter major on the medial side of the trochanter fossa (usually diagonal) and just below the femur head. As with the humerus the marks described above are consistent with a practice in which a knife was scraped down the bone to remove the meat after detachment of the tendons which hold the muscle to the proximal and distal ends of the femur.

Tibia

Cuts on the tibia were confined mostly to horizontal cuts across the dorsal ridge and vertical cuts to either side. These might have been the result of cutting through the flexor muscles of the hind leg. It is interesting that the heavy chipping, shaving and facetting apparent on the humerus and femur is not so much in evidence on the tibia. This may be due to the smoother outline of the tibia, or, alternatively, to the fact that the bone flares out into the proximal articulation so that a knife would be less likely to encounter obstacles. A few cuts also occurred on the posterior shaft below the proximal articulation, around the distal fusion point, and at the distal end of the muscular ridging on the posterior.

Summary

The available evidence seems to point to a butchery practice in which the meat was stripped off all these long bones prior to their fragmentation, but it is not clear whether this was whilst the limbs were still articulated, or had been separated into their constituent bones.

Epiphyseal fusion

Details of epiphyseal fusion have been included with the tables showing fragmentation of the long bones. Most animals represented in the sample seem to have survived beyond the age of 2.5 years although at least one animal was killed at around 18 months. However, the first main slaughtering peak seems to occur between three and four years of age, either before or at the time when the proximal humerus, distal and proximal femur, proximal tibia and distal radius fuse. The remaining animals seem to have survived beyond this age.

CONCLUSIONS

After completion of the analysis it was possible to determine with certainty to what extent cleavage had resulted from deliberate butchery. It is important to stress the positive nature of this approach. Without such a method only a negative conclusion can be drawn, such as "there seems to be no obvious reason for a high frequency of fracture in one restricted zone and .... butchery is likely to be responsible" (O'Connor 1982).

Nevertheless whilst the method is basically sound there were a number of problems that occurred during the analysis. The degree of fragmentation made it possible to match paired front and back, upper and lower and right and left fragments: so results may be distorted by inclusion of the same animal several times.

Secondly some bones have more distinctive features than others and some parts of the bones are more distinct than others, therefore more frequently recognised as types. However, this problem may be overcome by considering the shape of the unidentifiable fragments. If the shape and structure and size of a bone conditions the breakage pattern then it may eventually be possible to match the shapes of unidentified bone pieces to the shape left when all the recognisable type fragments are accounted for. Ultimately this may lead to an advance in our present methods of assessing the number of animals represented in the sample (by counting even shaft fragments which are of types that are mutually exclusive) and the degree of taphonomic loss.

Finally it was unfortunate that many fractures were classed as indeterminate because they could not be ascribed with confidence to deliberate or accidental breakage. In such cases it might have been more useful to describe the nature of the break; some grouping or "typing" may then have been possible. Clearly this sort of information would be more easily noted as a written rather than illustrated record, and thus analysis of the data might have to proceed with the aid of a computer data logging system. This could be accomplished fairly easily if the outline of each bone type shape was divided into numbered segments which could be allocated a code according to the nature of the fracture along its length. Indeed, the entire analytical system described in part 2 could be adapted without difficulty to enable storage and manipulation of the data by computer.

ACKNOWLEDGEMENTS

I wish to acknowledge the SERC Advanced Course Studentship which enabled me to take the M.A. course in Environmental Archaeology and Palaeoeconomy at the University of Sheffield. This paper is based upon work submitted as the dissertation component of that course. I would like to thank Terry O'Connor, firstly for helping obtain the faunal sample and secondly for so generously sharing some of this latest ideas about analytical techniques. I would also like to thank Jenny Mann of the Lincoln Archaeological Trust for allowing me access to "The Park" bones and providing me with all the background information she could find from old excavation records and the interim reports. Finally, I would like to thank the other people who work in the bone laboratory at Sheffield; Graeham Mountenay and Judy Cartledge, but most of all Gill Clark who through many hours of discussion has helped me formulate my ideas and streamline this final presentation.

REFERENCES

Aird, P.M. (1983) Butchering cattle in Roman Lincoln; some experiments in Methodology. Unpublished MA dissertation, University of Sheffield, Department of Prehistory and Archaeology.

Bahn, P.G. (1983) The case of the clumsy cave-bears, Nature 301, 565.

Behrensmeyer, A.K. (1978) Taphonomic and ecologic information from bone weathering, Paleobiology 4 (2), 150-162.

Binford, L.R. (1981) Bones: Ancient Men and Modern Myths. London.

Brain, C.K. (1981) The Hunters or the Hunted. London.

Colyer, C. (1975) Excavations at Lincoln 1970-72; the western defences of the lower town, Antiquaries Journal 55, 227-266.

Gifford, D. (1981) Taphonomy and Palaeoecology: A Critical Review of Archaeology's Sister Disciplines. In Advances in Archaeological Method and Theory 4, ed. M. Schiffer, pp. 365-438. New York: Academic Press.

Grant, A. (1975) The animal bones. In Excavations at Portchester Castle I Roman ed. B. Cunliffe, pp. 378-408 and 437-450, London:

Meadow, R.H. (1978) "Bone code" a system of numerical coding for faunal data from Middle Eastern sites. In Approaches to Faunal Analysis in the Middle East, ed. R.H. Meadow and M.A. Zeder. Harvard Peabody Museum Bulletin.

Miller, G.J. (1976) A study of cuts, grooves and other marks on recent and fossil bone: 2. In Lithic Technology: Making and Using Stone Tools, ed. E.H. Swanson, pp. 211-288. Chicago: Chicago University Press.

Noe-Nygaard, N. (1977) Butchery and marrow fracturing as a taphonomic factor in archaeological deposits, Paleobiology 3, 218-237.

O'Connor, T.P. (1982) Animal Bones from Flaxengate, Lincoln, London.

Ryder, M.L. (1969) Animal Bones in Archaeology, Oxford.

Sadek-Kooros, H. (1972) Primitive bone fracturing, a method of research, American Antiquity 37 (3), 369-382.

Sisson, S. (1976) The Anatomy of the Domestic Animals (revised J.D. Grossman) London.

	coded % of cut edges													
Bone type	1	1/2	2	2/3	3	3/4	4	?	Total	PF	PNF	PJF	DF	F?
1	6(1)	1	1	2(1)	2	0	0	0	12				11	1
2	8	1	0	1	1(1)	0	0	0	11				11	
3	1	0	1(1)	1(1)	3	0	1	0	7					7
4	4	0	0	1	2(2)	0	0	0	7					7
5	3	0	1	1	1	0	0	0	6					6
6	0	1	1	0	4(2)	1	0	0	7	1				6
7	0	0	0	1	7	0	2	5	15		4	1		10
8	13	0	0	5	1	0	0	0	19					10
9	0	0	0	1	2	0	0	0	3	1	2			
10	0	0	0	0	6(1)	1	2	1	10					10

Key: - ( ) Gnawing present

PF - proximal fused; PNF - proximal not fused; PJF - proximal just fused; DF - distal fused;

F? - indeterminate

TABLE 1: Bone types of the humerus showing % cut edges and canid breakage

	coded % of cut edges													
Bone type	1	1/2	2	2/3	3	3/4	4	?	Total	PF	PJF	DF	DNF	F?
1	1	0	0	0	9	0	2	0	12	12				
2	1	0	3	0	2(1)	1	0	2	9	8	1			
3	0	0	0	1	4	1	0	1	7			4	3	
4	0	0	0	3	2	0	1	0	6			5	1	
5	0	0	0	0	1	0	0	0	1	1				
6	0	0	0	0	2	0	2	3(1)	7	6			1	
7	0	0	0	1(1)	0	2(1)	0	0	3	2			1	
8	0	0	0	1	1	1	6	0	9				9	

Key: - ( ) Gnawing present

PF - proximal fused; PJF - proximal just fused; DF - distal fused; DNF - distal not fused;

F? - indeterminate

TABLE 2: Bone types of the radius and ulna showing % cut edges and canid breakage

coded % of cut edges

Bone type	1	1/2	2	2/3	3	3/4	4	?	Total	PF	PNF	PJF	DF	DNF	F?
Femur															
1	0	0	0	0	6	2	7	0	15				2		13
2	0	2(1)	0	3(1)	11(5)	1(1)	7(1)	0	24	1				1	22
3	0	3(1)	3(1)	4(2)	13(2)	2(1)	1	1	27	7	2				18
4/5	0	1	0	5	8	1	8	3	26				13	7	6
head	-	-	-	-	-	-	-	-	5	1	2				2
Tibia															
1a	0	1	0	1	1	1	0	0	4				4		
1b	0	0	0	1	8	0	2	1	12				11	1	
3a	0	0	0	0	9	0	1	1	11	9					2
3b	0	0	0	0	3	0	6	0	9	1	1				7
10	0	0	0	2	12	2	0	0	16						16

Key: - ( ) Gnawing present

PF - proximal fused; PNF - proximal not fused; PJF - proximal just fused; DF - distal fused;
DNF - distal not fused; F? - indeterminate

Table 3: Bone types of the femur and tibia showing % cut edges and canid breakage

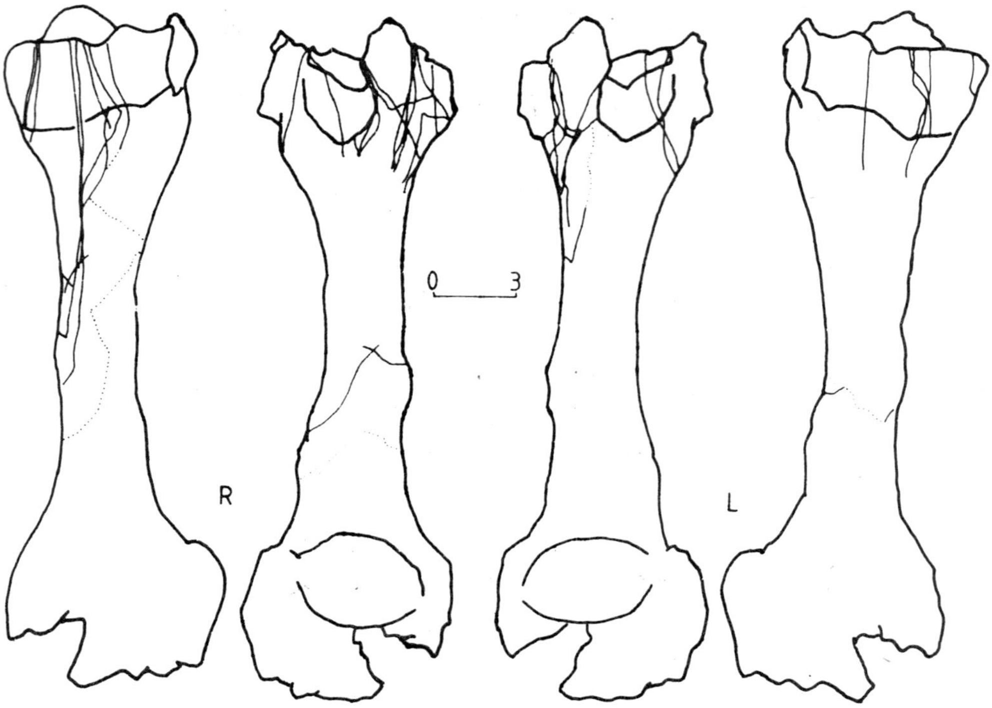

1. Type 1 humerus

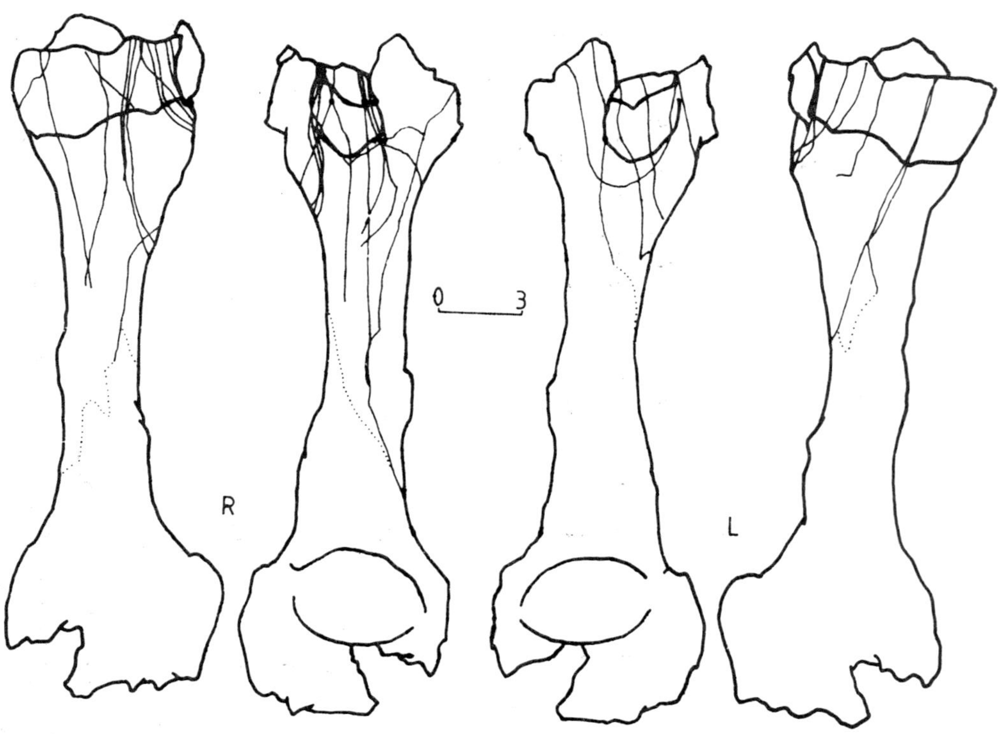

2. Type 2 humerus

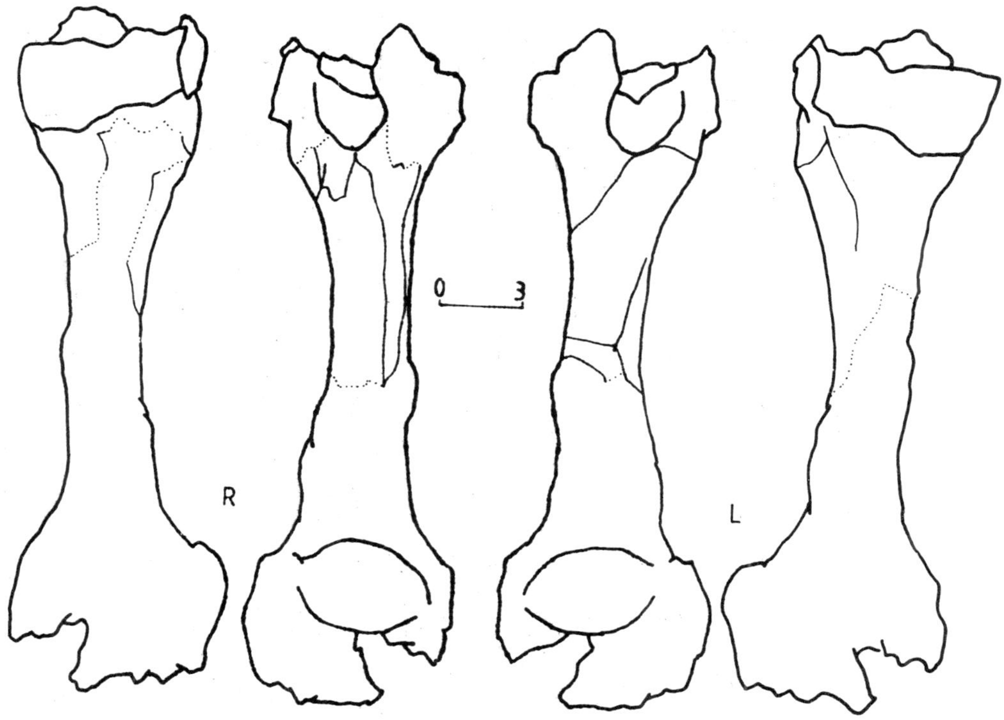

3. Type 3 humerus

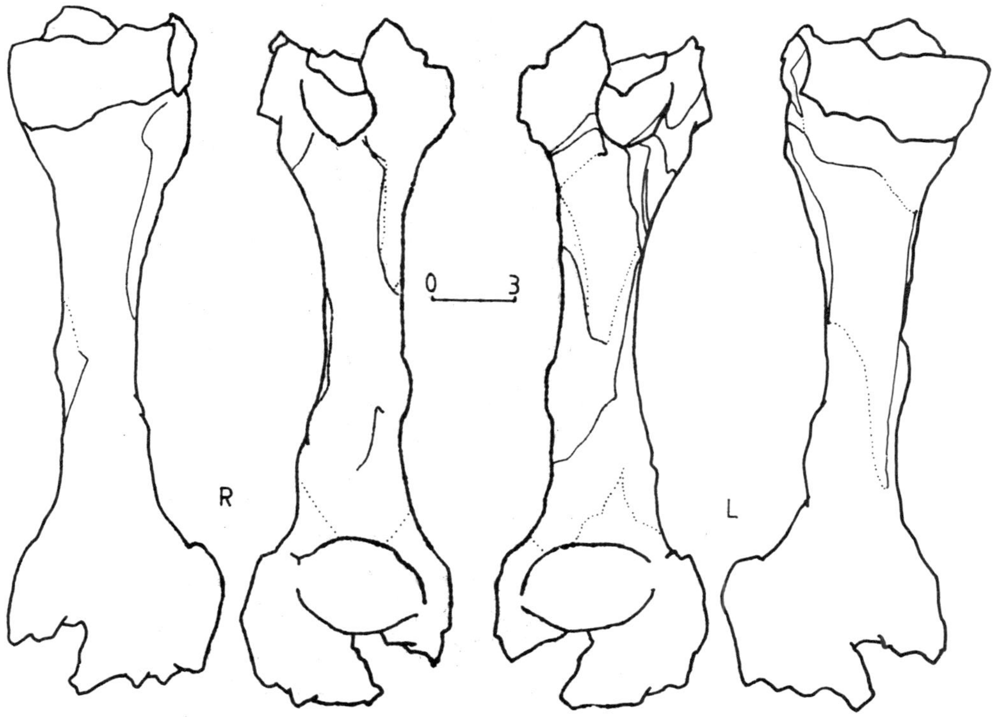

4. Type 4 humerus

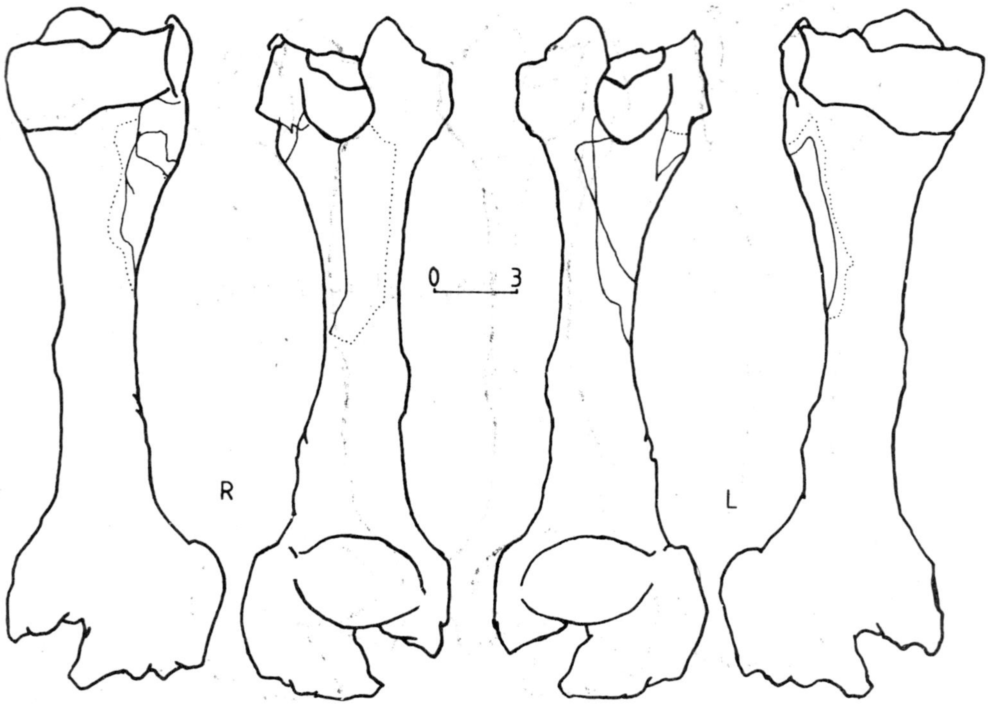

5. Type 5 humerus

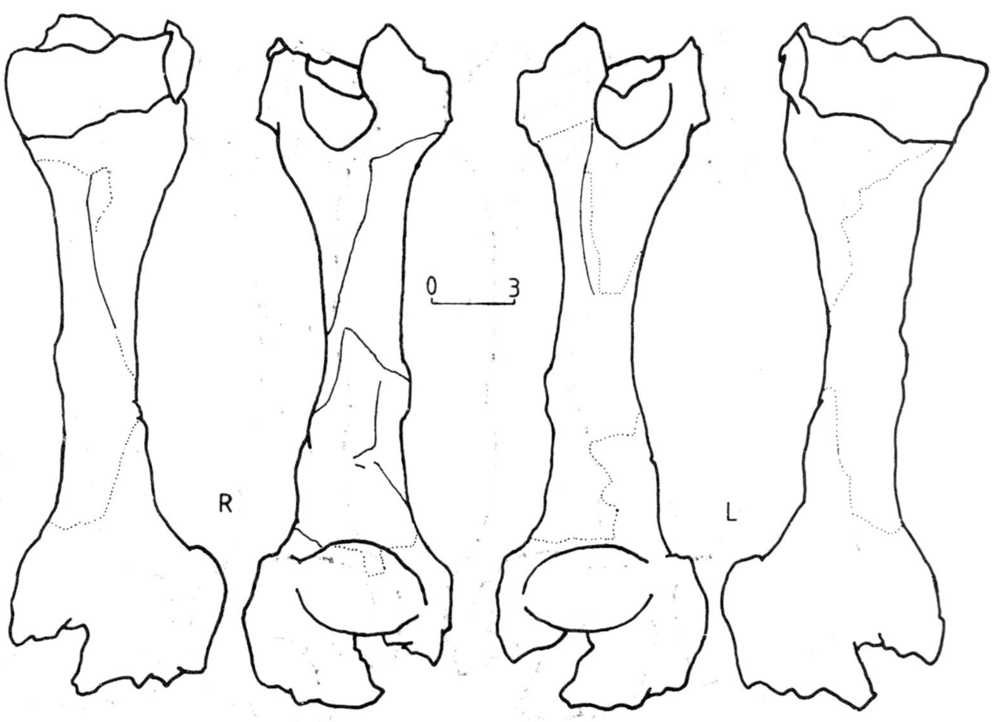

6. Type 6 humerus

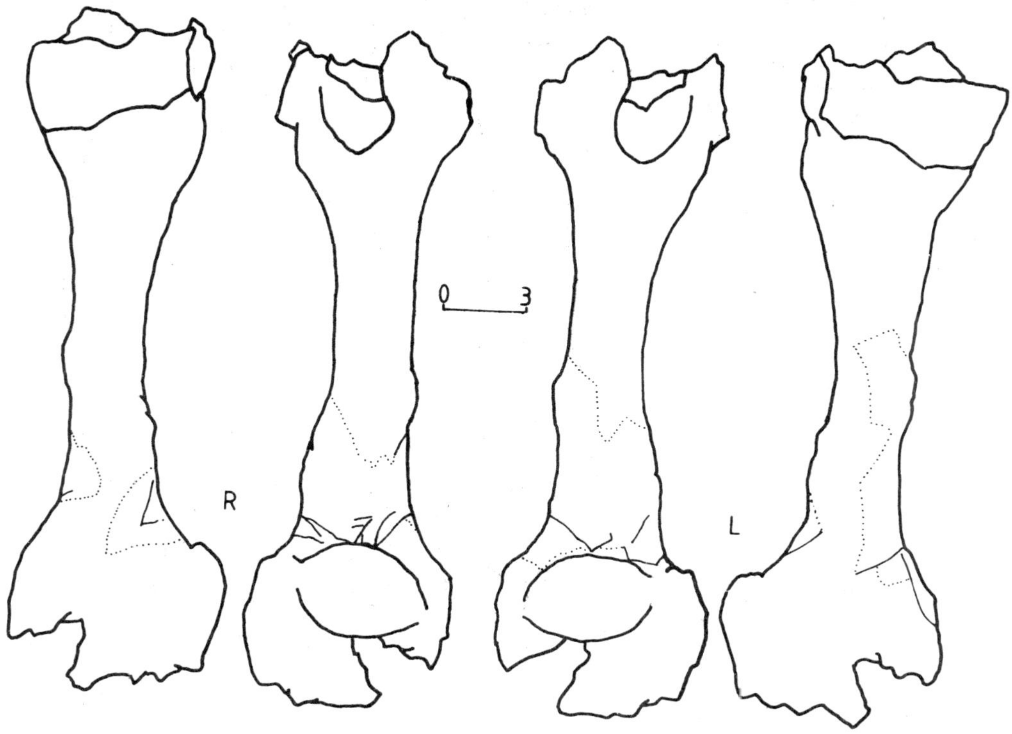

7. Type 7 humerus

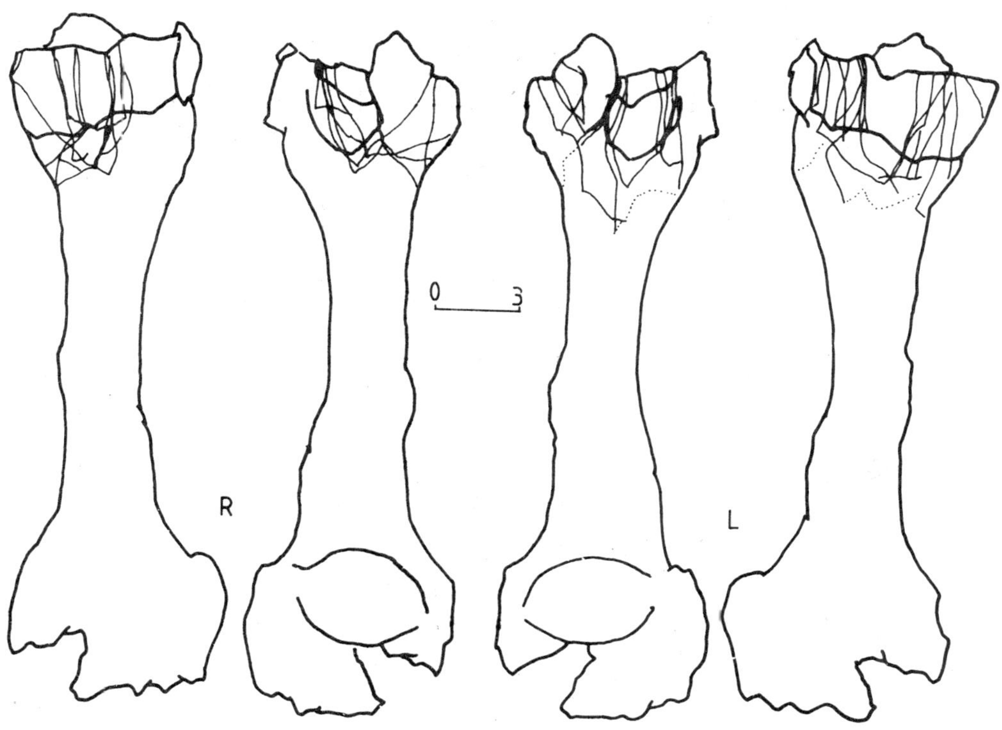

8. Type 8 humerus

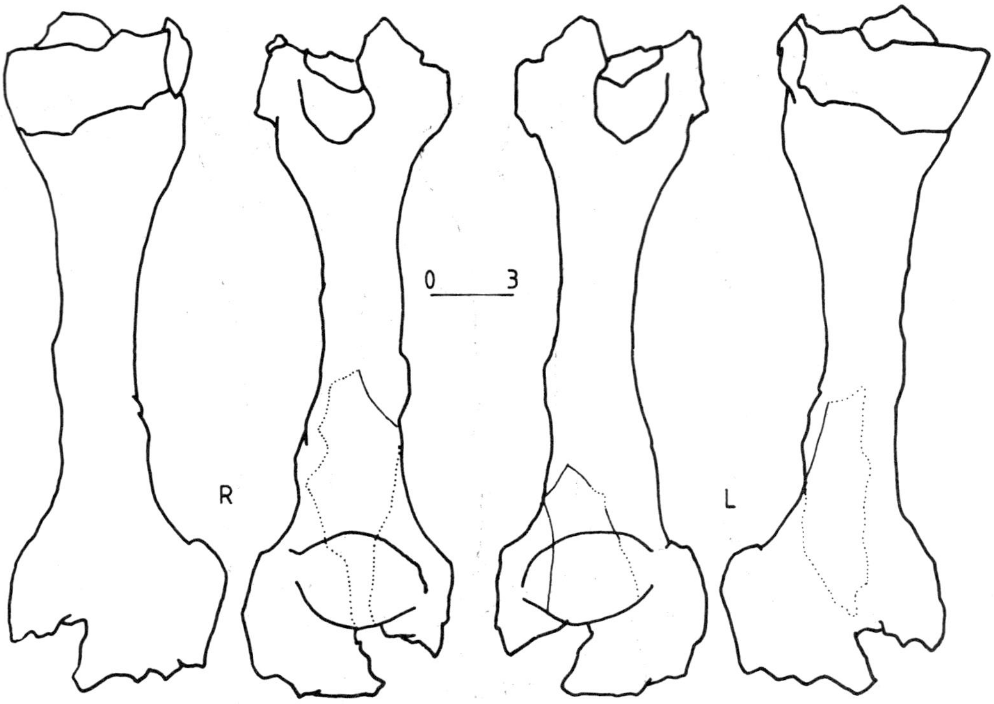

9. Type 9 humerus

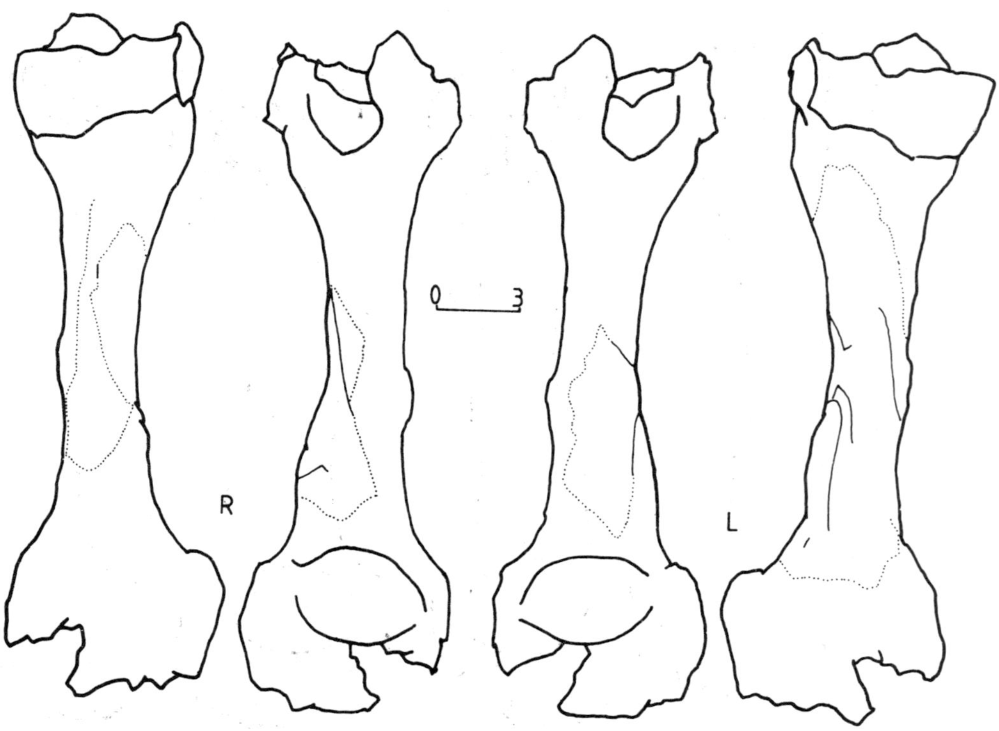

10. Type 10 humerus

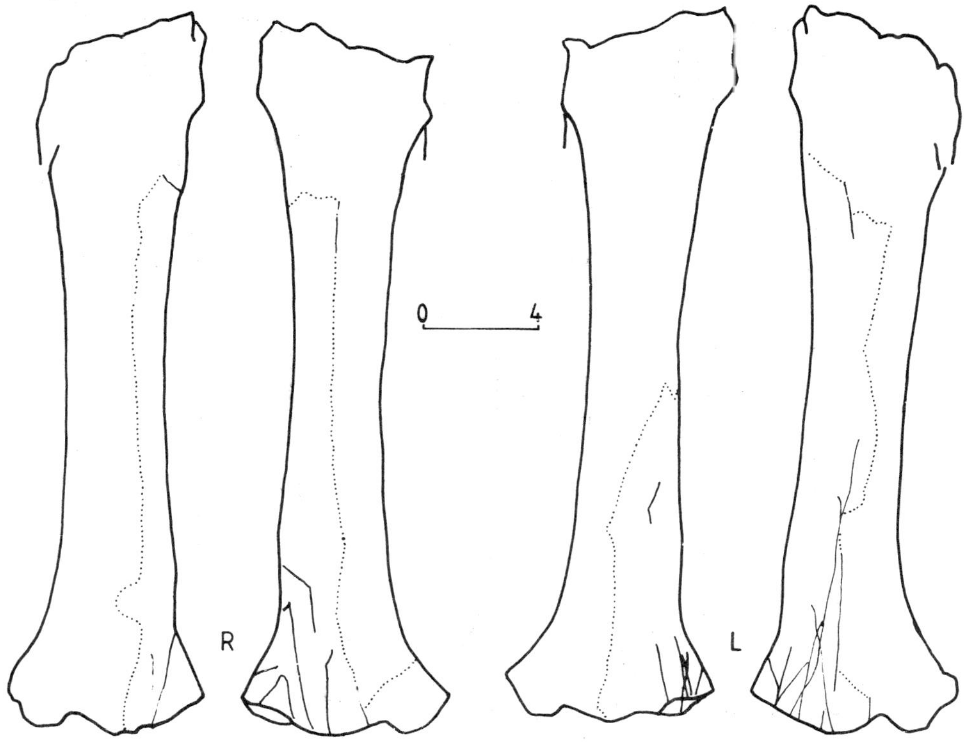

11. Type 1 radius and ulna

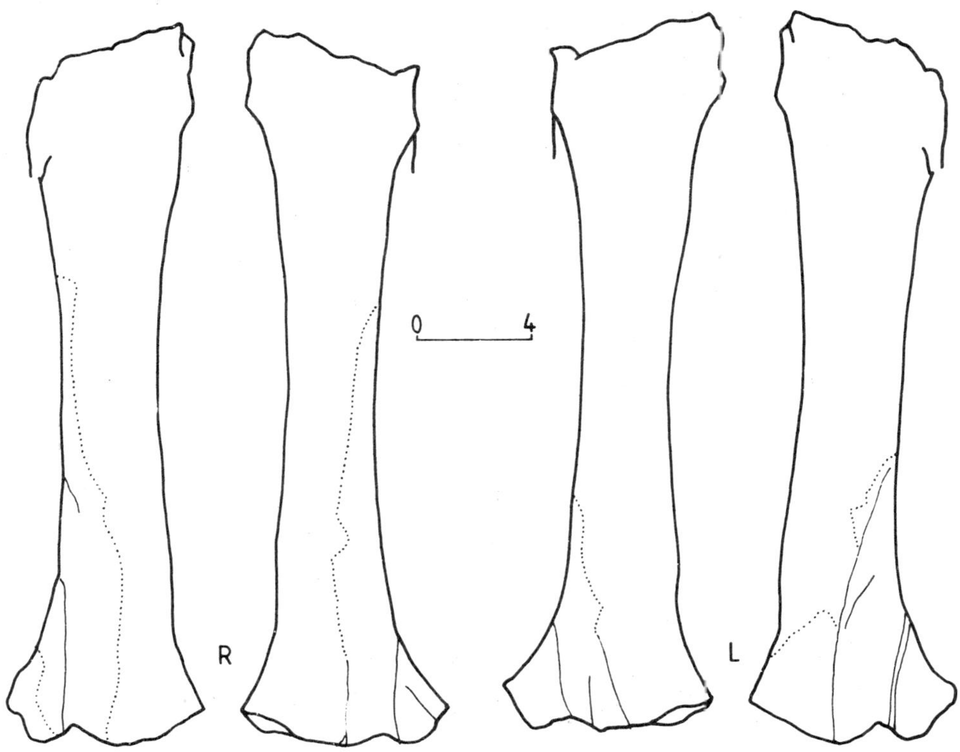

12. Type 2 radius and ulna

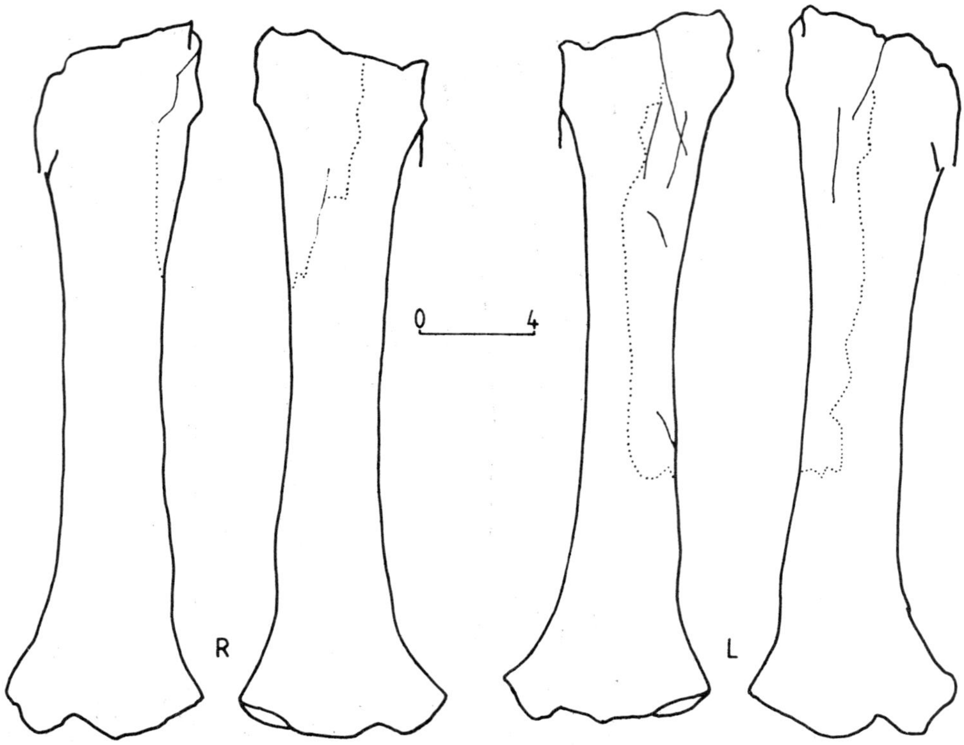

13. Type 3 radius and ulna

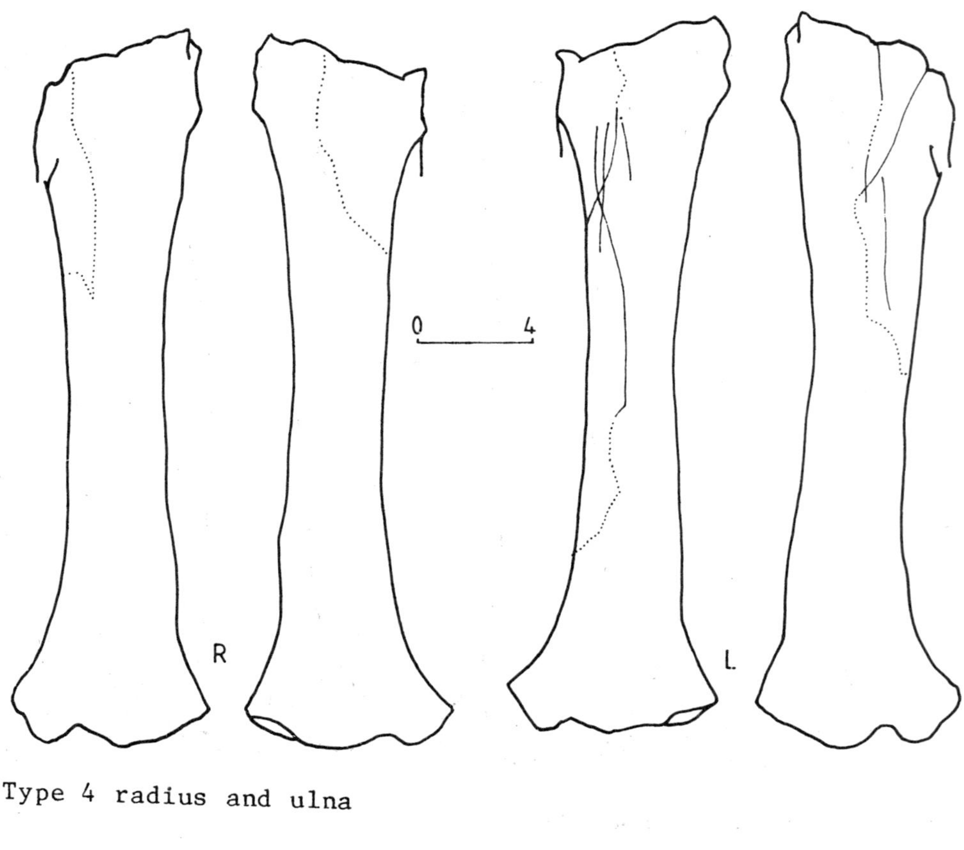

14. Type 4 radius and ulna

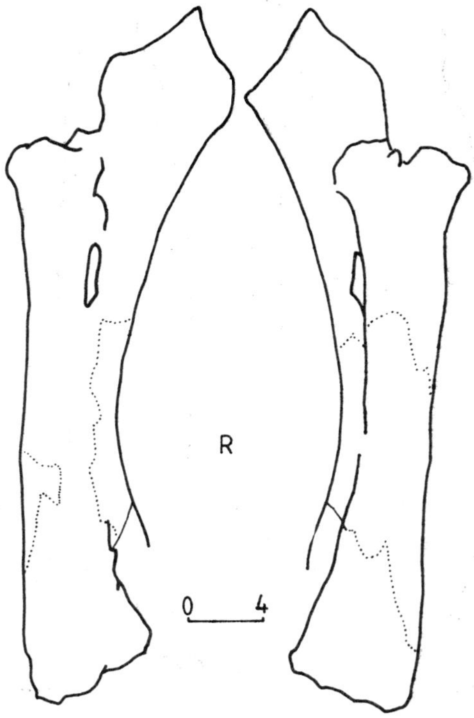

15. Type 5 radius and ulna

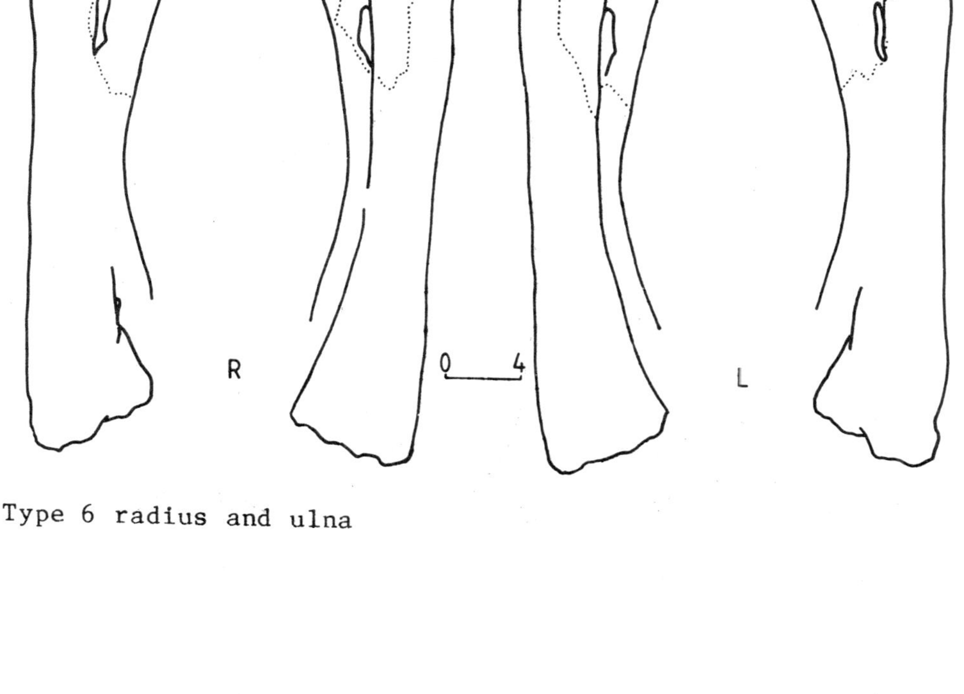

16. Type 6 radius and ulna

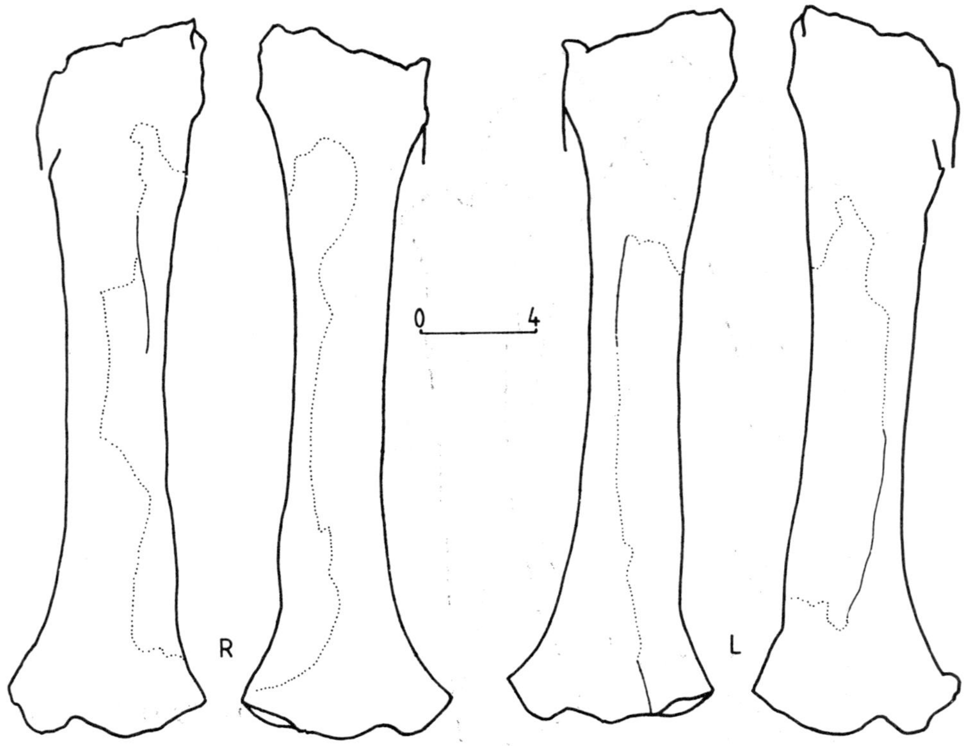

17. Type 7 radius and ulna

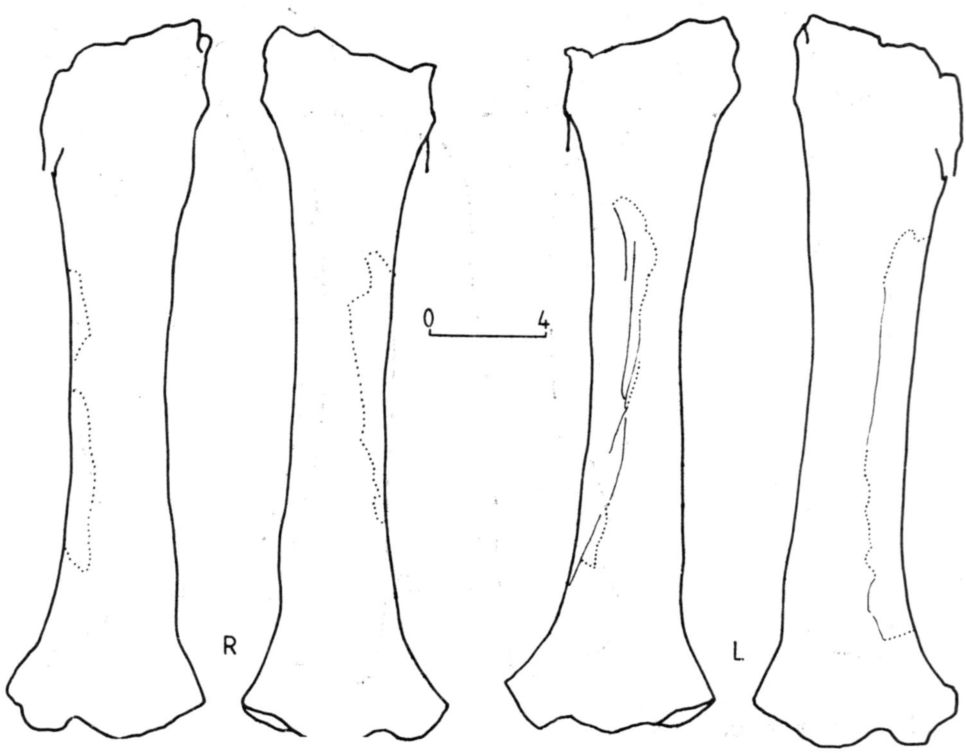

18. Type 8 radius and ulna

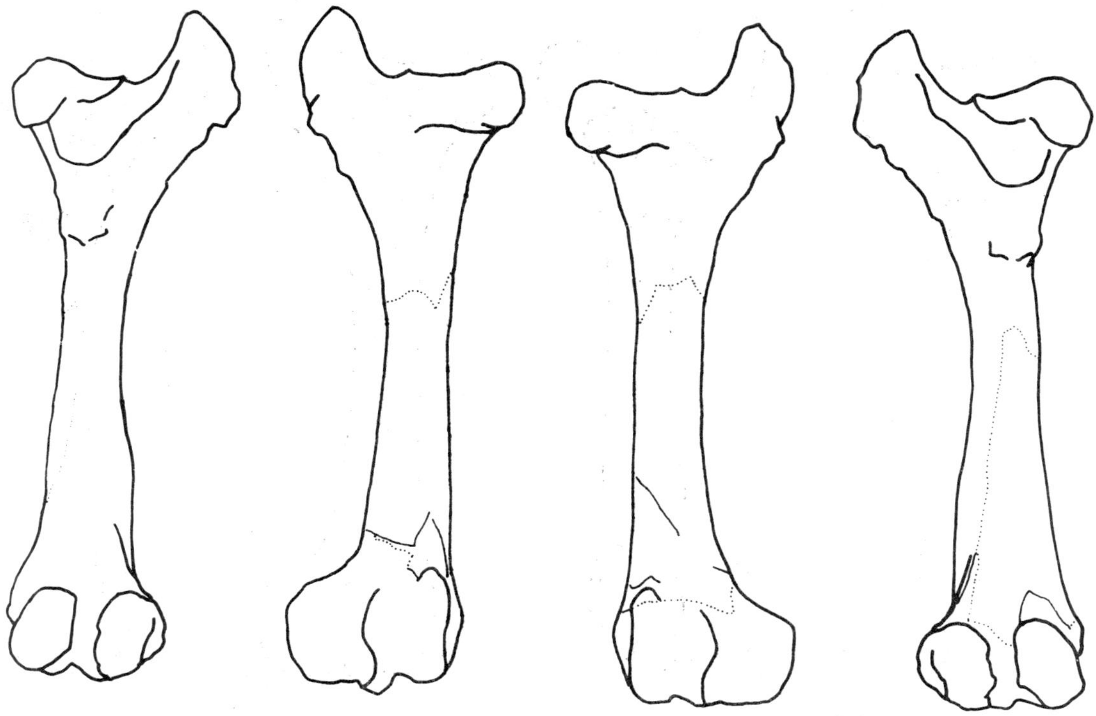

19. Type 1 femur

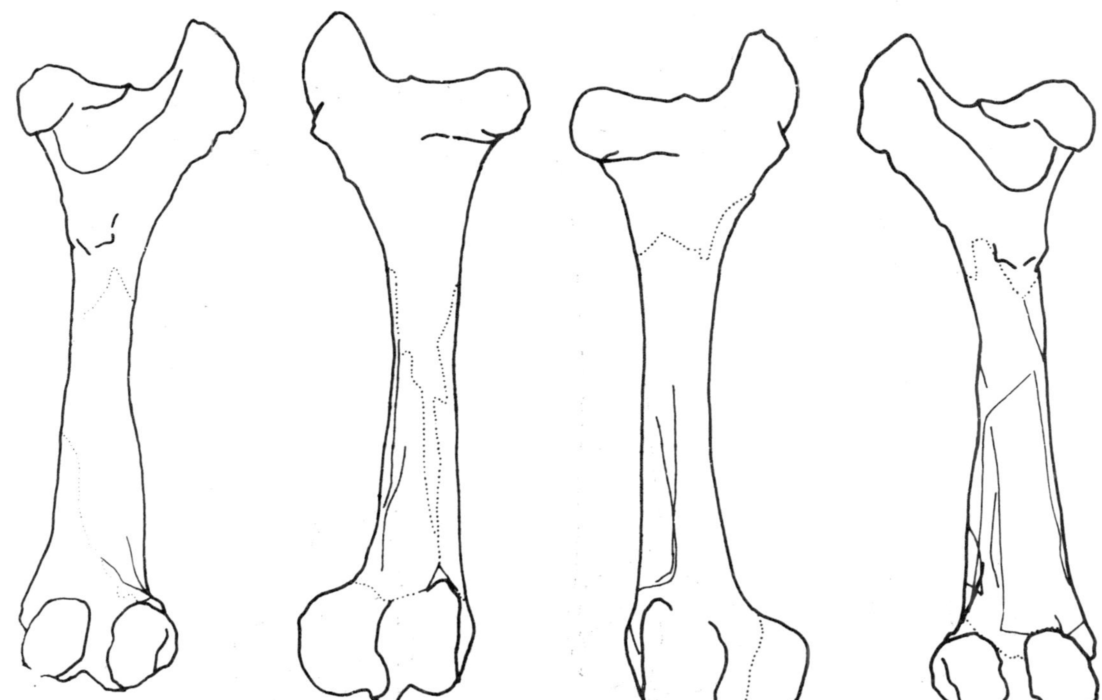

20. Type 2 femur

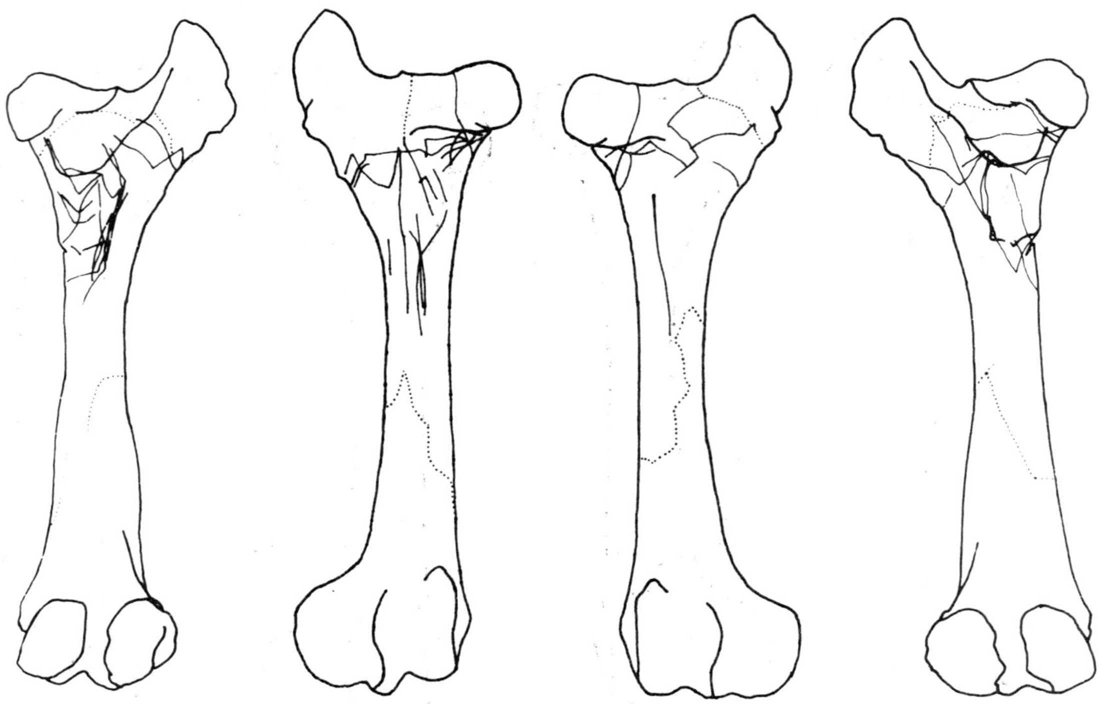

21. Type 3 femur

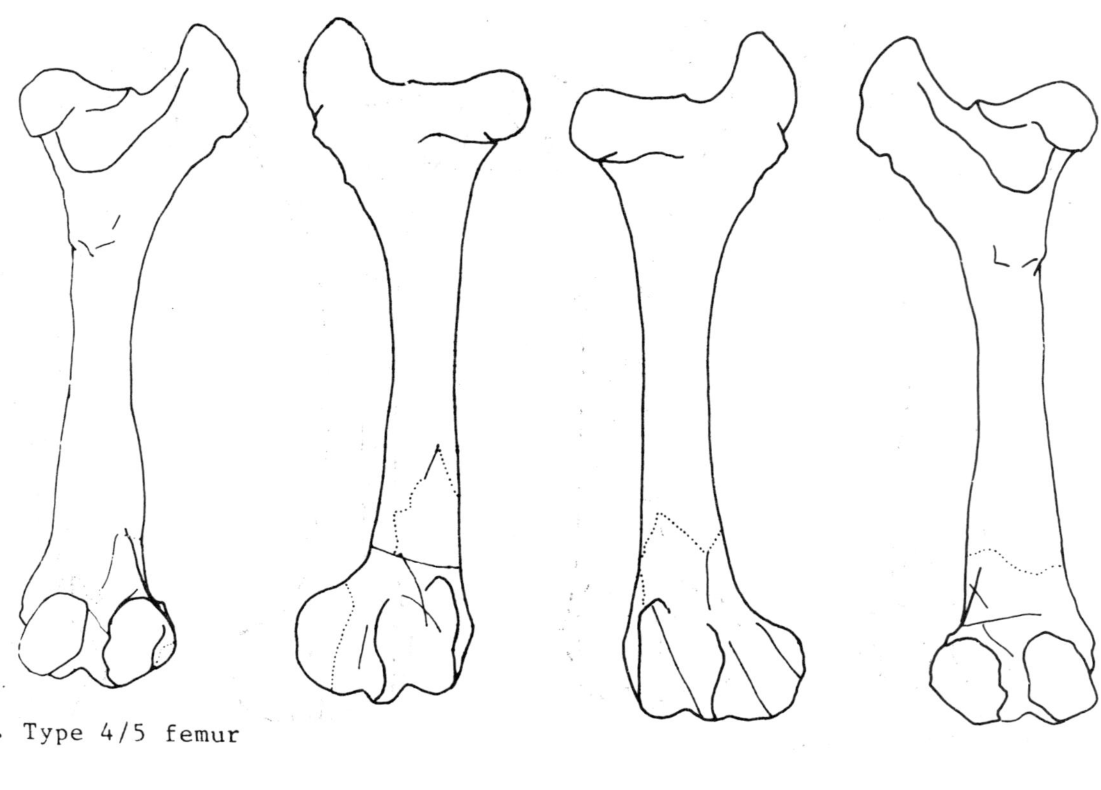

22. Type 4/5 femur

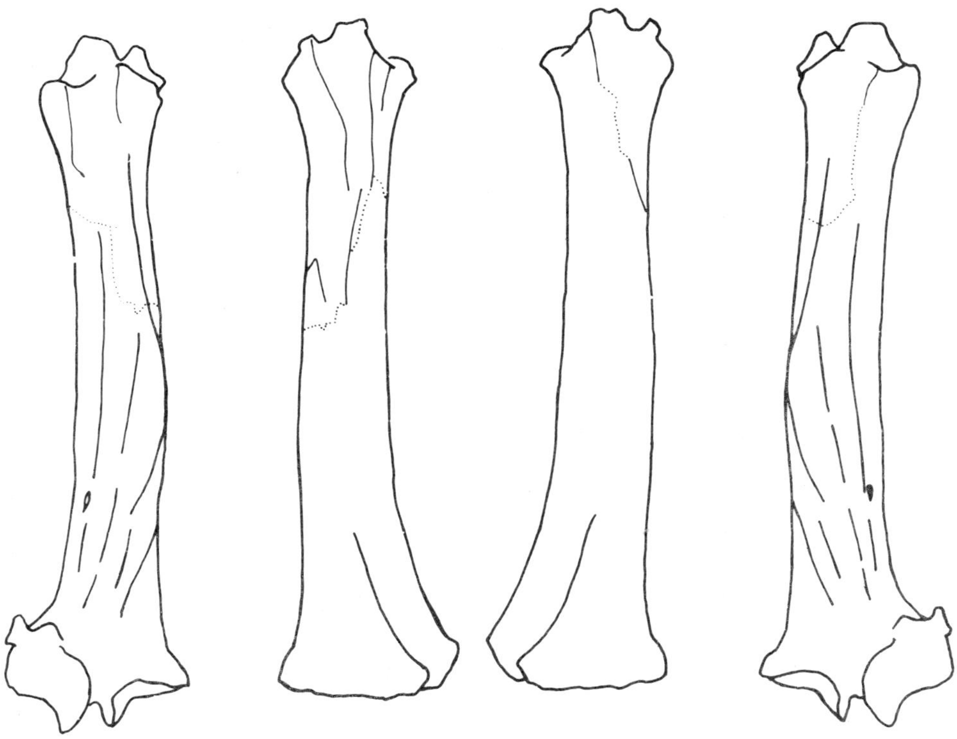

23. Type 1a tibia

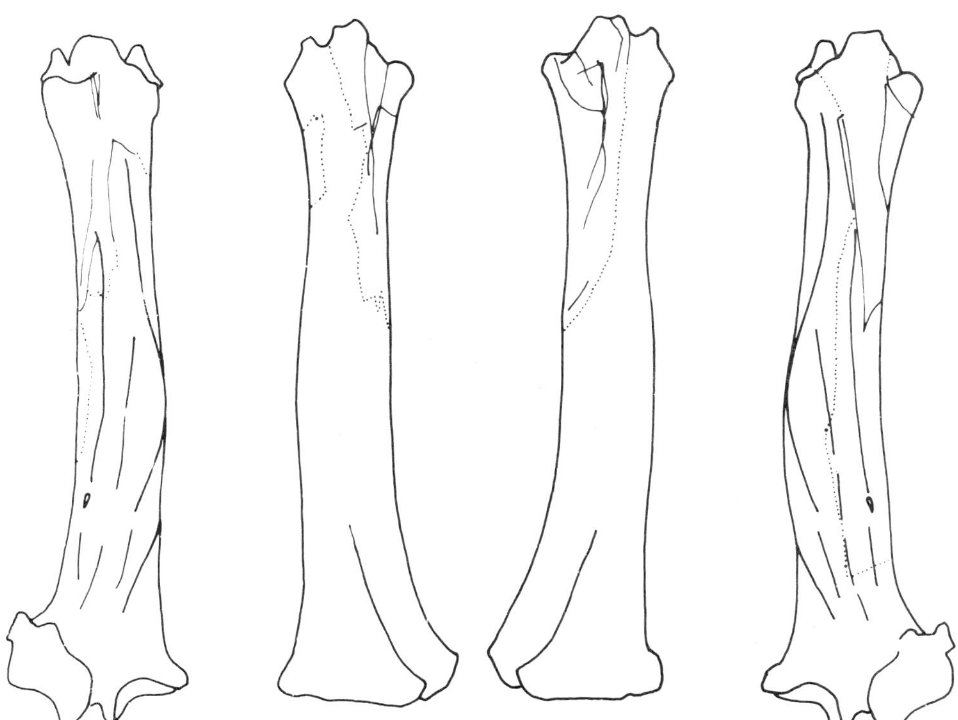

24. Type 1b tibia

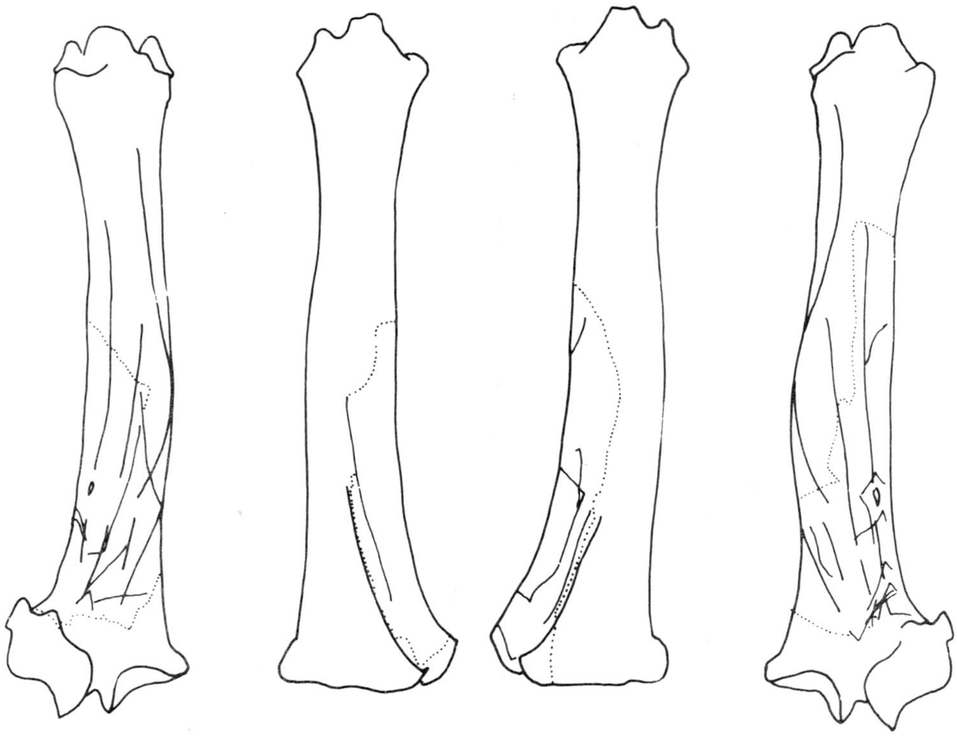

25. Type 3a tibia

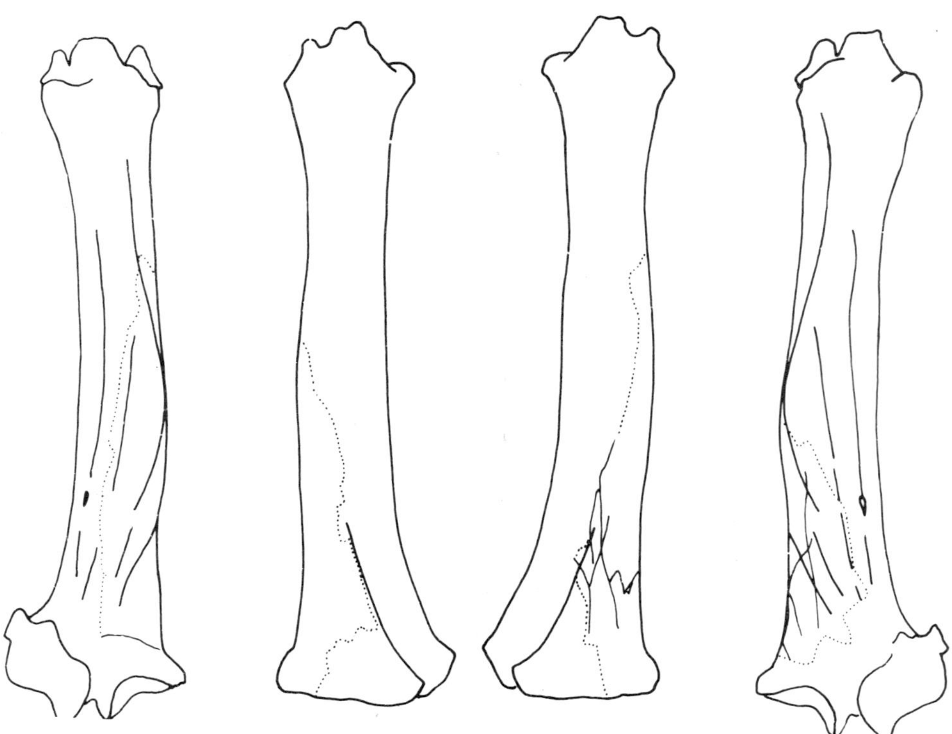

26. Type 3b tibia

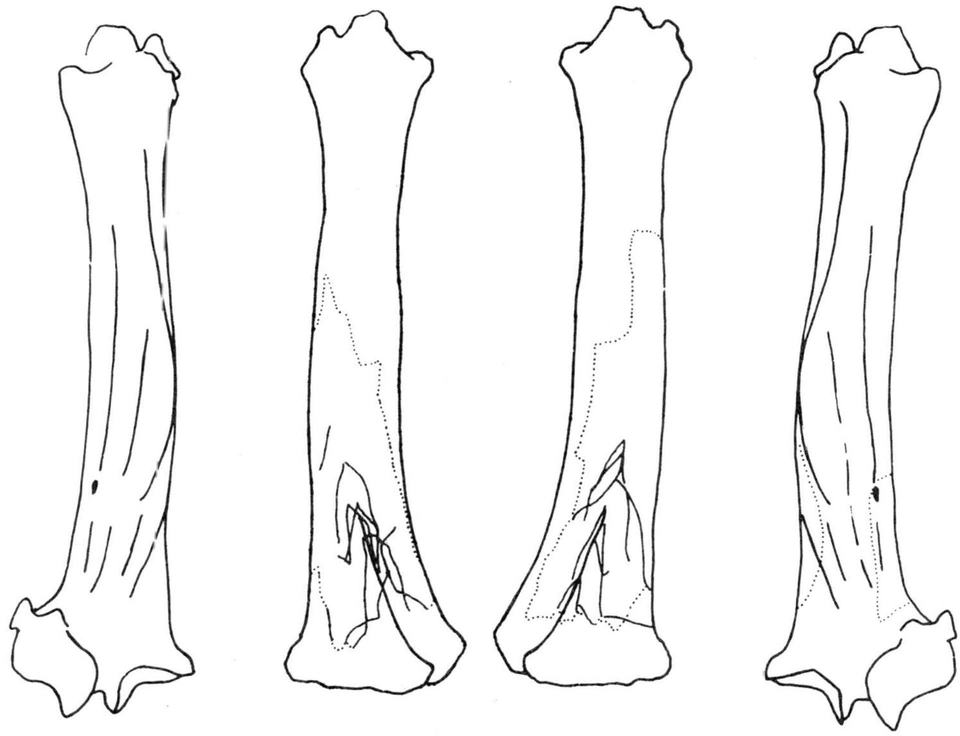

27. Type 10 tibia

ASSESSING VARIATIONS IN IRON AGE AND ROMAN BUTCHERY PRACTICES: THE NEED FOR QUANTIFICATION.

J.M. Maltby
Research Fellow, Department of Archaeology, University of Southampton.

ABSTRACT

This paper argues for the desirability of careful quantification of butchery evidence from archaeological faunal assemblages. Using the data from some Iron Age and Romano-British sites in southern England, aspects of cattle butchery are discussed. Specific examples of different types of butchery and fragmentation patterns are given to demonstrate some of the aims and problems of quantification. It is claimed that, provided detailed records are made of the location of butchery marks together with information about fragment size, gnawing and erosion damage, a much better understanding of the frequency and distribution of various butchery techniques can be obtained. In some periods, such as the Romano-British period, there was a great deal of variability in butchery practices. Quantification of such data would enable new insights to be formed into how carcases were utilised and their meat processed. The results may have wider implications for studies of the Romano-British period in general.

INTRODUCTION

The study of butchery practices employing the evidence of cut marks and fragmentation patterns on animal bones from British archaeological sites has been of a haphazard nature. A survey of published reports shows that in the majority of cases such evidence was ignored, or at best, given scant attention. For example, in his survey of bone assemblages mainly of Romano-British date, King (1978: 218-24) listed 116 site reports, of which only 26 mentioned butchery evidence at all. Only in a few instances have detailed summaries of the types of butchery marks observed been given (e.g. Wilson 1978; Maltby 1981; Thawley 1982a). In still fewer cases have the detailed results of fragmentation and butchery analysis been discussed (e.g. Wall 1980). This is surprising in view of the increased interest of archaeologists in faunal analysis in the last 20 years. In general, however, studies of butchery practices have been considered of less importance than the investigation of species abundance, ageing and metrical analysis.

However, several detailed ethnoachaeological studies of butchery have been carried out in recent years (e.g. Binford 1978; Yellen 1977), indicating the growing awareness of the importance of such studies for archaeozoologists. Several studies of archaeological bone assemblages of hunter groups have benefitted from such observations (Binford 1981; Speth 1983).

In Britain, gradually more archaeozoologists have begun to record butchery marks in detail. This is a welcome trend, but with the accumulation of this new body of data, the time has arrived to consider how such data should be analysed. It will be argued that careful quantification of butchery marks and fragmentation patterns is necessary to obtain a better understanding of how carcases were utilised, segmented and distributed. In addition, it is clear that in some periods there were variations in the treatment of carcases both within and between settlements. The types of butchery implement used (knife, cleaver, saw etc.) also varied and several types of cut marks may be found on bones from the same settlement or even in the same deposit. In such cases it is clear that careful quantification is required to investigate such interesting functional and cultural variability.

To illustrate the types of questions and problems facing archaeozoologists concerned with analysing butchery data, this paper will focus on aspects of cattle carcase utilisation in southern England during the Iron Age and Romano-British periods. The data will be used to examine some of the questions and problems to be faced in assessing such evidence.

THE QUANTIFICATION OF BUTCHERY MARKS

Is Quantification Necessary?

It first has to be shown that quantification of butchery data is needed. Certainly, if it can be shown that a uniform systematic procedure of butchery was practised, there may be no need to proceed beyond a thorough description of the marks and their causes (skinning, disarticulation, filleting etc.) and the types of implement used. Thus, in the case of many Iron Age sites, where fine sharp knife cuts were commonly produced during these processes (Wilson 1978:119-23; Maltby 1981), butchery techniques can be very consistent and can be interpreted as evidence for a regular butchery tradition. Such traditions can be contrasted with certain Romano-British butchery techniques. The use of choppers and cleavers was generally more common in the Romano-British period in southern England. In some cases concentrations of cattle bones displaying characteristic chop marks bear witness to systematic largescale butchery. Such assemblages have to date mainly been associated with urban sites such as Cirencester (Thawley 1982a: 217-20; 1982b: microfiche; Maltby in press 1), Colchester (Luff 1982: 101-4) and

Great Chesterford (Serjeantson n.d.). Clearly, there are occasions when discrete assemblages displaying consistent butchery need only be assessed qualitatively rather than quantitatively to establish which butchery tradition or process was in operation. However, in the Romano-British period such discrete butchery traditions are not universal. In Hampshire, for example, rural settlements in the early Romano-British period have produced a mixture of butchery marks produced with knives or choppers (e.g. at Winnall Down (SU 498304) Maltby n.d.1; Cowdery's Down (SU 658532) Maltby n.d.2; and Owslebury (SU 525246) Maltby in prep.) In such cases, in addition to simply stating that various butchery techniques were practised, quantitative assessments of the different types of butchery marks are needed, to establish their relative importance and any variations relating to their location within a settlement and their date.

Causes of Butchery Variation

Variations in the location and types of butchery marks and in the breakage of bones can be the result of a number of functional factors relating to how carcases were treated. Butchery techniques may have depended on whether meat was for immediate consumption or for storage by means of salting or smoking. The importance of marrow extraction and the boiling of some bones for broth could also have had a bearing on butchery strategies and subsequent fragmentation of the bones. The location of cut marks can be dependent upon whether the carcase was butchered on the ground or hung. The size and frequency of the butchery operations may have resulted in the use of different tools and butchery techniques. There is also the possibility that some butchery variability was related to cultural preferences or traditions.

The development of butchery studies needs to take two courses. The first is to obtain a better understanding of what the observed butchery marks and fragmentation represent. To which of the factors discussed above (and others) can the observed variations in butchery be related? Binford (1981) has made some progress through ethnoarchaeological studies of butchery practices along these lines. Ashdown & Evans (1981: 214-5) attempted to replicate chop marks found on cattle humeri and radii at Skeleton Green, Hertfordshire. More experimental work is urgently required, as by no means all types of butchery marks found on archaeological material have been explained. The second course is the need to develop reliable techniques to quantify butchery marks. The remainder of this paper will deal with some of the issues involved.

Problems of Quantification

Reliable quantification of butchery marks is faced with serious difficulties. In many instances these are the same problems encountered in the analysis of other aspects of faunal data. The detail of recording of butchery and fragmentation data varies between workers. Experience, skill, working (particularly lighting) conditions, time available and recording methods are all factors that could lead to incompatibility between results. Unfortunately such factors are extremely difficult to control for in quantitative comparisons and it mist be left to the judgement of the archaeozoologist to assess the reliability of making direct comparisons with other workers' data. Other biasing factors include the variation in the frequency of butchery marks produced by butchers of different skills and techniques. It is possible to butcher a carcase leaving very few marks on the bones. A skilled butcher, for example, is less likely to leave marks on the bones than an unskilled one. It is likely that butchers using choppers and cleavers to disarticulate a skeleton will leave marks more frequently than those using knives for the same purpose. Equal numbers of chop marks and knife cuts on specific bones need not mean that two alternative methods of butchery were employed with equal frequency. Such comparisons would tend to be biased towards chop marks. However, it should be possible to monitor the changes in the relative abundance of different types of butchery marks provided other biasing factors are taken into account.

One of the major problems is that the number of butchery observations is conditioned partly by the state of preservation of the assemblage. Weathering and scavenging of bones both destroy cut marks and increase fragmentation. Several reports on British faunal data have noted the percentage of bones which were butchered (e.g. Whitehouse & Whitehouse 1974;220; Thawley 1982b: microfiche 13). Such figures could be influenced by any of the variable factors discussed above. The effects of poor preservation, in particular, can have a dramatic effect upon the frequency of observed cut marks. Butchery observations on cattle bones at Winnall Down, Hampshire were in general found much more frequently on bones which had not been eroded or gnawed (Table 1). Consequently, much fewer records of butchery were made in the generally less well preserved assemblages from the Romano-British deposits than the Middle Iron Age assemblages. If relative frequencies of butchery observations are to be used as guidelines to the intensity of butchery, any calculations must take taphonomic factors into account.

SOME SPECIFIC EXAMPLES OF QUANTIFICATION PROBLEMS

Butchery marks near the proximal articulation of the metatarsus

Differential preservation may bias the frequency of observations of cut marks on different parts if the same bone. In the Iron Age samples from Hampshire, where disarticulation of the limb bones was usually achieved using a fine-bladed knife, most cut marks are found close to the articulations of the limb bones, the same areas most susceptible to canid gnawing (Binford 1981: 35-86; Payne this volume). The frequency of surviving knife cuts is therefore largely dependent upon the amount of subsequent scavenging.

For example, knife cuts and superficial chop marks are sometimes found close to the proximal articulation of cattle metatarsi. These were most probably produced during the disarticulation of these bones from the upper hindlimb. In assemblages of Iron Age and Romano-British date these marks have been found in varying numbers on all aspects (anterior, medial, lateral, posterior) of the bone. At the Iron Age settlement at Abingdon, Oxfordshire (SU 483973) most of the fine knife cuts were situated on the anterior of the metatarsi with others also on the medial and lateral aspects, as illustrated by Wilson (1978:122). These correspond with Binford's (1981:119-20) ethnographic observations of cut marks produced during dismemberment of the metatarsus from the upper hindlimb. The majority of knife cuts on cattle metatarsi in the Middle Iron Age assemblages at Old Down Farm (SU 356465)(Maltby 1981:151), Winnall Down (Maltby n.d.1) and Balksbury (SU 350445) (Maltby in prep.), in Hampshire were also located mainly on the anterior aspect. In contrast, on several Romano-British sites such knife cuts or superficial chop marks were more frequently located on the posterior projections of bone towards the medial and lateral aspects. Examples of such butchery have been found at Exeter (Maltby 1979: 11), Cirencester (Thawley 1982b; Maltby in press 1) and Silchester (Maltby n.d.3). In some assemblages these marks were mutually exclusive, on others cuts appeared in varying numbers on both the posterior and anterior aspects (although rarely if ever on the same specimen). Whatever the reasons for these variations, it is obviously desirable to obtain some quantitative assessment of the relative frequency of the different locations of the cutmarks. To do so, the possible effects of subsequent scavenging must be taken into account. Observations of gnawing damage on the proximal portions of cattle metapodia from archaeological assemblages indicate that the posterior portion of the articulation was usually attacked first. In some cases this has resulted in the destruction of the posterior half of the articulation surface (along with any evidence of butchery). Only in the more severe cases of gnawing is the anterior half also destroyed leaving only the shaft of the bone (see also Binford 1981: 35-86; Payne this volume for discussions of the effects of canid gnawing). Consequently, in an assemblage modified by scavengers, cut marks on the anterior of cattle metatarsi have a better chance of survival. Comparisons of the frequency of cut marks on the anterior and posterior aspects of these bones must take this into account. To do so, recording methods require details of exactly which parts of the proximal articulations have survived. The percentage occurrence of marks on either aspect can then be assessed more accurately. Unfortunately most faunal samples have not been recorded in sufficient detail for such calculations to be made, although recording methods on samples currently being examined from Owslebury (Maltby in prep.) and elsewhere should enable such analysis to take place.

Filleting marks on cattle scapulae

The same care with quantification is required with other types of butchery marks. A common modification of cattle scapulae on Romano-British sites is the removal of the lateral spine by chopping (or occasionally sawing) it along its length starting from towards the distal end (above the glenoid cavity). The process is illustrated by Marples (1974: 123). The reasons for such a procedure have received some attention in the literature. The general consensus has now favoured the theory that these marks were made during butchery rather than boneworking (Luff 1982:100). In which case, the longitudinal chopping is most likely to have been caused during the filleting of the shoulder meat. At Owslebury such chop marks tend to replace longitudinal knife cuts on scapulae, which were the usual evidence for filleting of these bones in the Iron Age (although one or two specimens have both types of cut mark on them - again experimental work would help to clarify the causes of these marks). However, the practice of chopping the spine was by no means consistent on Romano-British sites and it would be of interest to plot the relative frequencies of occurrence of such butchery on scapulae from different settlements. In some instances it was obviously a common procedure. Marples (1974:123) noted that 13 out of 26 scapulae at the fortress of Longthorpe (TL 158977) had such marks. This again raises the question of how such data should be quantified. Given the fragmentary nature of scapulae in many assemblages, it may be misleading to express the number of butchered specimens as a proportion of the total number of fragments, since not all the fragments would have consisted of the relevant part of the bone. For example, among the 143 scapulae fragments from layer 5 in the Tower Street excavations at Cirencester (1980/137 - T. Darvill pers.comm.), only 34 (23.8%) of the specimens bore evidence of chop marks associated with the removal of the spine. However, if one counts only the fragments possessing part of the spine, only 33 specimens did not have butchery recorded on them, increasing the percentage of butchery observations to 50.7%. In addition, 22 of the specimens without butchery marks consisted of very small fragments of the spine, many of which were broken in places where butchery marks may not have been made. Excluding these, the percentage of observed butchery records rises to 75.5%. This last figure gives a better reflection of the high degree of uniformity in butchery techniques observed on cattle scapulae in this extensive midden deposit (Maltby in prep.).

It is essential to make detailed records of fragmentation in order to assess butchery data accurately. Such recording has enabled direct comparisons to be made between the Cirencester data and the butchery patterns observed on cattle scapulae from the early Romano-British deposits at the rural settlement at Cowdery's Down, Hampshire (Maltby n.d.2). In this assemblage only two (6.5%) of 31 scapula fragments bore longitudinal chop marks on the spine. Omitting the specimens lacking the relevant part of the bone, 16 fragments did not have evidence of butchery (increasing the butchery percentage to 11.1%). Of these, 7 consisted of very small fragments of the area of the spine, which were broken at points where it was not clear whether they had been butchered or not. Omitting these, the percentage of scapulae with longitudinally chopped spines rises to 18.2%. This figure can be directly compared with the percentage obtained from the Cirencester data (75.5%) and it shows that, although the Cowdery's Down sample is quite small, it appears that this method of butchery was less frequently used there than at Cirencester. In this case, although the preservation of the Cowdery's Down bones was generally poorer than the Cirencester material, it is not thought that this had a significant bearing on the results. The example demonstrates that provided the recording methods of fragmentation and preservation are of sufficient detail, the frequency of specific butchery marks can be directly compared and the results used to monitor the frequency and distribution of butchery techniques.

Fragmentation of metapodia

Quantification of fragment size also requires careful consideration. Fragmentation can be the result of deliberate butchery, particularly for marrow, weathering, scavenging or trampling. Such ancient breaks need also to be distinguished from the modern breakages that can occur during excavation and transport of animal bones. In this instance some advances in data presentation have already been made. Several reports (e.g. Wall 1980) have illustrated the frequency of different size categories (1/2 a bone, 1/4 of a bone etc.) for each skeletal element using pie diagrams. Such observations do enable the overall picture of bone fragmentation to be appraised. Employing similar size divisions, variations in the degree of fragmentation of cattle metapodia have been observed between Iron Age sites in Hampshire and Romano-British assemblages (Table 2). The Iron Age samples from Winnall Down and Balksbury contained high percentages of complete bones and fragments consisting of c.3/4 of the complete bone. Conversely, the specialised butchery accumulations from Silchester and Cirencester produced no complete bones and were dominated by fragments consisting of only 1/4 of the bone or less. In this case the much greater degree of fragmentation in the Romano-British deposits was due to the deliberate breakage of the bones for marrow, since no gnawing and very little erosion was found on the bones. The metapodia of Middle Iron Age date from Winnall Down and Balksbury were deposited mainly in pits and the bones had survived generally in a good state of preservation. However, there was abundant evidence for the gnawing of metapodia in these deposits, which had destroyed one or both of the articulations. Many of the 3/4-fragments had been reduced to this size by such scavenging and had not previously been broken for their marrow content. In addition, the Early Iron Age and Romano-British assemblages from Winnall Down included a much greater proportion of eroded bones, which also increased fragmentation (Maltby in press 2). Nonetheless, the degree of fragmentation of the metapodia did not attain the very high levels witnessed at Silchester and Cirencester indicating that marrow extraction was less intensive at Winnall Down. This again demonstrates the variability of carcase utilisation on different Romano-British settlements.

Many more examples of variability of butchery marks and fragmentation patterns on cattle bones in these periods could be quoted. Provided that quantitative comparisons take differential preservation and other biasing factors into account, it should be possible to produce a coherent picture of carcase utilisation practices in the periods involved. The recognition of clear patterns of butchery variability on different settlements could add substantially to our knowledge of the scale and form of meat processing and redistribution systems in the Romano-British period in particular.

ASSESSMENTS OF THE DISTRIBUTION OF SKELETAL ELEMENTS

An important consequence of the recognition of variability in carcase utilisation is the need for more careful consideration of how to analyse and interpret the distribution of different anatomical elements within a settlement. Often, for convenience of analysis particularly with small samples, different skeletial elements are combined into categories that correspond to assumed divisions of the carcase (e.g. good or poor quality meat bones, marrow bones etc.).

Although the form and properties of animal carcases offer only a limited series of options for the butcher, it is dangerous to assume that the skeletal elements consistently fall into preconceived categories. In the analysis of animal bones from Exeter, Maltby (1979:7) arbitrarily separated limb bones of "high meat value" (scapula, humerus, pelvis, femur) from those of "moderate meat value" (radius, ulna, tibia). It is clear, however, from subsequent analysis (Maltby in press 2) that the proximal portions of sheep tibiae were in some periods often deliberately broken and separated from their distal halves. The proximal portions were regularly found in association with femora, whereas the distal portions were removed with the bones of the limb extremities. In other cases (e.g. on Iron Age sites in Hampshire) the breakage of sheep tibia was not a common butchery practice and the spatial separation of the proximal and distal portions was less likely to occur. Different butchery traditions may create quite different patterns of disposal of the same skeletal elements.

Similarly, Halstead, Hodder & Jones (1978: 121) suggested that bones of the trunk (ribs, vertebrae, pelves and scapulae) represented meat-bearing cuts and used variations in the relative proportions of these bones compared with skull bones and limb bones to attempt to distinguish between butchery, kitchen and table waste in the Iron Age and Romano-British assemblages at Wendons Ambo, Essex. The sample was small and the relevant butchery evidence may not have been found, but if the same techniques of butchery were practised as found commonly on Iron Age settlements in southern England, the flanks of the animals were often stripped away from the vertebral column and the meat filleted from the thoracic and lumbar vertebrae.

The sections of vertebrae were then sometimes dumped as a unit (Maltby 1981: 151; Wilson 1978: 119). If such a practice was commonplace at Wendons Ambo it would be better to regard such vertebrae as butchery rather than table waste.

It is only by paying regard to the butchery marks and fragmentation patterns of individual bones that a reliable understanding of the type of faunal assemblage being studied can be obtained. It has already been possible to demonstrate that largescale processing of cattle carcases in the Romano-British period can leave quite distinct spatially separated accumulations of particular bones (Maltby in press 1 & 2). Given a better understanding of the range of butchery practices, it may be possible to extend such studies to further our knowledge of carcase processing on a larger scale and thus make better sense of the derivation of the majority of faunal assemblages in archaeological deposits.

CONCLUSIONS

It is clear from the above discussion that successful interpretation of butchery data relies on detailed recording, not only of the location of the cuts but also of the part and the proportion of the bone present, and details of erosion, weathering and gnawing damage. Systems for the accurate recording of butchery exist, either through the use of diagrams (e.g. Wilson 1978: 119-23) or by means of descriptive coding systems (e.g. Jones et.al. n.d.) the latter system also has the facilities to record fragmentation and preservation criteria in sufficient detail for successful quantification of butchery data to be feasible. It has not been the intention of this paper to give a detailed description of the range and variety of butchery marks and carcase utilisation strategies on cattle in the periods under discussion. It is indeed premature for such an enterprise to be undertaken until further evidence has been collated. It is hoped, however, that the demonstration of some of the variability in butchery practices that existed will encourage a more rigorous attitude toward the recording of this data. It has been shown that very simple methods of quantification can, with care, be used to elucidate important information about butchery practices. More sophisticated statistical techniques may be needed once a better understanding of the range of variability has been obtained.

It can be argued that the need to investigate butchery practices is as important as the understanding of the effects of post-depositional taphonomic processes (particularly weathering and scavenging) that have occupied a much greater amount of archaeozoologists' time in recent years. For without it, assessments of carcase utilisation, activity patterning, the importance of different species to the diet, and ultimately the accurate interpretation of developments in the pastoral economy may be suspect.

In addition, if a clear pattern of butchery traditions can be monitored in the Romano-British period, for example, it should be possible to observe how those patterns relate to other aspects of archaeological data. Variations in butchery practices may in part be associated with distribution or trading systems, social traditions or economic status. Such topics have generally been considered to lie beyond the domain of archaeozoological studies, whereas they have sometimes exercised the minds of scholars studying other aspects of Roman archaeology (e.g. in studies of artefacts or settlement patterns). The potential that the development of reliable methods of quantifying butchery affords may therefore have ramifications extending far beyond the normal bounds of faunal studies.

ACKNOWLEDGEMENTS

I would like to thank Dale Serjeantson for permission to refer to unpublished information.

REFERENCES

Ashdown,R & Evans,C (1981). Animal bones. Part I. Mammalian bones. In (C. Partridge) Skeleton Green: a Late Iron Age and Romano-British Site. London: Britannia Monograph Series 2, 205-235.

Binford,L.R. (1978) Nunamiut Ethnoarchaeology. New York/London: Academic Press.

Binford,L.R. (1981). Bones: Ancient Men and Modern Myths. New York/ London: Academic Press.

Halstead,P., Hodder,I. & Jones,G. (1978). Behavioural archaeology and refuse patterns: a case study. Norwegian Archaeological Review 11, 118-131.

Jones,R.T. Wall, S.M. Locker,A.M., Coy,J.P. & Maltby,J.M. (n.d.) Computer-based osteometry: data capture user manual. Ancient Monuments Laboratory Report 3342.

King,A. (1978). A comparative survey of bone assemblages from Roman sites in Britain. University of London: Institute of Archaeology Bulletin 15, 207-232.

Luff,R.M. (1982). A Zooarchaeological Study of the Roman North-western Provinces. Oxford: British Archaeological Reports (International Series) 137.

Maltby,J.M. (1979). Faunal Studies and Urban Sites: the Animal Bones from Exeter 1971-1975. (Exeter Archaeological Reports 2)

Sheffield University: Department of Prehistory and Archaeology Monograph Series 1.

Maltby,J.M. (1981). The animal bones. In (S.M. Davies) Excavations at Old Down Farm, Andover: Part II Prehistoric and Roman. Proceedings of the Hampshire Field Club and Archaeological Society 37, 81-163.

Maltby,J.M. (in press 1). Animal Bones and the Romano-British economy. In (J. Clutton-Brock & C.Grigson,Eds) Animals and Archaeology: Volume 4: Husbandry and the Emergence of Breeds. Oxford: British Archaeological Reports (International Series).

Maltby,J.M. (in press 2). Piecing together the waste: interpreting patterns in faunal assemblage variability. In G.W.W. Barker & C.Gamble (Eds) Beyond Domestication... New York/London: Academic Press.

Maltby,J.M. (n.d.1) The animal bones from Winnall Down (M3 Motorway Archaeological Rescue Committee, Site R17), Hampshire. Ancient Monuments Laboratory Report 3453.

Maltby,J.M. (n.d.2). The animal bones from Cowdery's Down, Basingstoke, Hampshire. Ancient Monuments Laboratory Report 3875.

Maltby,J.M. (n.d.3). The animal bones from the 1974, 1975 and 1978 excavations at Silchester. Ancient Monuments Laboratory Report 3595.

Marples,B.J. (1974). Report on the animal bones. In (S.S. Frere & J.K. St. Joseph) The Roman fortress at Longthorpe.Britannia 5, 122-128.

Serjeantson,D. (n.d.). Animal bones from excavations in the Roman levels and the Anglo-Saxon cemetery at Great Chesterford, 1953-4. Ancient Monuments Laboratory Report 3995.

Speth,J.D. (1983). Bison Kills and Bone Counts: Decision Making by Ancient Hunters. Chicago: University of Chicago Press.

Thawley,C.R. (1982a). The animal remains. In (J.Wacher & A. McWhirr) Early Roman Occupation at Cirencester. Cirencester: Cirencester Excavations I, 211-227.

Thawley,C.R. (1982b). The animal bones. In (A. McWhirr, L. Viner & C.Wells) Romano-British Cemeteries at Cirencester Cirencester Excavations II, microfiche.

Wall,S.M. (1980). The animal bones from the excavation of the hospital of St. Mary of Ospringe. Archaeologia Cantiana 96, 227-266.

Whitehouse,R. & Whitehouse,D. (1974). The fauna. In S.C. Stanford,

Croft Ambrey. Hereford: privately printed, 215-221, 238-242.

Wilson,R. (1978). Methods and results of bone analysis/ General conclusions and discussion of the bone sample. In M. Parrington,(Ed.) The Excavation of an Iron Age Settlement, Bronze Age Ring-ditches and Roman Features at Ashville Trading Estate, Abingdon (Oxfordshire) 1974-76. (Oxfordshire Archaeological Unit Report 1) London: Council for British Archaeology Research Report 28, 110-126, 133-139.

Yellen,J. (1977). Cultural patterning in faunal remains: evidence from the !Kung Bushmen. In D.W. Ingersoll, J.E. Yellen & W.MacDonald,Eds. Experimental Archaeology. New York: Columbia University Press, 271-331.

	N	Eroded/Gnawed But.	NB.	%But.	Undamaged But.	NB.	%But.
Mandible							
EIA	74	2	34	5.6	8	30	21.1
MIA	106	0	46	-	12	48	20.0
R-B	92	5	69	6.8	1	17	5.5
Scapula							
EIA	34	2	17	10.5	7	8	46.7
MIA	51	1	14	7.1	16	20	44.4
R-B	45	1	38	2.6	5	1	83.3
Humerus							
EIA	41	5	22	18.5	6	8	42.9
MIA	55	2	26	7.1	12	15	44.4
R-B	46	3	36	7.7	3	4	42.9
Radius							
EIA	37	2	22	8.3	4	9	30.8
MIA	53	5	17	22.7	8	23	25.8
R-B	38	2	32	5.9	1	3	25.0
Os Coxae							
EIA	41	3	15	16.7	6	17	26.0
MIA	34	2	12	14.3	9	9	50.0
R-B	38	3	26	10.3	1	8	11.1
Metatarsus							
EIA	38	2	29	6.5	3	4	42.9
MIA	33	5	16	23.8	1	11	8.3
R-B	50	2	40	4.8	1	7	12.5

Key: N = number of fragments; But. = number of butchered specimens recorded; NB. = number of specimens with no butchery recorded; %But. = percentage of specimens with butchery recorded; EIA = Early Iron Age: MIA = Middle Iron Age; R-B = Early Romano-British.

TABLE 1: Frequency of Butchery Observations on Selected Cattle Bones from Successive Phases of Occupation at Winnall Down, Hampshire.

Assemblage		Metacarpus (%) C	3/4	1/2	1/4	Total
Winnall Down	EIA	39	35	4	22	23
	MIA	14	38	0	48	21
	R-B	9	22	4	65	23
Balksbury	MIA	21	32	11	36	63
Silchester 1978 1 26		0	5	5	90	22
Cirencester 1980/137 5		0	0	1	99	69

Assemblage		Metatarsus (%) C	3/4	1/2	1/4	Total
Winnall Down	EIA	25	39	0	36	28
	MIA	18	46	4	32	28
	R-B	5	21	7	67	43
Balksbury	MIA	20	27	12	41	56
Silchester 1978 1 26		0	4	18	78	28
Cirencester 1980/137 5		0	0	5	95	190

Key C = complete bone; $\frac{3}{4}$ = c. $\frac{3}{4}$ of bone present; $\frac{1}{2}$ = c. $\frac{1}{2}$ the bone present; $\frac{1}{4}$ = c. $\frac{1}{4}$ or less of the bone present; EIA = Early Iron Age, MIA = Middle Iron Age; R-B = Early Romano-British.

TABLE 2: Percentage of Different size Categories of Cattle Metapodia from Selected Iron Age and Romano-British Assemblages.

RUBY AND HOW MANY SQUIRRELS? THE DESTRUCTION OF BONES BY DOGS.

Sebastian Payne
Research Associate, Program in Classical Archaeology,
Indiana University, Bloomington, Indiana, USA.
9 Wilberforce Road, Cambridge, CB3 OEQ, Great Britain

Patrick J. Munson
Department of Anthropology, Indiana University,
Rawles Hall, Bloomington, IN 47405, USA.

ABSTRACT

A number of experiments were carried out to examine the destruction of bones by dogs, and the possible bias that this may cause in bone samples from archaeological sites. The combined effects of crushing and comminution by the teeth, and of corrosion by the digestive juices, were generally very destructive, affecting weaker parts of the skeleton more than stronger parts, and smaller animals (squirrels and cottontails) more than larger animals (goats). Bones and bone fragments that survived passage through the digestive system were usually visibly corroded. Minimum number estimates may be badly affected: in one experiment the heads and feet of 37 squirrels were fed to Ruby, and the minimum number of individuals calculated from the surviving bones was only 14, based on second phalanges, which survived better than any of the teeth.

INTRODUCTION

The likelihood that faunal samples from archaeological sites may be biased as a consequence of the destruction of bones by dogs and other scavengers has long been recognised by zoo-archaeologists. The work of Steenstrup (1862) on Danish Mesolithic midden sites is an early example of such awareness: Steenstrup recognised that the weaker parts of the skeleton were consistently under-represented in his animal bone samples, and concluded that dogs were probably responsible. This conclusion was supported by ethnographic and experimental observations: Steenstrup examined collections of bones from contemporary Eskimo middens, and fed bones to dogs to see what was destroyed. Steenstrup also recognised that the weaker parts of the skeleton were those which develop later, and suggested that the bones of younger animals were more liable to destruction than those of adults, and the the weaker bones of domestic and marine mammals were more liable to destruction than those of most wild land mammals.

Steenstrup's remarkable paper has been largely ignored by zoo-archaeologists, but the recent surge of interest in taphonomy has stimulated renewed work on the subject. Of particular importance is a series of papers by Brain (1967, 1969, 1981), who has used observations on bone samples from Hottentot villages and modern carnivore dens to re-examine Dart's 'osteodontokeratic' hypothesis, concluding that differences in the relative representation of different antelope skeletal parts at Makapansgat require no explanation in terms of selection and tool-use by Australopithecines, but instead reflect the biases that are to be expected in any bone sample that has been 'chewed-over' by carnivores. Steenstrup's suggestion that the bones of some species may be more liable to destruction than others has been taken further by Lyon (1970), who suggests, on the basis of ethnographic observations on the Wachipaeri in Peru, that dogs may totally destroy the bones of smaller mammals, birds and fishes, whose dietary and economic importance would thus be undetected by future zoo-archaeologists; and though Casteel (1971) argued in response that total destruction is unlikely, he agreed that it is important to recognise the partial and biased nature of the zoo-archaeological record.

In recent years, an increasing number of studies of carnivore kill-sites, carnivore den bone samples, dog pen bone samples and village refuse bone samples have been published; but there appear to have been surprisingly few fully controlled experimental studies with both precise information about what the carnivore started with, and consideration of all bones that were left and could be found by the archaeologist or palaeontologist of the future. Brain (1981), for instance, describing a number of feeding experiments, makes no mention of the bones that may have survived in the faeces after passage through the digestive system.

This paper reports briefly on a series of such experiments. One of the authors (PJM) was responsible for feeding the dogs and collecting the samples; the other (SP) for preparation, identification and analysis. The dogs concerned are two Redbone Coonhounds, both standing about 55 cm (22") at the shoulder and weighing 25-30 kg (55-65 lbs.). At the time of the first experiment, Ruby was about 4 years old; at the time of the other experiments Ruby was about 11 years old while Slick, her son, was about 4 years old.

THE EXPERIMENTS

A. RUBY 1: *Sciurus*.

During the course of one summer and autumn, Ruby was kept in a small enclosure and a record was kept of all the bones and parts of animals that were fed to her. Next spring, the bones that were visible on the surface of the enclosure were collected, and the enclosure was excavated to recover bones down to below the limits of any disturbance. The excavated earth was dry-screened with a 1/4" (*ca.* 6 mm) mesh; 25% of the earth was then water-sieved using window-screening with a mesh of about 1/16" (*ca.* 1.5 mm).

Among the bones fed to Ruby in the course of this experiment were the uncooked heads and feet (including distal radius, ulna, tibia and fibula) of 37 squirrels, and the cooked (some fried, others boiled) limb-bones of 15 of them. Some were fox squirrels (*Sciurus niger*), which usually average *ca.* 750g.; the rest were grey squirrels, which usually average *ca.* 550g.

Table 1 records the bones that were recovered and identified as *Sciurus*. After due multiplication to allow for sub-sampling (only 25% of the earth was water-sieved, and only a quarter of the 1/8"-1/16" fraction (*ca.* 3- .5mm) was sorted), the estimated total number of surviving bones (Table 1 column e) can be compared with the number that were fed to the dog, giving an estimate for the survival rate of different parts of the skeleton (Table 1 column f). As this and Figure 1 show, survival rates were generally low, the highest being for distal second and third phalanges (38%, 34%), incisors (28%) and calcanea (27%). Only 11-12% of cheek-teeth survived, and survival rates for the ends of the long-bones were surprisingly low. The number of squirrels would be seriously underestimated on the basis of the recovered bones (14, as compared with 37 originally fed to Ruby), and a zoo-archaeologist dealing with a sample of this kind from an archaeological context might, on the basis of the relative abundance of teeth and phalanges, have wrongly concluded that this was a 'processing' context, and that all the meat-bearing parts had been taken elsewhere. Without water sieving (many sites are still excavated without even token water-sieved samples) the number of squirrels would have been still more seriously underestimated: only 16 identified specimens were recovered by trowelling and dry-screening (Table 1 column b), of which 10 were incisors; and the relative abundance of squirrel phalanges would have gone undetected.

This experiment can reasonably be criticised on several grounds. First, most of the long-bones were cooked whereas the heads and feet were not, and it might be thought that the poor survival of the long-bones reflects this difference. Survival, however, was equally poor for the uncooked long-bone ends - around 1% (4/296) for distal radius, ulna, tibia and fibula taken as a group. A more serious criticism would be that the bones lay for some months, through an Indiana winter, at or near the surface of a rather acid soil, and thus bone destruction might have been caused by surface weathering and acid attack as well as by Ruby herself.

B. RUBY 4-9 AND SLICK 1-2.

A second series of experiments was therefore carried out in more controlled conditions. In each experiment, one animal was fed to the dog, rejected bones were monitored and collected once the dog showed no further interest for several days, before they could be trodden into the ground or buried (the dogs were tethered on short grass where any attempt to bury bones would have been detected, and in fact none were buried) and all faeces were collected each day until at least 3-4 days after the last bones were eaten or chewed. The faeces were agitated in water to break them up and macerated at 60° C with the addition of small amounts of Na_3PO_4, after which hair and other floating matter could be carefully decanted off, checking for any floating bones, until a clean bone sample was obtained.

B1 RUBY 4,6 and SLICK 1,2: Capra.

Two adult goats were used in these experiments. In the first pair of experiments (RUBY 4/SLICK 1), an 8-year-old female goat was skinned (apart from the head and feet) and roughly stripped of most of the meat. The stripped carcass was then disarticulated into units, and the skull and pelvis were split down the midline. The bones of the left side (left side of skull, mandible, forelimb and hindlimb) were fed to Ruby over a period of about 10 days (during which she also received her normal* processed food), while the bones of the right side, together with the axis and atlas vertebrae and about ten ribs, were fed to Slick. The remainder of the ribs and vertebrae and the sternum were fed to other dogs. In the second pair of experiments (RUBY 6/SLICK2), a 5-year-old female goat was similarly treated, except that only the limbs were fed and not the head and other parts, and the dogs were kept a little hungrier to see whether they would destroy more of the bones.

The results of these experiments are given in Table 2. The bones and bone fragments that were not swallowed by the dogs (Table 2 columns b and h) were generally in good condition, including complete metapodia as well as numbers of long-bone shaft splinters and bone chips (Plate 1). Some showed clear traces of gnawing: both mandibles, for instance, had lost ascending rami and much of the angle, and several of the long-bones showed clear tooth-marks and gnawing damage; but most of the bone fragments showed no definite traces of the activity of the dogs. Some of these 'rejected' pieces are fairly large - one radio-ulna is just over 20 cm long; but 90% were under 7 cm long, and nearly half were bone chips less than a centimetre in length (Table 5).

The very large number of bone fragments that were recovered from the faeces were all small: the largest was less than 5cm. long, and over 80% were less than a centimetre in length (Plate 2). The average weight of bone fragments that had passed through the dogs was 0.05g, as compared with 3.63 g in the case of the "rejected" bones.
The surfaces of the bone fragments from the faeces were typically in poor condition: many of the bones were very corroded - articular surfaces had sometimes been completely removed, exposing the cancellous bone below - and no tooth-marks remained. It was clear that two destructive processes had affected the bones: first they

* i.e. "boneless"

had been crushed and comminuted by the dogs' teeth and jaws, and then they had suffered corrosion by the digestive juices (Plates 3, 4). (A small group of bone fragments which had been vomited also showed considerable corrosion damage.) Only a small proportion of the resulting mass of tiny corroded bone chips could be identified, and, as indicated, some of the identifications are uncertain (Table 2 columns c and j). Some fragments were only identified in the light of the knowledge of what had been fed to the dogs; for this reason, in an attempt to provide counts that can more fairly be compared with archaeological data, Table 2 columns d and k treat the bones from each goat as a single sample, counting only those specimens which would have been identified to the 'sheep/goat' level in a normal archaeological sample. This has the effect of eliminating small fragments which can be identified to anatomical part but which might be from almost any mammal of similar size, such as small pieces of distal femur or proximal tibia. Counted in this way, the first goat was reduced by the dogs to 62 identified fragments (15 cranial and 47 postcranial), while the second goat, fed when the dogs were kept hungrier, was reduced to 29 postcranial fragments (the second goat's head was not given to the dogs). These totals are still further reduced by eliminating specimens of which less than half is preserved, as one might do to avoid the possibility of double-counting (Table 2 columns e and m).

One measure of bone survival is given by Table 2 columns o and p. Survival is high for the teeth (though one upper premolar has gone), high for early-fusing long-bone ends (e.g. distal humerus, proximal radius, distal tibia), moderate for metapodia, carpals and tarsals, and relatively low for late-fusing long-bone ends, scapula, pelvis and phalanges.

In Figure 1, the survival rates for selected elements are compared with those in Brain's Hottentot village sample and with those calculated for the Makapansgat sample. The overall pattern is clearly similar, despite the small size of the sample given by the present experiment. While survival rates for some parts of the skeleton appear to be comfortingly high, two notes of caution should be sounded. In the Makapansgat and Hottentot village samples, we have no way of telling how many animals were initially present; survival rates have been calculated on the basis of the minimum number of individuals given by the commonest element, and thus depend on the initial assumption that all specimens of the commonest element have survived. If not, true bone survival rates may in fact be much lower - these are relative survival rates rather than absolute rates. In the present experiment the rates given are absolute rates since we know how many animals we started with, but 1) some of the counted specimens are only fragments: no postcranial survival rate is above 75% if small fragments are eliminated (see Table 2 columns e and m), 2) a number of the faecal specimens have been so weakened by digestive corrosion that they would easily be destroyed by any further mechanical or chemical attack (e.g. trampling or soil corrosion), and 3) both goats were adults, with fully-fused postcranial skeletons; archaeological samples often include juvenile specimens, whose survival rates are probably much lower. In another experiment (RUBY 3) two juvenile *Capra* heads were fed to Ruby: the maxillae and mandibles were completely destroyed, and only about half the milk teeth survived.

B2. RUBY 5,7: *Sylvilagus*.

In these experiments, two adult cottontails (*Sylvilagus floridanus* (Allen), a lagomorph about the same size as a European wild rabbit, weighing 1-1.5kg) were fed to Ruby. Both were road casualties with damaged skulls (no teeth were missing), and were neither skinned nor gutted. Both were entirely consumed, and none of the bones were rejected (Tables 3,5). Fragmentation was extreme: each animal was reduced to about 300 bone scraps with an average weight of 0.01 g, all less than 3 cm long. The total weight of bone that survived (3.0 g, 4.3 g) is only 5-10% of the original weight of the skeleton (four complete skeletons in the Indiana University (Bloomington) Zooarchaeology collection weigh from 48.6 to 63.7 g). The degree of fragmentation is such that few of the bones would be likely to be recovered by excavation without water-sieving: all the identified specimens passed through an 8 mm mesh, and only a few were retained by a 4 mm mesh (Table 3 columns e,f,j and k).

Survival rates (Table 3 columns l, m) were generally low. The highest rate (50%) is given by the astragalus, while rates for the teeth (4-9%) are notably low. The relatively high survival rate for the distal third phalanges is caused by their protection within the keratin of the claw, which is relatively unaffected by digestion: several claws were found with the third phalanx still in place, but with its proximal end destroyed by digestive corrosion (Plate 5). Survival rates (Figure 1) are much lower in general than for *Capra*, and the pattern of skeletal part survival appears to be significantly different, even though samples are small. Teeth, in particular, survive poorly in *Sylvilagus* but well in *Capra*. This may reflect the relative sizes of the animals involved: Ruby could easily take a cottontail's head between her jaws and crunch it, while an adult goat maxilla or mandible is too large, and has to be gnawed from the edges.

B3 RUBY 8,9: *Sciurus*.

In these experiments, two squirrels were fed to Ruby. The first, a grey squirrel (RUBY 8), was neither skinned nor gutted; the second, a fox squirrel (RUBY 9), was skinned and gutted before being given to the dog. As with the cottontails, everything was swallowed (Tables 4,5).

In RUBY 8, bone destruction was even greater than for the two cottontails: only 57 fragments of bone survived which, with a total weight of 0.7 g, represent only 3-4% of the original weight of the skeleton (four skeletons of Sciurus carolinensis Gmelin in the IU(B) collection weigh from 18.1 to 26.1 g). The pattern of bone survival is similar to that in RUBY 1: most of the specimens which could be identified were teeth and phalanges, many of which clearly showed the effects of digestive corrosion, and again there were third phalanges whose distal ends had been protected by the claw.

In RUBY 9 bone destruction was less: 6.3 g of bone survived, which is around 15-20% of the original weight of the skeleton (three skeletons of Sciurus niger (Linnaeus) in the IU(B) collection weigh from 33.4 to 43.0 g), and many more specimens could be identified. Again, many of the bones showed the effects of digestive corrosion, especially the long-bone ends. The reasons for the differences between the results of the two experiments are not clear - one possibility is that the dog bolted the second squirrel more quickly, with less chewing and bone breakage, while another is that its digestive juices were for some reason less corrosive.

The pattern of survival shown by these squirrels and by RUBY 1 (Figure 1) again clearly differs from that shown by Capra, and survival is generally poorer. Survival rates are low for teeth and jaws, as in Sylvilagus, but relatively higher for phalanges. As squirrel feet are relatively small, it may be that they are often bolted whole, with little or no chewing, and this may give the phalanges some measure of protection.

DISCUSSION

We would stress that these experiments are only a small beginning: larger samples are needed, using a variety of dogs, before one can feel any real confidence in the extent to which the patterning observed can be expected to be repeated, and can the results have any predictive value. Many other variables also need to be examined - for instance the relative survival of bones of older and younger individuals, and the effects of different kinds of cookery on bone survival.

Limited though they are, however, these experiments add to the body of evidence that indicates that the combined effects of dogs' teeth and digestions may destroy large numbers of bones, and may be a very significant factor in the biasing of archaeological samples. As has been previously pointed out (Steenstrup 1862, Brain 1982), harder and more resistant parts of the skeleton can be expected to survive better, to which should probably also be added the effects of the relative size of the dog and the bone or animal involved. Greater destruction is to be expected for bones that are small enough to be crunched and at the same time worth crunching, such as goat phalanges and rabbit-sized long-bones, while survival is probably better for bones that are both large and strong, such as adult cow bones and adult goat mandibles, and for bones that are small, strong and have little to offer such as squirrel phalanges. Probably all that can be done to compensate for this destruction is to be aware of the problem in interpreting faunal data; though some indication of the probable seriousness of such biasing in a sample may be given by the frequency of dog bones, of bones showing traces of dog-gnawing, and of bone fragments affected by digestive corrosion. (Digestive corrosion can be distinguished from soil corrosion by the fact that it will only affect bones of a certain size - the size that a dog can comfortably swallow. The common presence of such corroded fragments may in fact provide one source of evidence for early dog domestication - though scavenging wild carnivores will also leave scats on sites; cave sites in particular are often used as pens by wild carnivores.) ~~Deer phalanges which appear to have passed through dogs are identified and illustrated by Guiday (1971, Figure 7k) in a report on bones from a 17th-century Indian village in West Virginia.~~

At the same time it is encouraging that in all these experiments some identifiable bones survived, and this was equally true of another experiment in which two fishes were fed to Ruby (RUBY 4A) (Jones 1984). This then indicates that while dogs may seriously bias archaeological samples, they are unlikely to destroy all trace of any significant vertebrate food resource as long as the excavations include areas where dogs defecated, and as long as reasonable samples are water-sieved.

ACKNOWLEDGEMENTS

We are grateful to W.R. Adams, Director of the Zoo-Archaeology Laboratory, Indiana University, Bloomington, for kindly allowing us to use his laboratory and comparative collection.

REFERENCES

Brain, C.K. 1967. Hottentot food remains and their bearing on the interpretation of fossil bone assemblages. Scientific Papers of the Namib Desert Research Station, 32, 1-11.

Brain, C.K., 1969. The contribution of Namib Desert Hottentots to an understanding of australopithecine bone accumulations. Scientific Papers of the Namib Desert Research Station, 39, 13-22.

Brain, C.K. 1981. The hunters or the hunted? An introduction to African cave taphonomy. University of Chicago Press.

Casteel, R.W. 1971. Differential bone destruction: some comments. American Antiquity, 35, 466-469.

Guilday, J.E. 1971. Biological and archaeological analysis of bones from a 17th century Indian village (46 PU 31), Putnam County, West Virginia. Report of Archaeological Investigations 4, West Virginia Geological and Economic Survey, Morgantown, West Virginia, USA.

Jones, A.K.G. 1984. Some effects of the mammalian digestive system on fish bones. In Desse-Berset, N. (ed), 2nd Fish Osteoarchaeology meeting. C.N.R.S. Centre de Recherches Archaeologiques. Notes et Monographies Techniques 16, Paris; 61 - 65.

Lyon, P.J. 1970. Differential bone destruction: an ethnographic example. American Antiquity, 35, 213-215.

Steenstrup, J.J. 1862. Et Blik paa Natur- og Oldforskningens Forstudier til Besvarelsen af Sporgsmaalet om Menneskeslaegtens tidligste Optraeden i Europa. Inbydelsesskrift til Kjøbenhavns Universitets Aarsfest til erindring om Kirkens Reformation. Kjøbenhavn.

TABLE 1: RUBY 1: Sciurus

Part of skeleton		Number fed	Recovered by trowelling and coarse dry-sieving to ¼"	Recovered by wet-sieving to 1/8" (25% sample)	Recovered by wet-sieving to 1/16" (6.25% sample)	Estimated total present: b + 4c + 16d	Survival rate: e/a
		a	b	c	d	e	f
Maxilla		74	0	0	0	0	0%
Mandible		74	(1)	(1)	0	(5)	(7%)
Cheek-teeth		592-666*	0	6	3	72	11-12%
Incisors		148	10	8	0	42	28%
Scapula		30 cooked	0	0	0	0	0%
Humerus	p	30 cooked	0	0	0	0	0%
	d	30 cooked	0	?1	0	?4	0-13%
Radius	p	30 cooked	0	0	0	0	0%
	d	74	0	0	0	0	0%
Ulna	p	30 cooked	0	0	0	0	0%
	d	74	0	0	0	0	0%
Innominate		30 cooked	0	0	0	0	0%
Femur	p	30 cooked	0	0	0	0	0%
	d	30 cooked	0	0	0	0	0%
Patella		30 cooked	0	0	0	0	0%
Tibia	p	30 cooked	0	0	0	0	0%
	d	74	0	0	0	0	0%
Fibula	p	30 cooked	0	0	0	0	0%
	d	74	0	1	0	4	5%
Astragalus		74	1	2	0	9	12%
Calcaneum		74	4	4	0	20	27%
Metapodia	p	666	0	3 } 1	0	16	2%
	d	666	0	7 }	1	48	7%
Phalanx 1	p	740	0	3+?1 } 6	?1	36+?20	5-8%
	d	740	0	8 }	5	136	18%
Phalanx 2	p	666	0	0 } 7	1 } 3	92	14%
	d	666	0	4 }	10 }	252	38%
Phalanx 3	p	740	0	0 } 23	0 } 1	108	15%
	d	(740)	0	(3) }	(8) }	(248)	(34%)

* Some of the Sciurus were S. niger, which has one upper premolar, while others were S. carolinensis, which has two.

Note: In Tables 1-4, the following are included in the counts: teeth when at least half the tooth is present, maxilla and mandibles when containing at least one countable tooth, and the listed postcranial parts when at least part of the relevant articulation (or its fusion surface) is present. Counts are also given in parentheses, but not included in column totals, for edentate jaws, teeth which are still in the jaw, and distal third phalanges.

TABLE 2: RUBY 4, 6 and SLICK 1, 2: <u>Capra</u>

	RUBY 4 and SLICK 1						RUBY 6 and SLICK 2							RUBY… a… SLICK…
	Number fed	Rejected	In faeces	Total if archaeological sample	Total of which more than half preserved	Total held by 8 mm mesh	Number fed	Rejected	In vomit	In faeces	Total if archaeological sample	Total of which more than half preserved	Total held by 8 mm mesh	Poten… archaeo… survi… (d+k)/…
	a	b	c	d	e	f	g	h	i	j	k	m	n	o
Part of skeleton														
Maxilla	2	3	0	3	2	3	0							3/2
Maxillary teeth	12	(9)+1	1	(9)+2	(9)+2	(9)+2	0							(9)+2/12
Mandible	2	2	0	2	2	2	0							2/2
Incisors	8	8	0	8	8	0	0							8/8
Mandibular cheek-teeth	12	(12)	0	(12)	(12)	(12)	0							(12)/12
Scapula	2	0	0	0	0	0	2	0	0	0	0	0	0	0/4
Humerus p	2	0	0	0	0	0	2	0	0	0	0	0	0	0/4
d	2	2	0	2	2	2	2	1	0	?1	1	1	1	3/4
Radius p	2	0 }1	0	0 }1	1	0 }1	2	1 }1	0	0	1 }1	0 }1	1 }1	4/4
Ulna p	2	0 }1	0	0 }1	0 }1	0 }1	2	0	0	1+?1	?2	0	?2	3+?2/4
Radius d	2	0	?2	0	0	0	2	0	0	3+?1	1+?2	0	1+?2	2+?2/4
Ulna d	2	0	0	0	0	0	2	0	0	0	0	0	0	1/4
Carpals	12	0	10	9+?1	9+?1	9+?1	12	0	0	7+?1	6+?1	6+?1	6+?1	15+?2/24
Innominate	2	1	2	1+?1	1	1+?1	2	0	0	4	1+?2	0	1+?1	2+?3/4
Femur p	2	0	2	?2	0	?2	2	0	1	0	1	0	1	1+?2/4
d	2	0	1	0	0	0	2	0	?3	?1	0	0	0	0/4
Patella	2	0	2	2	2	2	2	0	1	0	1	1	1	3/4
Tibia p	2	0	2	0	0	0	2	0	0	?5	0	0	0	0/4
d	2	1	1	2	2	2	2	1	2	1	2+?1	1	2+?1	4+?1/4
Astragalus	2	2	0	2	2	2	2	0	0	1	?1	0	?1	2+?1/4
Calcaneum	2	1	1	2	1	2	2	0	0	1	?1	0	?1	2+?1/4
Other tarsals*	8	4	2+?1	6	6	5	8	0	0	5	2+?3	2+?2	2+?1	8+?3/16
Metapodia p	4	2 }2	1+?1	2+?1 }2	0 }2	2 }2	4	2 }1	0	0	1+?1 }1	0 }1	1+?1 }1	6+?2/8
d	4	0	3	2+?1	1	2+?1	4	0	0	0	0	0	0	5+?1/8
Proximal sesamoids	16	4	2	6	6	2	16	0	0	9+?1	5+?4	5+?4	0	11+?4/32
Phalanx 1 p	8	0 }2	2+?1	?2 }2	?1 }2	?2 }2	8	0	0	2+?1 }1	2 }1	2 }1	2 }1	5+?2/16
d	8	0	1+?2	?1	0	?1	8	0	0	1+?1	?1	0	0	3+?2/16
Phalanx 2 p	8	0	3 }?1	1	1	1	8	0	0	2+?1	1+?1	1+?1	1+?1	2+?1/16
d	8	0	?1	0	0	0	8	0	0	1+?3	1	1	1	1/16
Distal sesamoids	8	0	2	2	2	0	8	0	0	1	?1	?1	0	2+?1/16
Phalanx 3 p	8	0	2+?1	2+?1	2	2	8	0	0	0	0	0	0	2+?1/16
d	(8)	0	0	0	0	0	(8)	0	0	0	0	0	0	(0/16)
Total identified frs.		37	40+?10	62+?10	57+?2	47+?8		7	4+?3	40+?17	29+?21	23+?9	24+?12	
Total bone frs.		165	4822	4987				102	35	3021	3158			
Total bone weight (g)		701.3	252.3	953.6				267.5	14.4	147.7	429.6			

* including lateral malleolus

Note: Sometimes, and especially in columns o and p, the figures appear to suggest that more bones were recovered and identified than were originally fed to the dogs. This is because, as in many bone reports, the counts include <u>all</u> identified fragments, and one broken bor… may thus be counted more than once. Columns e and m give counts after eliminating small fragments.

TABLE 3: RUBY 5, 7: *Sylvilagus*

	Number in skeleton	RUBY 5: Rejected	RUBY 5: In faeces	RUBY 5: Total if archaeological sample	RUBY 5: Total held by 8 mm mesh	RUBY 5: Total held by 4 mm mesh	RUBY 7: Rejected	RUBY 7: In faeces	RUBY 7: Total if archaeological sample	RUBY 7: Total held by 8 mm mesh	RUBY 7: Total held by 4 mm mesh	RUBY 5 and RUBY 7: Potential archaeological survival: (d+i)/2a	
	a	b	c	d	e	f	g	h	i	j	k	l	m
Part of skeleton													
Maxilla	2	0	0	0	0	0	0	0	0	0	0	0/4	0%
Maxillary teeth	16	0	2	2	0	0	0	1	1	0	0	3/32	9%
Mandible	2	0	0	0	0	0	0	0	0	0	0	0/4	0%
Mandibular teeth	12	0	1	1	0	0	0	0	0	0	0	1/24	4%
Scapula	2	0	2	1+?1	0	1+?1	0	?1	0	0	0	1+?1/4	25-50%
Humerus p	2	0	0	0	0	0	0	1	?1	0	?1	?1/4	0-25%
d	2	0	?1	0	0	0	0	0	0	0	0	0/4	0%
Radius p	2	0	1	?1	0	0	0	1	1	0	0	1+?1/4	25-50%
d	2	0	0	0	0	0	0	0	0	0	0	0/4	0%
Ulna p	2	0	1	0	0	0	0	1	1	0	0	1/4	25%
d	2	0	0	0	0	0	0	0	0	0	0	0/4	0%
Carpals	20	0	?1	0	0	0	0	0	0	0	0	0/40	0%
Innominate	2	0	0	0	0	0	0	0	0	0	0	0/4	0%
Femur p	2	0	0	0	0	0	0	0	0	0	0	0/4	0%
d	2	0	0	0	0	0	0	0	0	0	0	0/4	0%
Patella	2	0	1	0	0	0	0	1	?1	0	0	?1/4	0-25%
Tibia p	2	0	0	0	0	0	0	0	0	0	0	0/4	0%
d	2	0	1	?1	0	0	0	0	0	0	0	?1/4	0-25%
Fibula p	2	0	0	0	0	0	0	0	0	0	0	0/4	0%
Astragalus	2	0	1+?1	1	0	1	0	1	1	0	1	2/4	50%
Calcaneum	2	0	0	0	0	0	0	1	?1	0	0	?1/4	0-25%
Other tarsals	12	0	0	0	0	0	0	1	1	0	0	1/24	4%
Metapodia p	16	0	0	0	0	0	0	?1	0	0	0	0/32	0%
d	16	0	6	6	0	0	0	1	1	0	0	7/32	22%
Phalanx 1 p	16	0	0	0	0	0	0	2	1+?1	0	0	1+?1/32	3-6%
d	16	0	1	0	0	0	0	0	0	0	0	0/32	0%
Phalanx 2 p	18	0	0 } 5	0 } 4+?1	0	0	0	0	0	0	0	4+?1/36	11-14%
d	18	0	2 }	1 }	0	0	0	0	0	0	0	5+?1/36	14-17%
Phalanx 3 p	18	0	0 } 1	0 } 1	0	0	0	0 } 2	0 } 2	0	0	3/36	8%
d	(18)	0	(2) }	(2) }	0	0	0	(5) }	(5) }	0	0	(10/36)	(28)%
Total identified frs.		0	25+?3	17+?4	0	2+?1	0	13+?2	9+?4	0	1+?1		
Total bone frs.		0	339	339	0	12	0	272	272	2	26		
Total bone weight (g)		0	3.0				0	4.3					

TABLE 4: RUBY 8, 9: Sciurus

	Number in skeleton	RUBY 8: Sciurus carolinensis					RUBY 9: Sciurus niger					RUBY 8 and RUBY 9	
		Rejected	In faeces	Total if archaeological sample	Total held by 8 mm mesh	Total held by 4 mm mesh	Rejected	In faeces	Total if archaeological sample	Total held by 8 mm mesh	Total held by 4 mm mesh	Potential archaeological survival: (d+i)/2a	
	a	b	c	d	e	f	g	h	i	j	k	l	m
Part of skeleton													
Maxilla	2	0	0	0	0	0	0	2	2	0	(2)	2/4	50%
Maxillary incisor	2	0	2	2	0	0	0	1	1	0	0	3/4	75%
Maxillary cheek-teeth	8/10*	0	2	2	0	0	0	1+(7)	1+(7)	0	0	3+(7)/18	56%
Mandible	2	0	0	0	0	0	0	1	1	0	(1)	1/4	25%
Mandibular incisor	2	0	2	2	0	0	0	0	0	0	0	2/4	50%
Mandibular cheek-teeth	8	0	0	0	0	0	0	2+(1)	2+(1)	0	0	2+(1)/16	19%
Scapula	2	0	0	0	0	0	0	2	1	0	1	1/4	25%
Humerus p	2	0	0	0	0	0	0	1	1	1	1	1/4	25%
d	2	0	1+?1	?1	0	0	0	2	1+?1	1	1+?1	1+?2/4	25-75%
Radius p	2	0	1	?1	0	0	0	2	1+?1	0	1+?1	1+?2/4	25-75%
d	2	0	0	0	0	0	0	2	2	0	0	2/4	50%
Ulna p	2	0	0	0	0	0	0	1	1	0	1	1/4	25%
d	2	0	0	0	0	0	0	1	?1	0	0	?1/4	0-25%
Carpals	18	0	0	0	0	0	0	4+?1	4	0	0	4/36	11%
Innominate	2	0	0	0	0	0	0	1	1	0	1	1/4	25%
Femur p	2	0	0	0	0	0	0	1	?1	0	?1	?1/4	0-25%
d	2	0	?1	0	0	0	0	1	?1	?1	?1	?1/4	0-25%
Patella	2	0	0	0	0	0	0	0	0	0	0	0/4	0%
Tibia p	2	0	0	0	0	0	0	1	0	0	0	0/4	0%
d	2	0	0	0	0	0	0	1	1	0	1	1/4	25%
Fibula p	2	0	0	0	0	0	0	0	0	0	0	0/4	0%
d	2	0	0	0	0	0	0	0	0	0	0	0/4	0%
Astragalus	2	0	0	0	0	0	0	1	1	0	1	1/4	25%
Calcaneum	2	0	0	0	0	0	0	3	?1+?1	0	1	1+?1/4	25-50%
Other tarsals	14	0	0	0	0	0	0	9+?1	8	0	0	8/28	29%
Metapodia p	18	0	0	0	0	0	0	10 } 1	8+?2 } 1	0	0	9+?2/36	25-31%
d	18	0	0	0	0	0	0	14 }	10+?4 }	0	0	11+?4/36	31-42%
Phalanx 1 p	20	0	0	0	0	0	0	4 } 2	1+?1 } 2	0	0	3+?1/40	8-10%
d	20	0	3	3	0	0	0	5 }	4 }	0	0	9/40	23%
Phalanx 2 p	18	0	0 } 1	0 } 1	0	0	0	1 } 4	1 } 4	0	0	6/36	17%
d	18	0	3 }	3 }	0	0	0	7 }	7 }	0	0	15/36	42%
Phalanx 3 p	20	0	0 } 7	0 } 7	0	0	0	0 } 16	0 } 15	0	0	22/40	55%
d	(20)	0	(2) }	(2) }	0	0	0	(2) }	(2) }	0	0	(26/40)	(65)%
Total identified frs.		0	22+?2	20+?2	0	0	0	104+?2	83+?13	2+?1	9+?4		
Total bone frs.		0	57	57	0	0	0	441	441	3	36		
Total bone weight (g)		0	0.7		0	0	0	6.3					

* 8 in S. niger, 10 in S. carolinensis

TABLE 5: Fragmentation data: <u>Capra</u> (RUBY 4, 6 and SLICK 1, 2), <u>Sylvilagus</u> (RUBY 5, 7) and <u>Sciurus</u> (RUBY 8, 9)

	Size class: maximum length (cm) 0–1	1–2	2–3	3–4	4–5	5–6	6–7	7–8	8–9	9–10	10–11	11–12	12–13	13–14	14–15	15–16	16–17	17–18	18–19	19–20	20–21	21–22	Total	Total weight (g)	Average weight (g)
Capra																									
Rejected -Identified	1	6	10	4	2	2	3	1	1	1	1	3	3	2	1	.	.	2	.	.	1	.	44		
-All fragments	111	37	28	25	16	13	10	4	6	2	1	3	3	3	1	.	1	2	.	.	1	.	267	968.8	3.63
In vomit -Identified	0	2	1+?2	1+?1	.	.	.	.	.	.	.	.	.	.	.	.	.	.	.	.	.	.	4+?3		
-All fragments	2	20	8	5	.	.	.	.	.	.	.	.	.	.	.	.	.	.	.	.	.	.	35	14.4	0.41
In faeces -Identified	3+?1	46+?18	26+?7	5+?1	.	.	.	.	.	.	.	.	.	.	.	.	.	.	.	.	.	.	80+?27		
-All fragments	6466	1127	221	28	1	.	.	.	.	.	.	.	.	.	.	.	.	.	.	.	.	.	7843	400.0	0.05
Total -Identified	4+?1	54+?18	37+?9	10+?2	2	2	3	1	1	1	1	3	3	2	1	.	.	2	.	.	1	.	128+?30		
-All fragments	6579	1184	257	58	17	13	10	4	6	2	1	3	3	3	1	.	1	2	.	.	1	.	8145	1383.2	0.17
-Identified in archaeological context	1+?3	35+?17	26+?8	8+?1	2+?1	1	2+?1	1	1	1	1	3	3	2	1	.	.	2	.	.	1	.	91+?31		
Sylvilagus																									
In faeces -Identified	34+?4	3+?1	1	.	.	.	.	.	.	.	.	.	.	.	.	.	.	.	.	.	.	.	38+?5		
-All fragments	563	44	4	.	.	.	.	.	.	.	.	.	.	.	.	.	.	.	.	.	.	.	611	7.3	0.01
-Identified in archaeological context	22+?8	3	1	.	.	.	.	.	.	.	.	.	.	.	.	.	.	.	.	.	.	.	26+?8		
Sciurus																									
In faeces -Identified	92+?4	32	2	.	.	.	.	.	.	.	.	.	.	.	.	.	.	.	.	.	.	.	126+?4		
-All fragments	429	67	2	.	.	.	.	.	.	.	.	.	.	.	.	.	.	.	.	.	.	.	498	7.0	0.01
-Identified in archaeological context	73+?12	28+?3	2	.	.	.	.	.	.	.	.	.	.	.	.	.	.	.	.	.	.	.	103+?15		

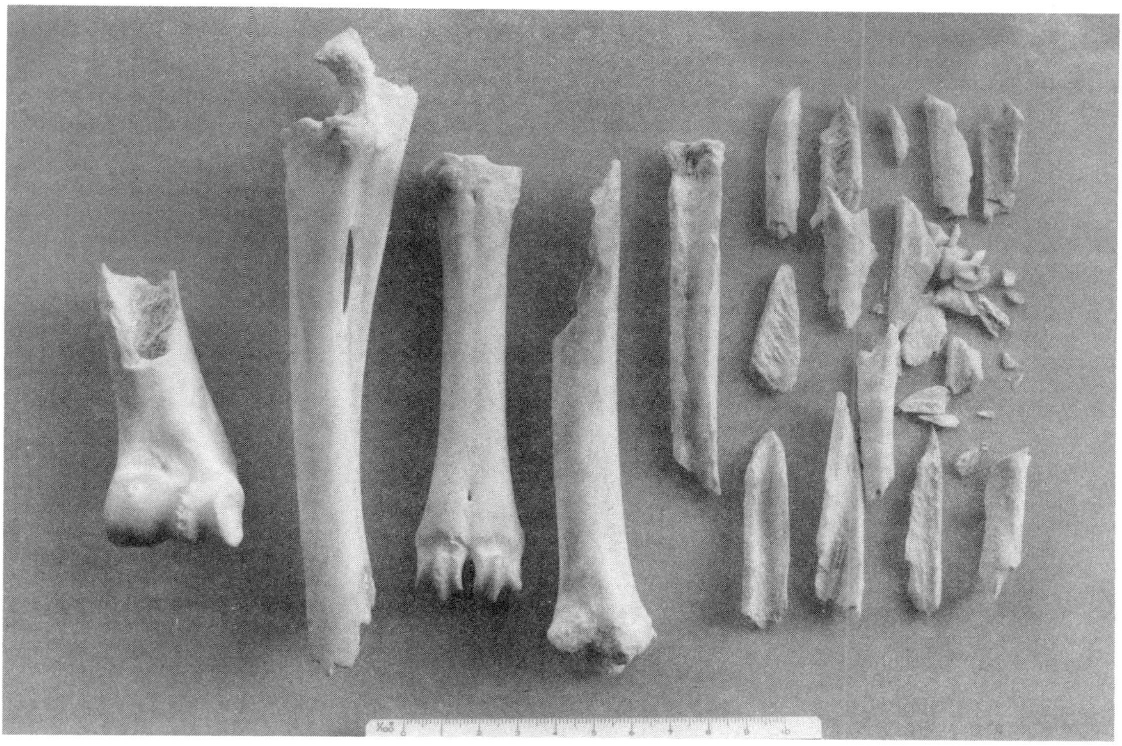

Plate 1.

Plate 2.

Plates 1 and 2: RUBY 6: bones 'rejected' (Plate 1) and recovered from faeces (Plate 2) after Ruby was given a <u>Capra</u> forelimb and hindlimb.

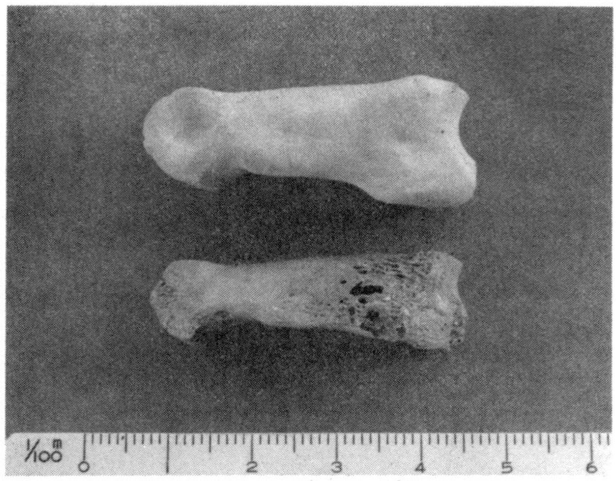

Plate 3: The effects of digestive corrosion on a *Capra* first phalanx (bottom: RUBY 6, from faeces) as compared with one that was 'rejected' (top: RUBY 4).

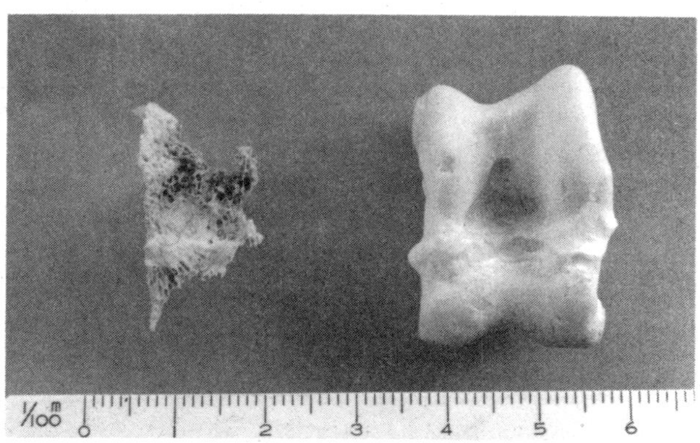

Plate 4: The effects of digestive corrosion on a *Capra* astragalus (left: SLICK 2, from faeces) as compared with one that was 'rejected' (right; RUBY 4).

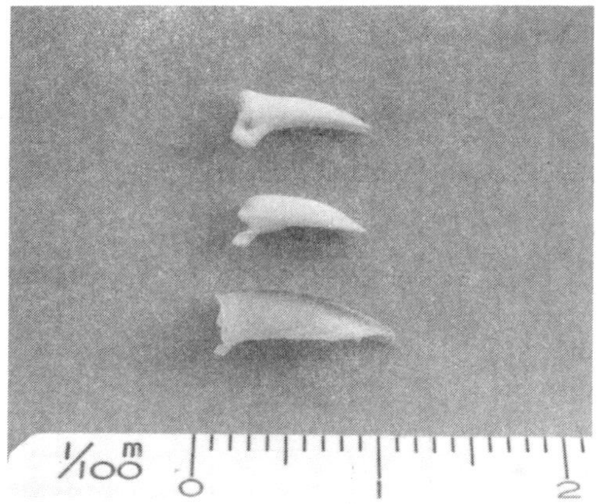

Plate 5: The effects of digestive corrosion on Sylvilagus third phalanges. Top: unaffected; middle: specimen with corroded proximal end (on left); bottom; claw sheath containing and protecting distal part of phalanx. (All from RUBY 7).

Plate 6: RUBY 8; all that was left after Ruby was given a whole squirrel (all from faeces).

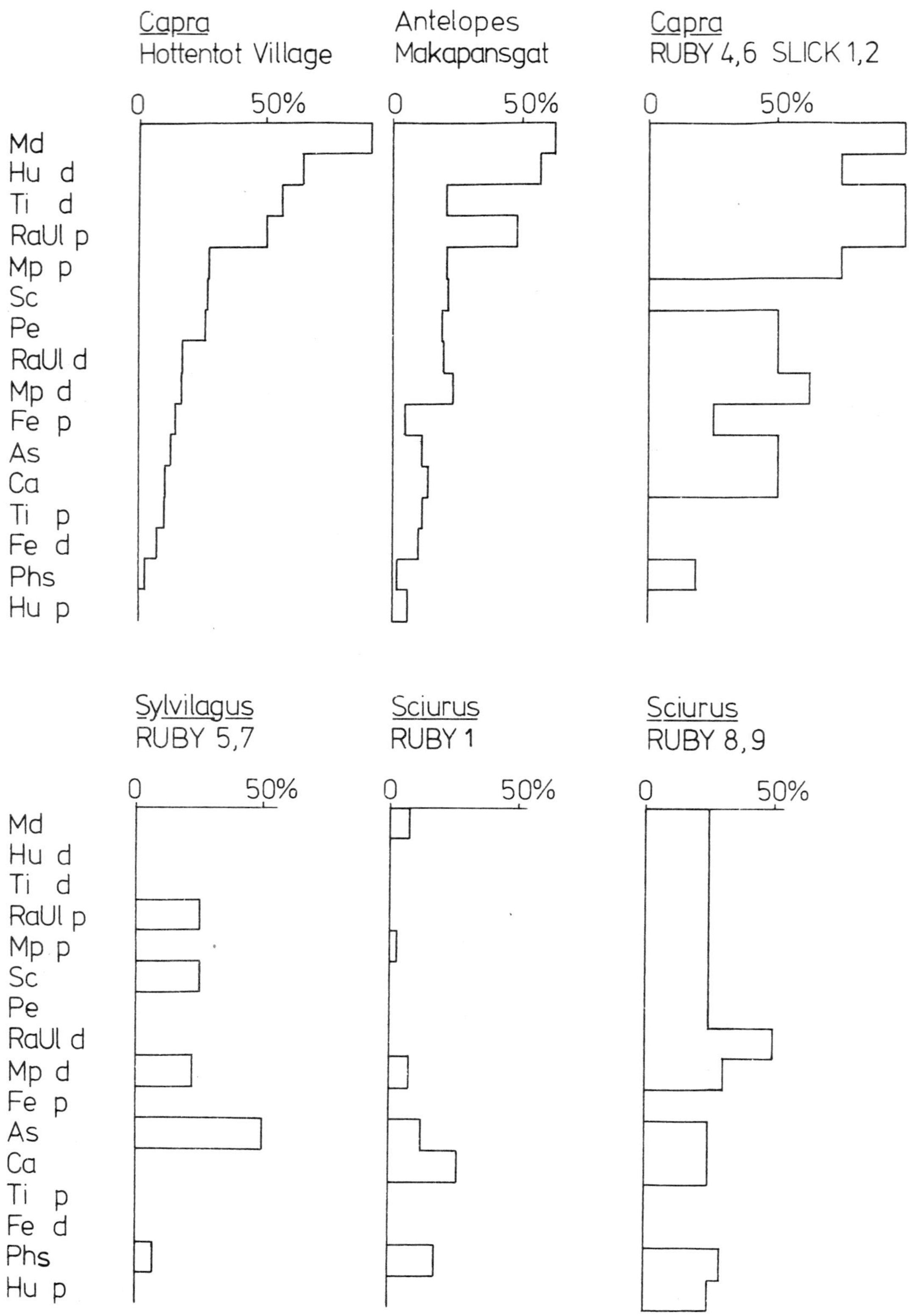

Figure 1. Survival rates for different parts of the skeleton in dog-feeding experiments, compared with data given by Brain (1981) for *Capra* from Hottentot village refuse and for antelopes from Makapansgat.

A METHODOLOGY FOR RECORDING THE PATHOLOGY AND OTHER ANOMALIES OF UNGULATE MANDIBLES FROM ARCHAEOLOGICAL SITES.

BY BRUCE LEVITAN

South Western England Faunal Analysis Project

City Museum and Art Gallery

Bristol

ABSTRACT

Teeth have an intimate contact with the external environment, so there is greater potential in reconstructing details about health and nutrition from teeth and jaws than from other bones. A key aspect in any such study is palaeopathology; this has been recognised by bone analysts, and oral pathology is sometimes quite well described in bone reports. However, little has been done to study this aspect in detail. The aim of this paper is to fill this gap in the analysis of bones from archaeological sites by providing a methodology for recording pathology of the mandibles of ungulates.

The methodology is a modified and improved version of an original devised by the author as undergraduate research. It is based on a sample of 1855 ovicaprid mandibles from the first millennium AD site of West Hill, Uley, Gloucestershire. The anomalies encountered in this sample are described and discussed with reference to modern veterinary studies where possible, and the method of recording them is described and illustrated. The paper concludes with three examples of results obtained from the West Hill assemblage using this methodology.

INTRODUCTION

Mandibles and teeth survive well in the archaeological record in comparison with other bones. Teeth are unique in that they are the only bones with a direct and intimate contact with the external environment. The mouth is an important zone in vertebrate anatomy because it is the entrance for food, the region of mastication, and as such is the area of greatest access for toxins, parasites and other alien material. Thus oral pathology is more common than other forms of bone pathology, and mandibles are more sensitive to change than other bones. Somewhat surprisingly, little attempt has been made to study this topic in detail, despite the obvious potential. One problem has been the lack of a non-subjective and well defined method of recording the anomalies, and this paper seeks to provide such a system. The author made a study of a group of sheep mandibles from the site of Wicken Bonhunt, Essex (TL 334502) as an undergraduate dissertation (Levitan, 1977). This formed the basis for a much more detailed study of a larger group of sheep and goat mandibles from the first millennium AD ritual site of West Hill, Uley, Gloucestershire (ST 792998). The present methodology bears some resemblance to the earlier study, but surpasses and replaces it. Although the methodology has been devised specifically for ungulates, it could be modified to include other groups of vertebrates.

A METHODOLOGY FOR RECORDING MANDIBULAR PATHOLOGY OF UNGULATES

The Anomalies

The recording sheet, Figure 1, is used for scoring the anomalies. Bracketed abbreviations following each sub-heading below refer to the relevant sections of the recording sheet. Figure 2 is a specimen sheet filled out for an imaginary mandible. Many of the features described below and 'recorded' in Figure 2 are illustrated in Figure 3.

i) tooth presence/absence (P/A)

This section may be used to record wear stages and eruption status of teeth where they are present: deciduous (milk) teeth are recorded in the D row, and permanent (adult) teeth are recorded in the M (mature) row. If eruption/wear are recorded elsewhere, presence may be indicated by placing a cross (+) in the relevant box. Absence is recorded in the P row for post-mortem loss, and in the A row for ante-mortem loss. Some authorities designate the deciduous cheek teeth as molars, using the conventions ml, m2, etc, but the more logical system of deciduous premolars (p2, p3 or pm2, pm3, etc) being replaced by permanent premolars (P2, P3 or Pm2, Pm3, etc) is followed here. Molars may be abbreviated using the conventions M1, M2, etc.

ii) fragmentation

Fragmentation is recorded diagrammatically in the P/A section. Thus the mandible illustrated in Figure 3 indicates the zones of breakage recorded in Figure 2. The zone of breakage is indicated by drawing a zig-zag line. Where the break is fresh, the line may be labelled F. When the mandible is severed due to butchery, a straight line is used instead of a zig-zag one.

iii) attrition (ATT)

Dental attrition is due to the opposing forces of growth and wear and is related to physiological, environmental and individual factors. Different species and different diets produce characteristic wear patterns on the occlusal (biting) surface of the teeth. Individual variations occur within a species, and it is important to establish the range of these variations, to define a normal range, and thus identify the anomalies. Sometimes environmental factors may be extremely important, as in the case of ingested soil. The role that this plays in tooth wear has been investigated in a series of studies which have shown that attrition can be extreme where much soil is ingested and supplementary feeding is not carried out. Such considerations have an important bearing on the tooth wear methodologies employed by bone analysts to estimate age, (Healy and Ludwig, 1965; Ludwig, Healy and Cutress, 1966; Healy, Cutress and Mitchie 1967). Even the relatively subtle effects of pasture juices have a part to play in dental attrition due to the varying effects produced on dentine (Cutress and Healy, 1965).

Anomalies of attrition can occur in two ways: anomalies in crown height, and intra-dental attrition. The latter occurs when an adjacent tooth rubs against its neighbour and wears away the enamel sheath of the tooth. It is often a symptom of overcrowding, and recording its frequency and occurrence has an important bearing on this topic. It is scored in the IDA row with an arrow pointing to the right (→) to indicate attrition at the mesial end, and an arrow pointing to the left (←) for attrition of the distal end. It is scored only when the enamel has worn away completely (Plate 1). Overcrowding in modern sheep is recorded by Colyer, particularly in Panir, Rocky Mountain Bighorn and Moufflon (Colyer, 1936: 376).

The upper line of the ATT section (CH) is used for scoring of anomalous crown height. This is achieved by making a diagrammatic sketch of the tooth row. Where a minority of teeth have a lesser height than the rest of the row, a downwards pointing arrow is appended (↓). Conversely, where a minority have a greater crown height, the arrow is reversed (↑). Examples of both are illustrated in Plates 2 and 6. Oligodentia may occur when opposing teeth are lost, the surviving antagonist showing elongation and displacement. Such factors are in any case fairly common on a less severe sense as the dental arcades seldom coincide directly, and may cause conditions such as hook formation on the most mesial and most distal teeth. Restricted chewing practices - what ever the reason - may allow ridges in herbivore teeth to become accentuated and to lead to 'shear mouth' (Jubb and Kennedy, 1963:2). 'Weave mouth' and 'step mouth' occur from different rates of wear in which successive teeth in an arcade wear at different rates (Plate 2). Developmental dystrophies decrease wear resistance. Deficiencies in essential elements such as calcium and phosphorus soften the teeth in growing individuals (Franklin, 1950:28) and there is also severe retardation of enamel and dentine development. Anomalous eruption sequences may be due in some cases to malnutrition in early life, and wear patterns may also be affected (Jensen, 1974:209-210).

iv) periodontal disease (PER)

Various names are used for this condition: periodontitis, pyorrhoea, paradontal disease and periodontal disease. The specific cause of ovine periodontal disease is unknown, but it is presumably due to a mixed species of oral bacteria which penetrate the periodontal membrane. The course of the disease is well known and defined (Jensen, 1974:211). It is characterised by inflammation of the periodontal membrane (Armstrong,1960:429). Introduction of bacteria may be by means of food deposits (dental calculus) which become lodged between the teeth and under the gums, leading in the first instance to infection of the gingival membrane (gingivitis). It should be noted that this is very much the traditional veterinary view, but in fact it may be that calculus has a less important primary role (see below). This stage is followed by recession of the gum and aggravation which leads to infection of the periodontal membrane (Jensen, 1974:212). The bony alveolus may then begin to recede, eventually swelling and enlarging (Plate 4). The tooth loosens and fistulae of the ventral margins may occur in the mandible. In severe cases the tooth may be lost and eventually resorption of the alveolar cavity may occur (Plate 7) (Jubb and Kennedy 1963:5). In modern flocks infection rates may be between two and eleven percent (Hart and MacKinnon, 1958:120; Armstrong, 1960:431).

Scoring for this anomaly is in the form of five severity stages: Stage 1 - characterised by recession of the alveolar margin, especially lingually, but no widening of the alveolus. This is perhaps the most difficult stage to recognise as alveolar recession may be very slight. (Plate 3).

Stage 2 - the alveolus has widened out and the tooth has become loose. Further recession of the margins may occur, and the external surface of the mandible is sometimes pitted and 'porous' in appearance. Teeth may be lost post-mortem at this stage due to the loosening described above and care must be exercised in separating this from ante-mortem tooth loss. (Plate 4).

Stage 3 - the tooth has been shed ante-mortem. The alveolar cavity will be characterised by first stages of infilling and new bone formation - this has not occurred where the tooth has been shed post-mortem. (Plate 5).

Stage 4 - the infilling of the alveolar cavity well advanced but not complete, and a cavity of sorts will still be present. (Plate 6).

Stage 5 - the new bone formation is nearly complete or has been completed. There will be no cavity present, but the surface of the mandible may be pitted and 'porous' in appearance. In some cases infilling may be so successful that no superficial sign of the cavity will be present. (Plate 7)

These severity scores are recorded in the lower line of the PER section. The upper line may be used for notes or sketches.

v) abscess (ABS)

Abscess of the alveolus or pulp cavity may result from a number of infections, some of which may be caused by injury, others from eruption anomalies (especially in piglets) and others from periodontal disease. Whatever the cause, the gross effect is likely to produce similar conditions in the bone, so that it may not be possible to assign a cause in archaeological specimens. In pulpitis, which affects the pulp cavity, the inflammation is usually acute but may be chronic. Only mild forms may heal, most forms terminating in necrosis (Jubb and Kennedy, 1963:4). This may be caused by, or be the cause of periodontal disease! Early stages of abscess will be extremely difficult to recognise unless detailed X-ray coverage is used (Baker and Brothwell, 1980:154). Where X-ray coverage is possible, the following scoring system may be used. Stage 1 - low grade infection: there is obvious disturbance to the pulp cavity and/or alveolus, but this is not discernable from the exterior (X-ray must be used). Stage 2 - medium grade infection: the disturbance has spread to the extent that the condition is obvious externally; there is severe root disfigurement, but the tooth is retained (Plate 6 second molar). Stage 3 - high grade infection: the tooth has been shed ante-mortem (Plate 6 first molar). The scoring of this condition is very problematic because abscess and periodontal disease often go hand in hand. Furthermore, unless X-ray coverage is possible, only the more severe cases will be recorded, and the quantification will thus be biassed. As for periodontal disease, it is scored in the lower line (of the ABS section), the upper line being reserved for notes/sketches.

vi) supernumerary teeth, absent teeth, tooth rotation (S/A.R)

The presence of extra teeth (and extra cusps), absences in teeth, and tooth rotation may all arise from a number of causes including genetic and congenital defects, developmental conditions, disease, etc. Colyer discussed the presence of extra teeth in modern breeds and found 'fourth' molars to be present in two of a sample of 44 domestic sheep. A similar sample of wild sheep had no extra teeth. He also noted that extra incisors sometimes occur (Colyer 1936:120). Such cases probably arise as congenital conditions. If extra teeth do occur, they are recorded with the letter S, followed by a 'designation', e.g. S(M4), in the S/A row.

Teeth sometimes bear extra cusps or pillars ('bovine' pillars). They may occur in sheep and goats (Plate 15), and the presence of 'bovine' pillars on the deciduous fourth premolar is a (not entirely reliable) indication of goat (Plate 10). This can be related to the cusp patterning on the deciduous third premolar which has characteristic sheep and goat patterning (Payne pers. comm.). Note that the mandible in Plate 15c has a 'goat' p4, but a 'sheep' p3! 'Bovine' pillars are recorded with the letter B immediately beneath the R line, under the relevant tooth. Deciduous fourth premolars sometimes have two pillars situated between the main columns of the tooth (Plate 10). In such cases the mesial pillar is labelled B1, and the distal B2.

Absence of teeth is only recorded in this section where the tooth has never been present. In some cases Stage 5 periodontal disease may give the appearance of true absence, but X-rays generally show the infilled alveolus as a 'ghost' of less dense bone. Where true absence has occurred, the letter A is inserted for the appropriate tooth in the S/A row. Such anomalies are most likely to be congenital, and absence of the permanent second premolar appears to be fairly common in ungulates (Andrews and Noddle 1975).

Occasionally teeth may lack cusps or pillars, and the third molar appears to be the tooth most commonly affected in such cases, where the third cusp is absent or ill-formed. This condition is scored with the letter C followed by a designation of the absent cusp, e.g. for third molar described above: C3. This is placed in the S/A line.

Tooth rotation or malalignment is recorded in the R line. Where this occurs it is generally the result of overcrowding, though it may also be due to congenital defects or developmental disorders, etc. (where the tooth erupts with the wrong alignment, or even in the wrong location). It could also be due to disease, trauma, etc. Rotation was observed in modern sheep, but seemed confined mainly to maxillary teeth (Colyer 1936:377). In archaeological specimens, mandibular premolars are sometimes rotated, particularly the second premolar (Plate 8), whilst displacement or malalignment may occur in almost any tooth. Teeth may be rotated lingually or labially (according to the angle of alignment in relation to the mesial edge - thus the first molar (a) in Figure 3 is rotated labially since the mesial end is aligned outwards, and the fourth premolar (b) is rotated lingually, with the mesial end aligned inwards) up to an angle of 90 - I have never seen greater rotation than this. Displacement can be lingual or labial, mesial or distal. The various possibilities are illustrated in Figure 3, which also illustrates the symbols used to record them.

vii) hypoplasia (HYP)

This occurs in the enamel, generally as horizontal banding, and seems to be related to vitamin D deficiency which also leads to impaired calcification of dentine (Baker and Brothwell 1980:143). The defects are most clearly seen on the incisors. They are caused when the ameloblasts involved in enamel production are inactivated.

This is irreversable, so any damage caused is permanent. However, subsequently formed parts of the tooth may be unaffected - hence the banding (Jubb and Kennedy 1963:2). Detailed studies of hypoplasia thus have the potential of indicating periods of ill health or malnutrition during the growth of an individual. It has been found in dogs, pigs and man, and may sometimes occur in herbivores. It is scored with a cross (+) in the LN line for occurrence on the lingual surface, and in the LB line for occurrance labial surface.

viii) dental calculus (CAL)

Dental calculus or tartar is common in old dogs and in horses, cattle and ovicaprids. Apparently it is uncommon in other (domestic) species (Jubb and Kennedy 1963:4) though one would expect it to occur in other herbivores. It is a composite mass of bacteria, fungi, food particles, desquamated epithelium and leucocytes. It is not a dental tissue, but it is closely linked with health and with functioning of the teeth. In ovicaprids and cattle it is recognisable as a black deposit with a metallic sheen. Build-up of deposits may be extremely heavy, and it can be instrumental in periodontal infection, as has been shown, when deposited on the cementum in the gingival sulcus (Jensen 1974:212); but the importance of calculus as a causal agent in periodontal disease should not be taken as fact until more conclusive evidence is available. In fact it would appear from the archaeological evidence that calculus acts as a bacterial reservoir once infection has started, but may not actually cause the disease (Hillson, pers. comm.). In scoring for calculus, the upper line is taken to represent the lingual (LN) aspect of the tooth row, and the lower line to represent the labial (LB) aspect. The extent of the deposit can be sketched in diagrammatically, and the heaviness of the deposit is shown by blacking in for a heavy deposit, and hatching for a light deposit. Where cleaning of the teeth has been too assiduous, all or part of the deposit may have been removed, and although it is sometimes possible to see where the deposit had been and to show its extent, this is a problem which should be given more careful consideration. Plate 1 illustrates a mandible with a heavy deposit of calculus. Beyond noting its presence and extent, it may also be collected in order to extract and identify some of the plant constituents (Armitage 1975).

ix)pigmentation (PIG)

'Normal enamel is never discoloured' (Jubb and Kennedy 1963:4), thus any discolouration of the enamel in the living animal will be anomalous. Such discolouration is due to pigmentation of the dentine which then shows through the semi-transparent enamel. Pigmentation of the dentine may be caused by pulpal haemorrhages, an inflammation (when it is reddish-brown), putrid pulpitis (when it is greyish-green) or icterus (when it is yellowish). So called 'pink tooth' in cattle and pigs is a congenital porphyria - this becomes darker brown in adults (Jubb and Kennedy 1963:4). Its presence in archaeological specimens (Plate 9) can be more problematic since pigmentation of the dentine may occur by chemical staining, particularly in urban waterlogged contexts when the dentine may be stained black. In such cases there is no way of separating artificial and pathological pigmentation. Where such staining is not present - for example where pigmentation only occurs on some teeth from a single deposit or type of deposit - it is recorded as present using a cross (+). This is placed in the upper line where secondary dentine is affected (2ary), and in the lower line when primary dentine is pigmented (1ary).

x) pulp cavity exposure (PCE)

This condition appears to be rare in archaeological specimens. It is due to excessive dental attrition, and occurrence will depend on such factors as softness of dentine, ingestion of soil and malnutrition (Jensen 1974:209-210). In a normal tooth, as soon as wear exposes the dentine, new secondary (or substitution) dentine is deposited to protect the pulp cavity because dentine wears more rapidly than enamel. The secondary dentine is often easily distinguishable where pigmentation has occurred (Plate 9). In time the secondary dentine may completely fill the pulp cavity and cause the death of the tooth. Where there has been extremely rapid wear, or insufficient deposition of secondary dentine, the tooth may wear down to expose the pulp cavity (Plate 9). This is scored using a twofold system. A cross (+) is entered where secondary dentine is present but not completely filling the cavity, and a star (*) is entered when no secondary dentine is present. The PCE section is taken to represent a diagrammatic view of the occlusal surface of the teeth. The upper line represents the labial side in a left-hand jaw, and the lower line the lingual side. This is reversed for a right-hand jaw. Thus if the mesial end of the labial surface of a left-hand tooth is affected, the symbol is placed on the right-hand side of the upper line in the relevant tooth box.

xi) butchery and gnawing/chewing (B.G.)

Scoring for these is not the province of oral pathology, and butchery at least is adequately covered in most bone reports. Inclusion here is in order to provide a complete record of the mandible assemblage. Notes and/or sketches for butchery are recorded in the upper line, and records about chewing/gnawing are placed in the lower line.

xii) nutrient foramena (FOR)

In sheep and goats there is generally one large nutrient (or mental) foramen on the labial side of the mandible, close to the incisors. Occasionally one or more extra, smaller foramena are present, a discontinuous trait which may be very variable, possibly even within a single population. It is worthwhile recording because there is potential in discovering population variability. The foramena may be located mesially to the major foramen, or as far distally as beneath the fourth premolar. They are almost always much smaller, being 1-2mm in diameter. Rarely they are of similar size to the major foramen. They are probably congenital in origin, though sometimes they may form as exit canals for internal infection. In the latter case, associated pathology should make this evident. At present little is known about this phenomenon but it is possible that it may be related to different types, blood lines or even breeds. Until research on modern specimens has resolved this, the problem must remain. Detailed recording of this anomaly will provide both a corpus of information for future work, and a measure - albeit unqualified - of variability within the archaeological sample. Plates 2 and 3 illustrate mandibles with extra nutrient foramena. It is recorded by drawing a circle in the diastema box to represent the major foramen and an arrowed line (or lines) radiating from the circle, and terminating with an arrowhead representing the location of the extra foramen (Figures 2 and 3). Where the major foramen is not present due to fragmentation, the arrowed line is shown emerging from a zig-zag line which indicates the location of the break.

xiii) disturbance of ventral margin (VMD)

Occasionally disturbances of the ventral margin of the mandible occur in the form of small lumps which may be roughened. The most common location is beneath the fourth premolar in juvenile mandibles, through other locations have been observed. These are formed as a reaction to the developing permanent tooth within the alveolus. The largest of the permanent premolars is the fourth, and this accounts for the reason why this is the commonest location of the disturbance, (Plate 10). Presumably it is caused by retarded shedding of the deciduous tooth, advanced development of the permanent tooth, or simply too small a ramal width (or depth) to accommodate the developing tooth. This causes the roots of the tooth bud to come into contact with the ventral portion of the ramus, and sometimes even penetrate it. It may be related to congenital conditions, or due to developmental defects (e.g. malnutrition) which may not always be obvious. A simple protuberance is recorded by appending an inverted triangle immediately below the FOR section under the relevant tooth. Where penetration has occurred, an arrow is added to the apex of the triangle (Figure 2).

xiv) other anomalies (Other Anom)

The section at the base of the sheet is reserved for anomalies not covered in the above. They are discussed below in order of frequency of occurrence for West Hill, though in no case (except the first) are they commonly occurring. Appended to these are a few anomalies mentioned in the veterinary literature which did not appear at West Hill, though some did so for other sites. The list is not meant to be exhaustive. They may be recorded diagrammatically or descriptively.

a) Developmental deposits and anomalies. The most common are thin, 'platey' deposits of extra bone near the alveolar margins, generally but not exclusively on the labial side of the mandible (Plate 11). These appear to be age related in that they occur most commonly on juvenile mandibles. Presumably they are related to disturbances caused during tooth development and (especially) eruption. Sometimes the 'platey' deposits are accompanied by pitted 'porous' margins, and the latter may also occur in isolation.

b) 'Traction epiphyses'. In aged animals, bony prominences are sometimes formed around muscle attachment points (Silver 1969:284). One such point is at the angle of the horizontal and ascending rami on the labial surface. Where this is unusually well formed it is probably worth noting.

c) Developmental deposits of teeth. Baker and Brothwell (1980:150) illustrate examples of abnormal deposits on bovid teeth from Westbridge Priory. Similar examples were found at West Hill on ovicaprid teeth (Plate 12).

d) Effects of trauma. These include fractures, reaction to superficial damage, etc. An unusual example is illustrated in Plate 16, possibly a reaction to muzzling.

e) Actinobacillosis. This condition is commonly referred to as 'lumpy jaw'. In sheep the condition is caused by the bacteria Actinobacillus ligniersi, in cattle it is a reaction to the 'ray fungus' Actinomyces bovis, (Baker and Brothwell 1980:150). The aetiology and pathogenesis of this condition in sheep is described by Jensen (1974:213-215). One specimen from West Hill is possibly an example of this condition (Plate 14).

f) Miscellaneous phenomena. These include examples such as the compound tooth illustrated in Plate 13, depressions in teeth and mandibles, etc.

g) Other anomalies. Caries are undoubtedly common in some species, but apparently are very rare in ungulates (Baker and Brothwell 1980:146). Dentigerous cysts are reported to be quite common in modern sheep (Colyer 1936:693), but are extremely rare in archaeological specimens. A mandible from the medieval site of Exe Bridge, Exeter may be an example of this (Plate 17). Neoplastic lesions include adomantinoma, amelobastoma and odontoma (Jubb and Kennedy 1963:7). In sheep they occur in both mandibles and maxillae, but are rare (Marsh 1965:302-304). Fluorine poisoning produces a chalky appearance and mottled discolouration in teeth (Franklin 1950:6) and molars may be extensively malformed, deciduous teeth seem unaffected. Grass contamination has been responsible for fluorine poisoning in modern sheep (Marsh 1965:380). A number of other anomalies are discussed in more accessible papers and need not be repeated here (Baker and Brothwell 1980; Brothwell 1981).

This concludes the description of the anomalies and of the methodology. Each recording sheet represents one mandible, and where a serious study of the mandibles from a site is being undertaken, it is suggested that the sheets be filled in for all mandibles, whether or not they have any anomalies. Such a record will then form a complete and self contained corpus of information about the mandible assemblage which can be analysed in depth and at length if necessary. The methodology is a manual system, but it is not inconceivable that it could be transformed into a computer based system.

ANALYSIS OF THE WEST HILL MANDIBLES

The methodology was used to record the West Hill mandibles, a sample numbering 1855. Results from this assemblage will appear in full in the bone report from this site (Levitan in prep). This paper, therefore, is not the place to present or discuss these results, but the final part will summarise a selection of the results in order to provide a practical example of how the system works.

periodontal disease

The West Hill assemblage contained almost equal proportions of mandibles with deciduous and permanent dentition: 936 permanent (50.5%), 919 deciduous (49.5%). Table 1 shows that periodontal disease was almost entirely confined to permanent teeth. Frequency of occurrence ranged from 0.3% in deciduous second premolars to 17.5% in permanent fourth premolars. Table 1 indicates clearly that occurrences were centred on the first molar - permanent fourth premolar, and in fact the frequency in the latter is more than twice that of any other tooth. This is an interesting result because it provides a clue as to the initial cause of infection. Since it is centred on the fourth premolar, but is very rare in deciduous teeth, it would seem logical to conclude that most cases of infection do not occur until the permanent teeth are present. It would appear from this that infection might take place while the permanent premolars are erupting - particularly the fourth, which is the site of maximum disruption due to the size of the teeth involved. Perhaps calculus also has a part to play, but the fact that it is more or less as common on deciduous teeth as permanent (see below) indicates that perhaps the eruption event is the major catalyst. Table 1 also shows that severe infection was rare - never more than 1.5% of teeth were lost due to the disease (Stages 3-5). This result may, however, reflect the special nature of the site, and this aspect is discussed more fully in the report (Levitan in prep).

calculus

Table 2 provides a summary of the frequency and occurrence of calculus deposits. No attempt has been made to analyse heaviness of deposit because of the post excavation cleaning problem described above. The table shows that with the exception of the second premolars, frequencies were fairly consistent: 14.2-18.8% on the lingual side, and 11.6-16.8% on the labial surface. Secondly, calculus was always more common on the lingual side. A subtle pattern can be made out from the frequencies, which show peaks for the fourth premolar and first molar, a result which tends to underline and corroborate the periodontal disease results. The frequencies do seem a little low, however - particularly in the light of veterinary information - and undoubtedly the cleaning problem has caused a bias in this respect, which tends to throw some suspicion on the results as a whole.

intra-dental attrition

Intra-dental attrition was very common, with frequencies as high as 88.8% (Table 3). Logically, the second premolars and third molar escaped with the least amount of attrition - less than 15%. In the case of the third molar this is because it is the last tooth to erupt, so undergoes a shorter period of attrition than the other teeth. The second premolars are small, nearly vestigial teeth, so the surface area involved is small and the relatively shallow roots allow for movement away from anatagonistic forces distally. All the other teeth have in excess of 25% attrition on at least one interface. It is interesting to note that for the premolars greatest attrition is distally, and for the molars it is mesially. Such a pattern may be related to the eruption sequence and morphology of the teeth. Thus the molars each erupt with a tooth already present mesially, but none distally, and they present a greater surface area at their mesial end. Conversely, the premolars present their greatest surface area distally, and the permanent fourth premolar - to exaggerate a point - often erupts at an angle extremely antagonistic to the first molar (Plate 18)!

These three examples have been given only brief attention, but hopefully they illustrate that interesting and important information can be recovered from mandibles, and that the above methodology is a useful way of recording and retrieving this information. It is hoped that this paper will lead to the wider adoption of this methodology, thus providing the basis for compiling a corpus of information about oral palaeopathology and a greater understanding of health, nutrition and husbandry in ancient times.

ACKNOWLEDGEMENTS

The excavation and post excavation programmes for West Hill are being financed by the Historic Buildings & Monuments Commission for England The British Museum, The Society of Antiquaries of London and The Bristol and Gloucester Archaeological Society. All the finds, including the bones, have been donated by the landowner, Major C.A. Goldingham, to the British Museum.

REFERENCES

Andrews, A. and Noddle, B. (1975) Absence of Premolar Teeth from Ruminant Mandibles found at Archaeological Sites, Journal of Archaeological Science 2, 137-144.

Armitage, P.L. (1975) The Extraction and Identification of Opal Phytoliths from the Teeth of Ungulates, Journal of Archaeological Science 2, 187-197.

Armstrong, M.C. (1960) Paradontal Disease of Sheep in South Canterbury, New Zealand Journal of Agriculture 100, 429-431.

Baker, J. and Brothwell, D. (1980) Animal Diseases in Archaeology London: Academic Press.

Brothwell, D.R. (1981) Disease as an Environmental Paramater, In The Environment of Man: the Iron Age to the Anglo-Saxon Period, Eds. Jones, M. and Dimbleby, G., 231-247. Oxford: British Archaeological Reports, British Series 87.

Colyer, F. (1936) Variation and Diseases of Teeth of Animals. London: Bale and Danielsson.

Cutress, T.W. and Healey, W.B. (1965) Wear of Sheep's Teeth II. Effects of Pasture Juices on Dentine, New Zealand Journal of Agricultural Research 8, 753-762.

Franklin, M.C. (1950) The Influence of Diet on Dental Development in the Sheep, Commonwealth Science and Industrial Research Organisation, Australia Bulletin 252.

Hart, E.K. and MacKinnon, M.M. (1958) Enzootic Paradontal Disease in Adult Sheep, New Zealand Veterinary Journal 6, 118-123.

Healy, W.B., Cutress, T.W. and Mitchie, C. (1967) Wear of Sheep's Teeth IV. Reduction of Soil Ingestion and Tooth Wear by Supplementary Feeding, New Zealand Journal of Agricultural Research 10, 201-209.

Healy, W.B. and Ludwig, T.G. (1965) Wear of Sheep's Teeth. I. The Role of Ingested Soil, New Zealand Journal of Agricultural Research 8, 737-752.

Jensen, R. (1974) Diseases of Sheep. New York: Lea and Febiger.

Jubb, K. and Kennedy, P. (1963) Pathology of Domestic Animals. 2 vols. London: Academic Press.

Levitan, B. (1977) Pathological Anomalies in Sheep Mandibles: a Methodological Approach. University of London, Institute of Archaeology, unpublished undergraduate dissertation.

Levitan, B. in prep. The animal bones from the first millennium AD ritual complex at West Hill, Uley, Gloucestershire.

Ludwig, T.G., Healy, W.B. and Cutress, T.W. (1966) Wear of Sheep's Teeth III. Seasonal Variation in Wear and Ingested Soil, New Zealand Journal of Agricultural Research 9, 157-164.

Marsh, H. (1965) Newsom's Sheep Diseases 3rd edn. Montana: Veterinary Research Laboratory.

Silver, I.A. (1969) The Ageing of Domestic Animals, In Science in Archaeology 2nd edn., Eds. E. Higgs and D. Brothwell, pp 283-302, London: Thames and Hudson.

Mandibular Pathology Species:

Site: Ref: Mand. No: Right/Left

Ascending Ramus	M3	M2	M1	P4	P3	P2	Diast.	I4/C	I3	I2	I1	
A												P/A
P												
M												
D												
CH												ATT
IDA												
												PER
												ABS
												S/A
												R
LN												HYP
LB												
LN												CAL
LB												
2ary												PIG
1ary												
												PCE
												B
												G
												FOR
												VMD Other Anom
A.R.	M3	M2	M1	P4	P3	P2	Diast.	I4/C	I3	I2	I1	

Figure 1. Recording sheet for Ungulate mandibular pathology.

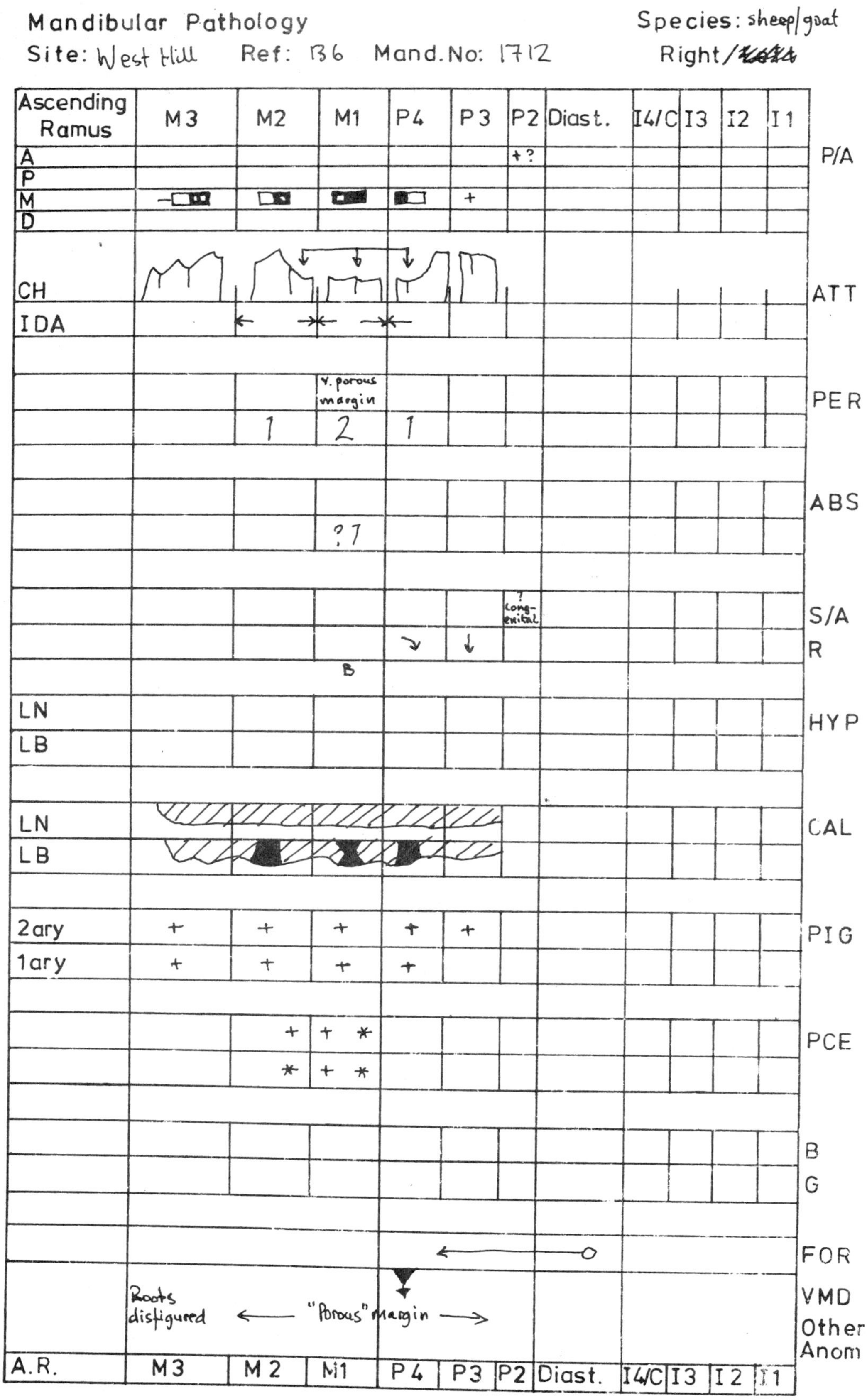

Figure 2. Recording sheet for Ungulate mandibular pathology filled out to illustrate scoring system.

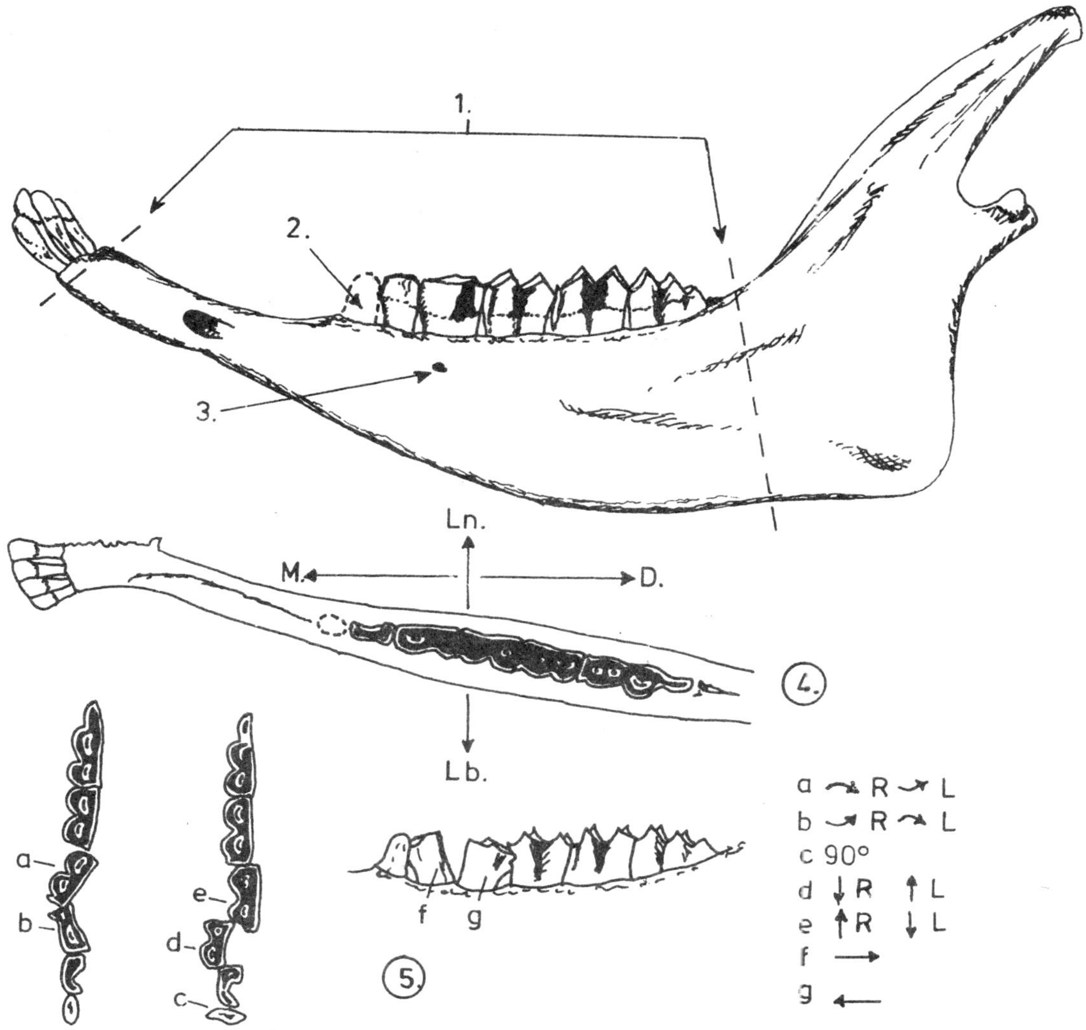

Figure 3. Definitions of orientations and symbols described in the text.

Key: 1. zone of breakage in Figure 2.
2. absent second premolar.
3. extra nutrient foramen.
4. orientation of teeth and mandible.
5. scoring system for rotation and displacement:

a. labial rotation	d. lingual displacement
b. lingual rotation	e. labial displacement
c. 90° rotation	f. mesial displacement
	g. distal displacement

score	p2		p3		p4		P2		P3		P4		M1		M2		M3	
	n	%	n	%	n	%	n	%	n	%	n	%	n	%	n	%	n	%
1	3	0.3	3	0.3	5	0.6	12	1.3	35	3.7	127	13.6	105	6.0	41	3.0	15	1.6
2	-	-	2	0.2	2	0.2	1	0.1	12	1.3	23	2.4	27	1.6	16	1.2	9	1.0
3	-	-	-	-	-	-	4	0.4	3	0.3	9	1.0	6	0.3	1	0.1	-	-
4	-	-	-	-	-	-	1	0.1	-	-	5	0.5	1	0.1	-	-	-	-
5	-	-	-	-	-	-	2	0.2	1	0.1	-	-	-	-	-	-	-	-
total	3	0.3	5	0.5	7	0.8	20	2.1	51	5.4	164	17.5	139	8.0	58	4.3	24	2.6

Table 1. Occurrence and Frequency of Periodontal Disease in Ovicaprid Mandibles from West Hill.

aspect	p2		p3		p4		P2		P3		P4		M1		M2		M3	
	n	%	n	%	n	%	n	%	n	%	n	%	n	%	n	%	n	%
lingual	21	8.9	92	14.4	115	17.1	28	10.7	125	18.8	142	18.1	263	17.5	196	17.2	81	15.3
labial	13	5.5	74	11.6	105	15.6	22	8.4	90	13.6	132	16.8	253	16.8	168	14.7	68	12.8

Table 2. Occurrence and Frequency of Calculus Deposits on Ovicaprid Teeth from West Hill.

location	p2		p3		p4		P2		P3		P4		M1		M2		M3	
	n	%	n	%	n	%	n	%	n	%	n	%	n	%	n	%	n	%
mesial	-	-	37	5.8	488	63.3	-	-	10	1.5	100	12.7	937	62.4	287	25.2	68	12.8
distal	23	9.7	568	88.8	389	50.9	12	4.6	174	26.2	299	38.1	508	33.8	144	12.6	-	-

Table 3. Occurrence and Frequency of Intra-dental Attrition on Ovicaprid Teeth from West Hill.

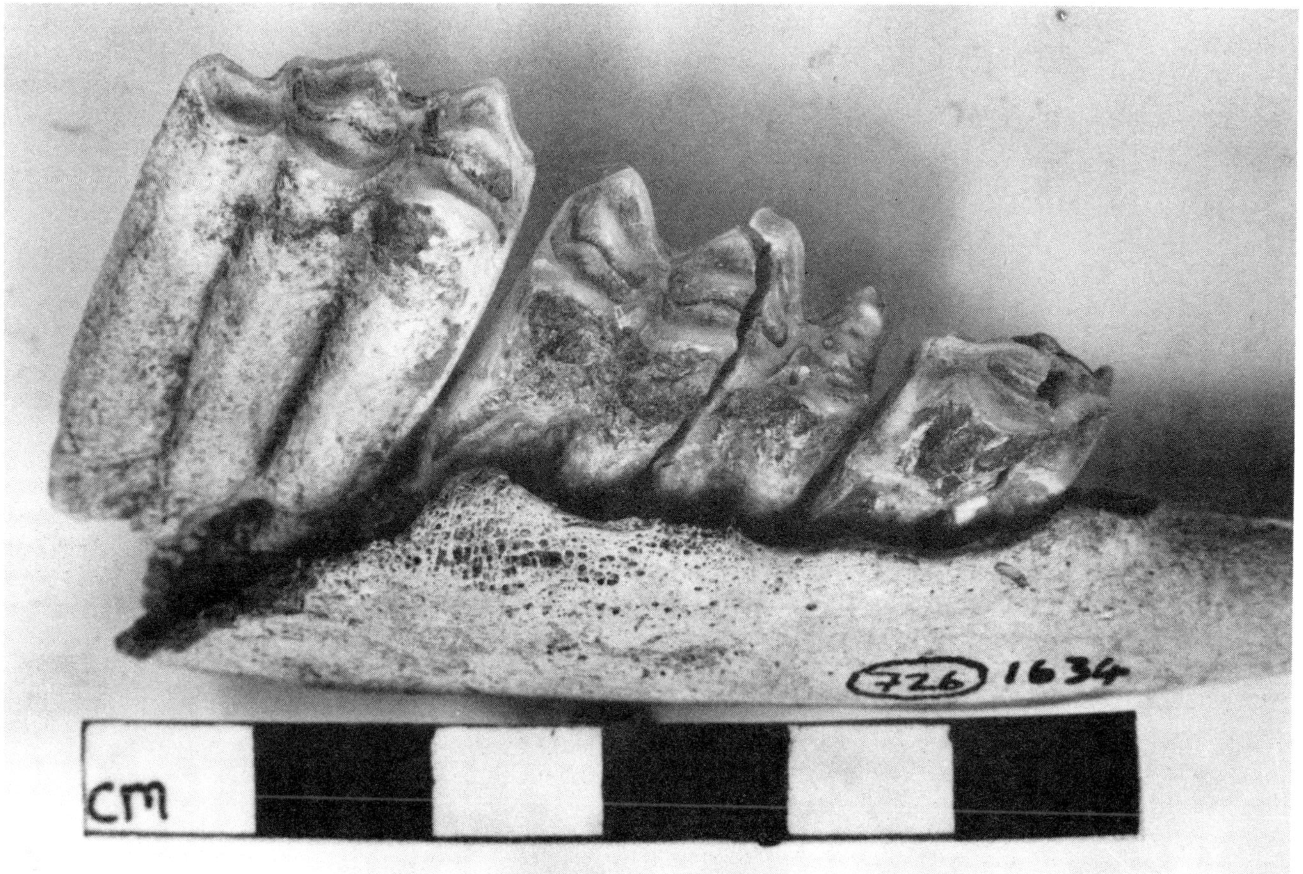

Plate 1. Intra-dental attrition on first and second molars. Calculus deposits.

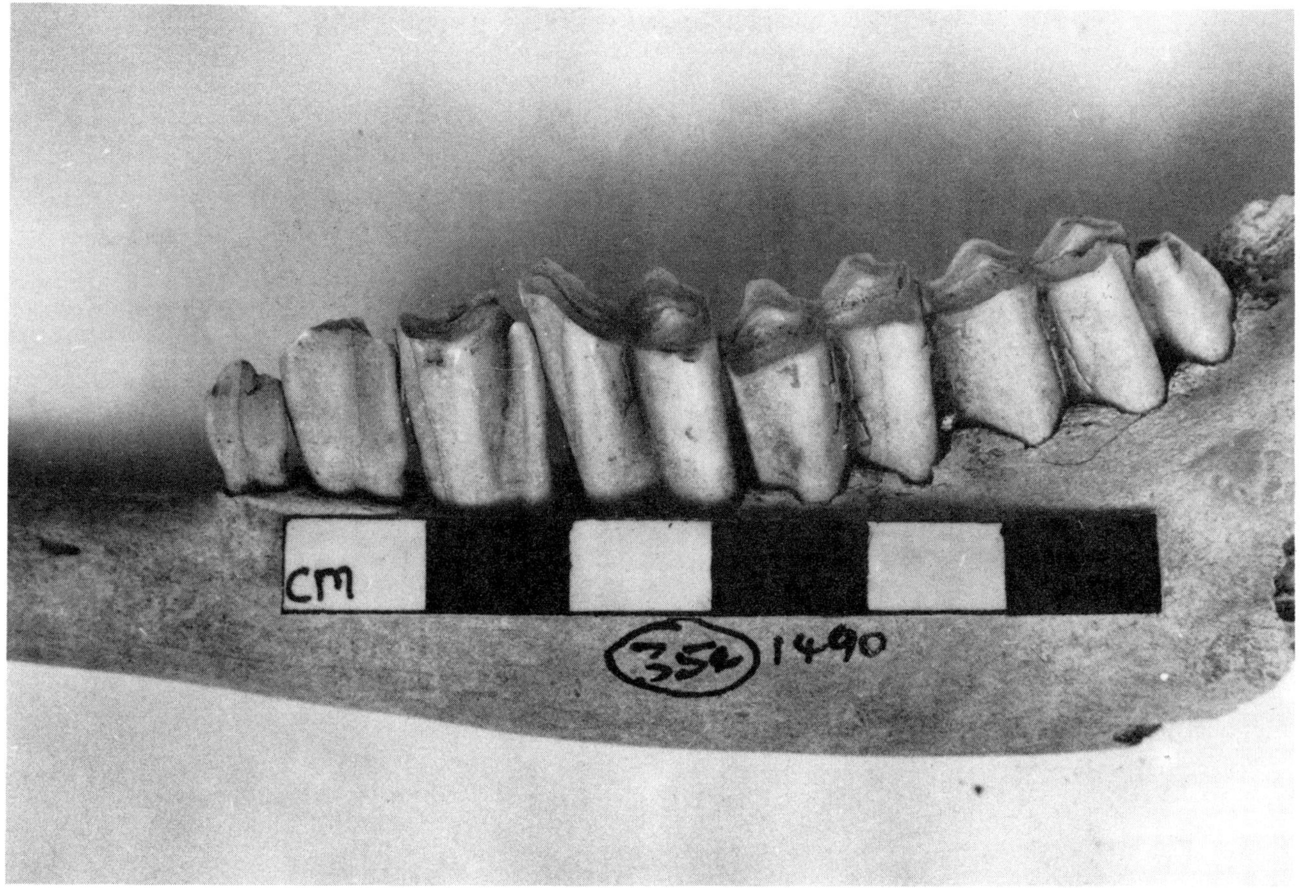

Plate 2. Anomalous crown height on first molar. Extra nutrient foramen.

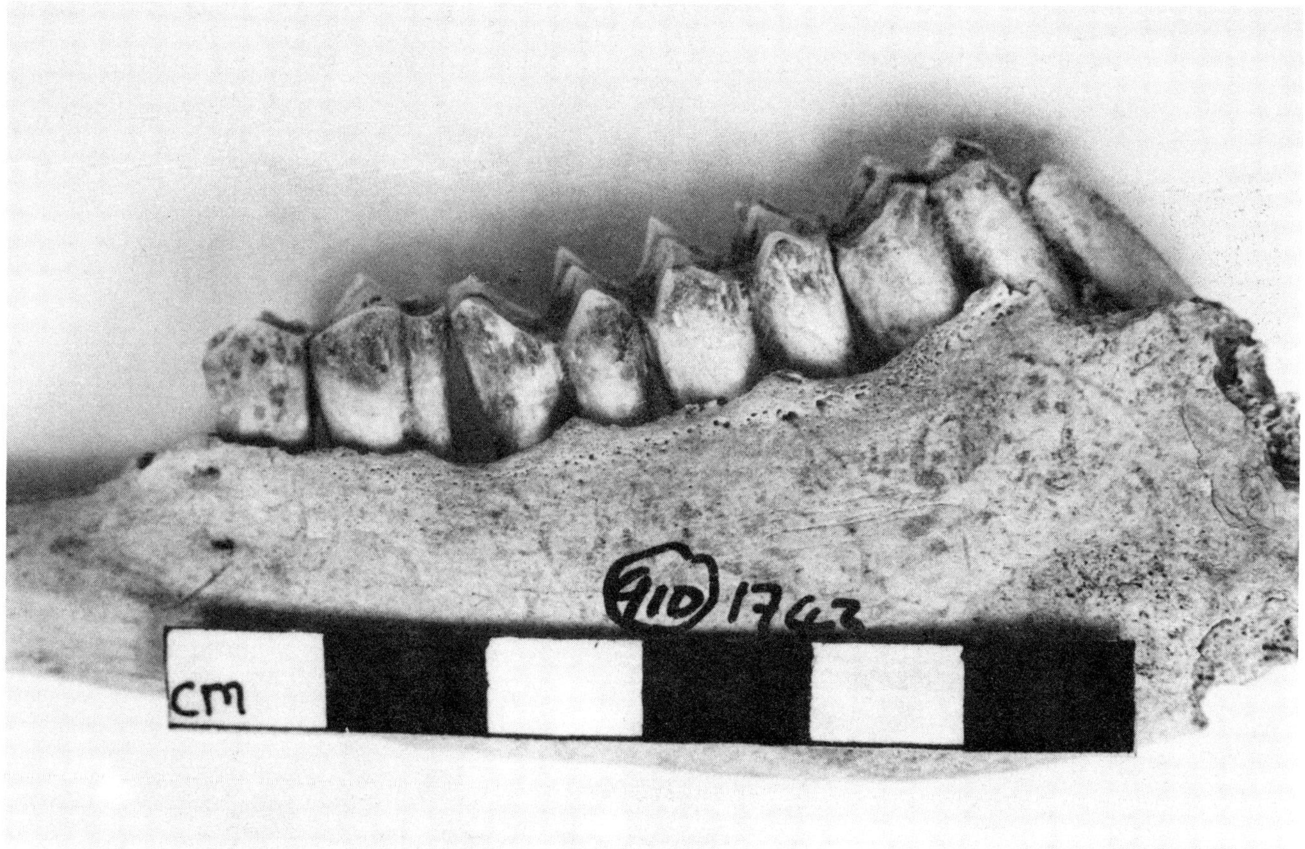

Plate 3. Stage 1 periodontal disease affecting first molar and fourth premolar. Extra nutrient foramen.

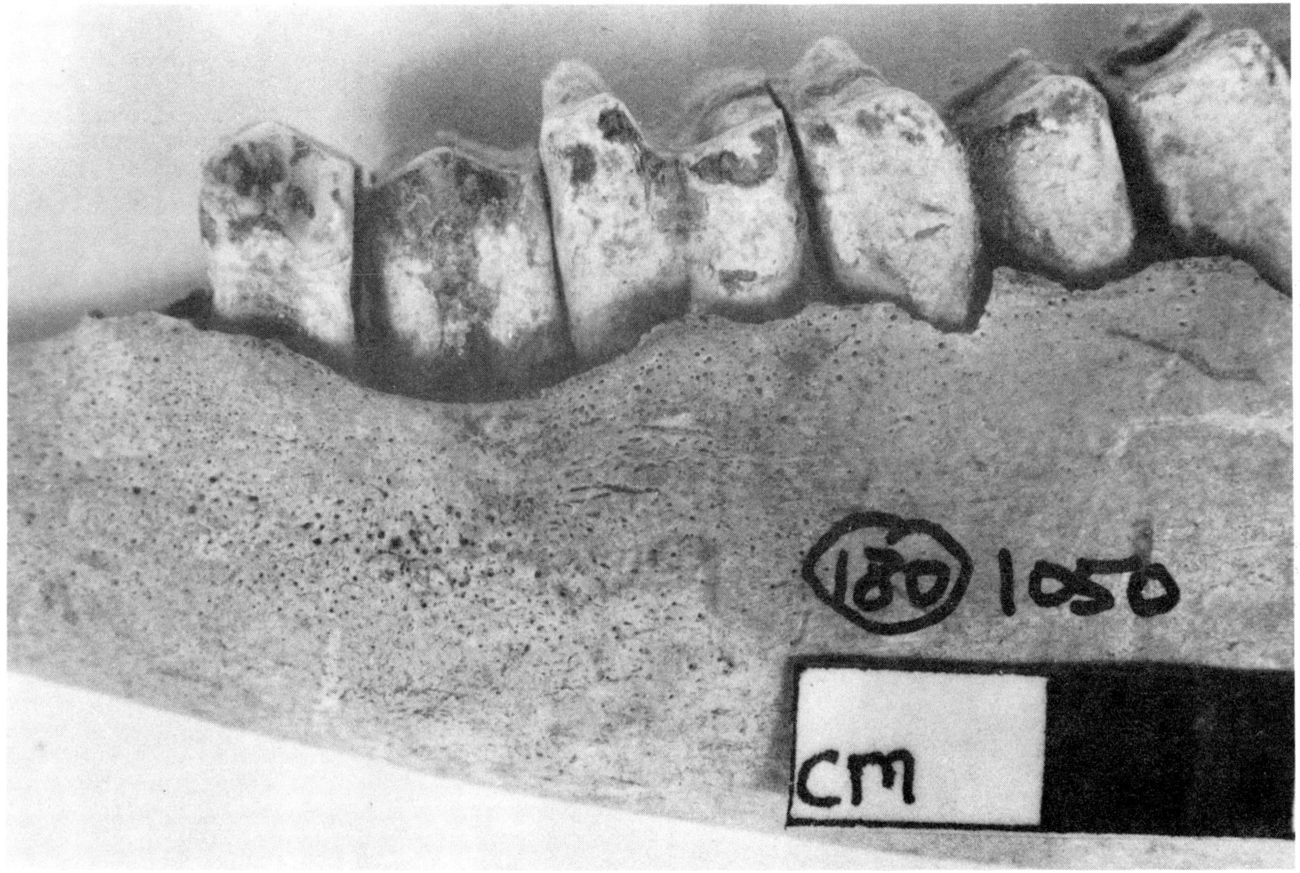

Plate 4. Stage 2 periodontal disease affecting fourth premolar. Third premolar and first molar have Stage 1 periodontal disease.

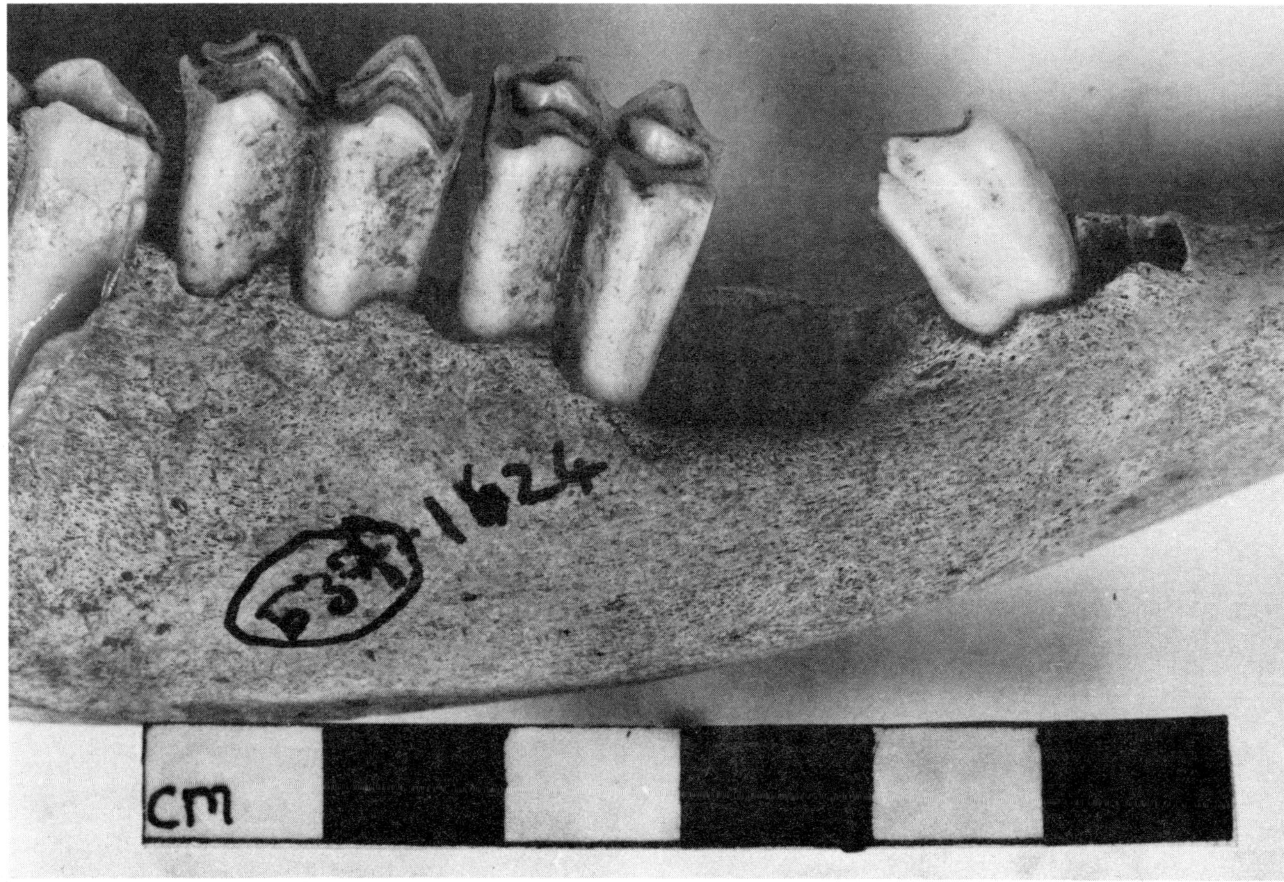

Plate 5. Stage 3 periodontal disease: ante-mortem loss of fourth premolar.

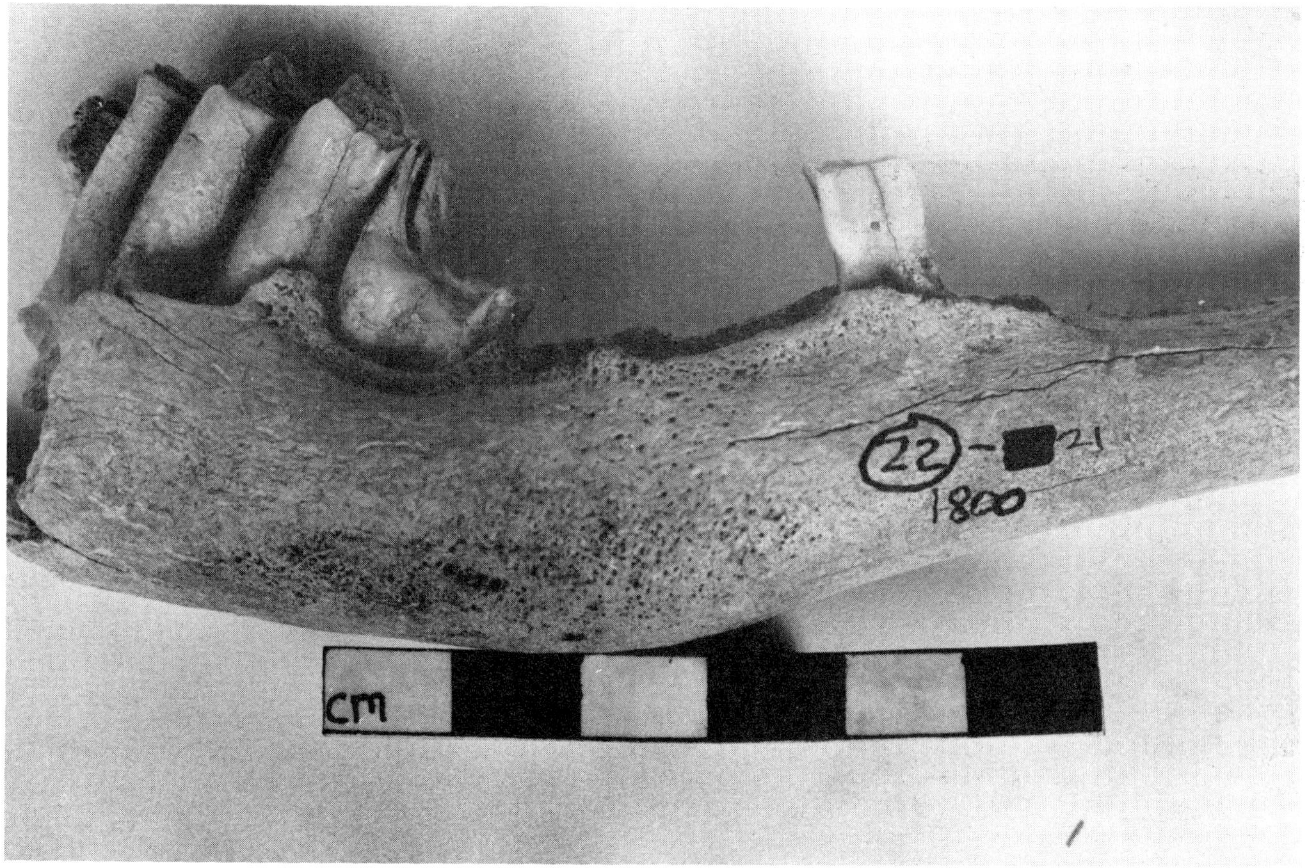

Plate 6. Stage 4 periodontal disease and Stage 3 abscess: loss of fourth premolar and first molar with considerable reaction in alveolar cavities. A-labial aspect,

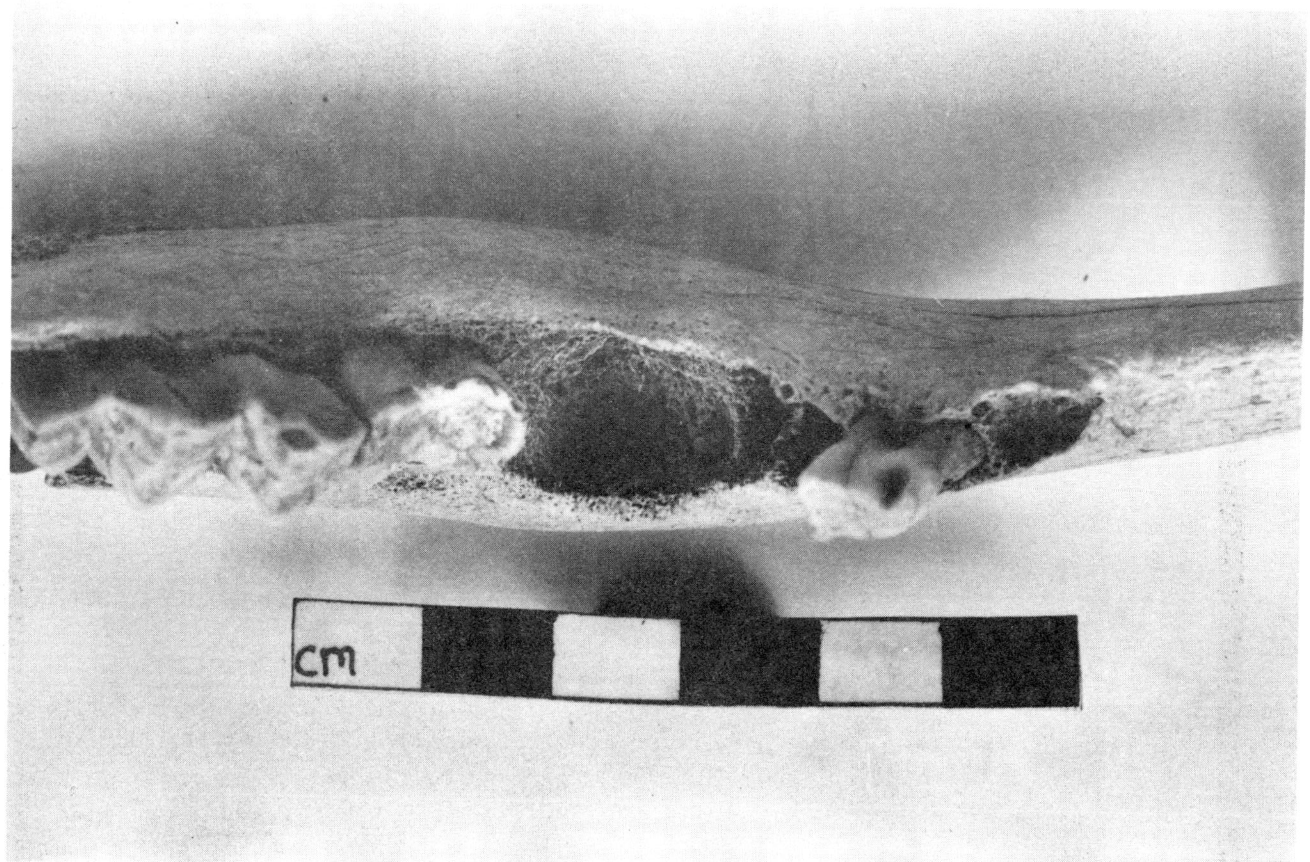

B-occlusal aspect,

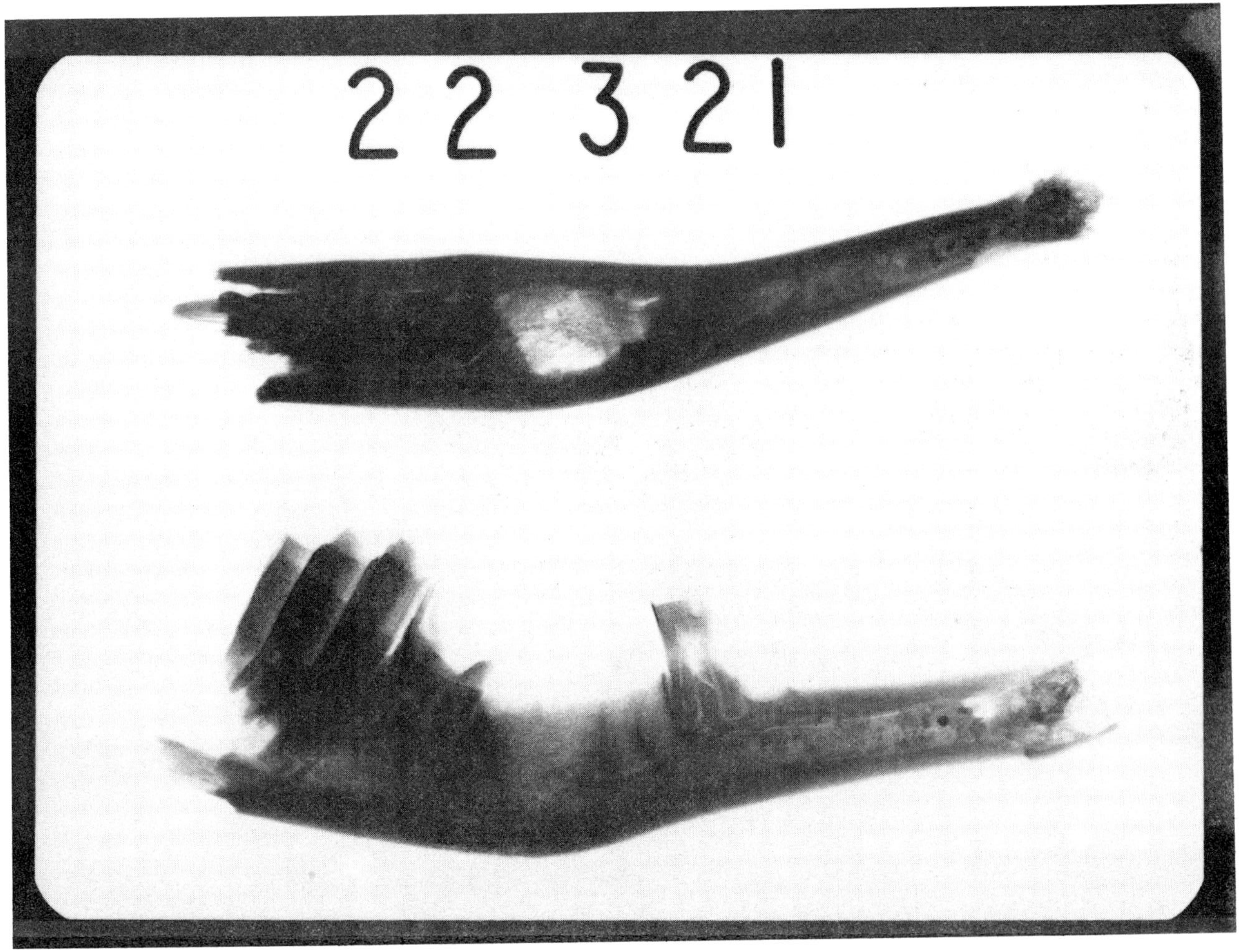

C- X-ray.

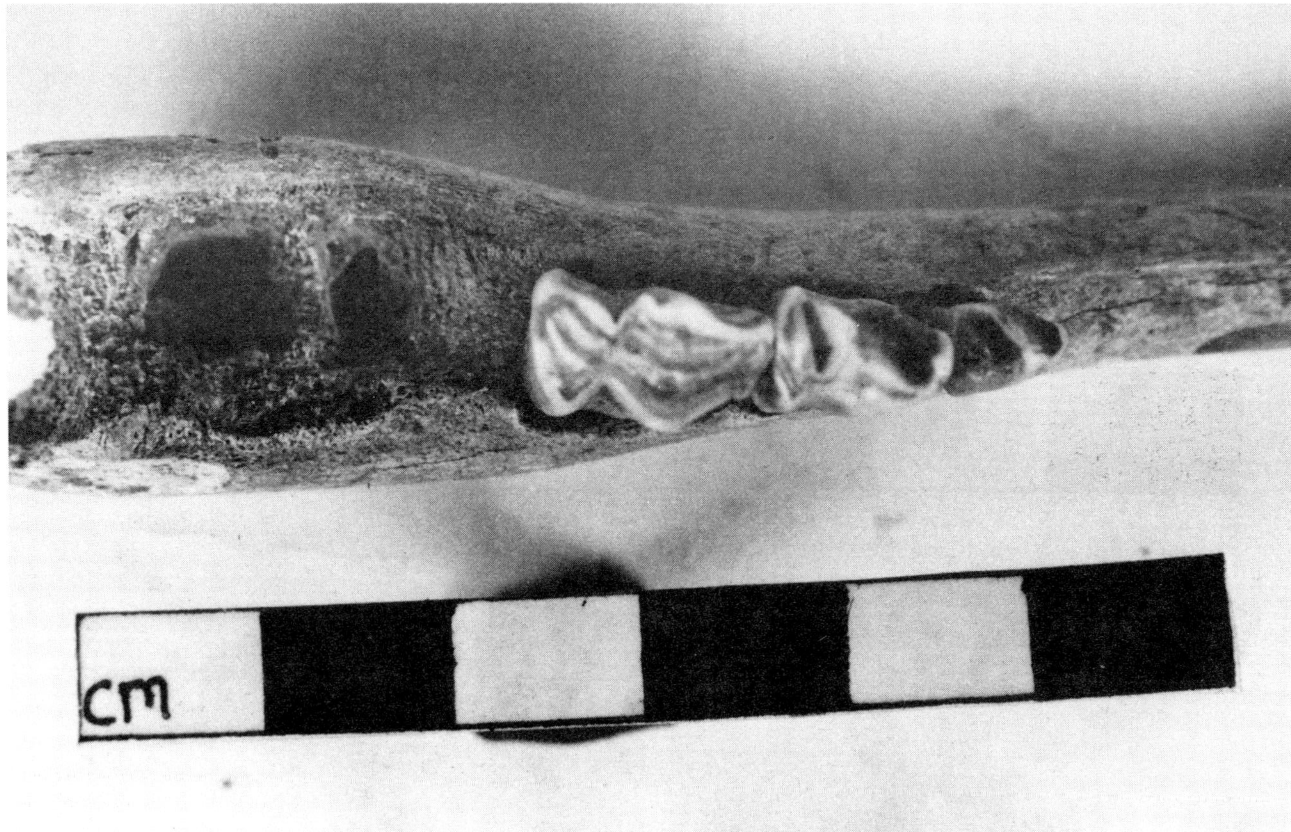

Plate 7. Stage 5 periodontal disease: first molar.

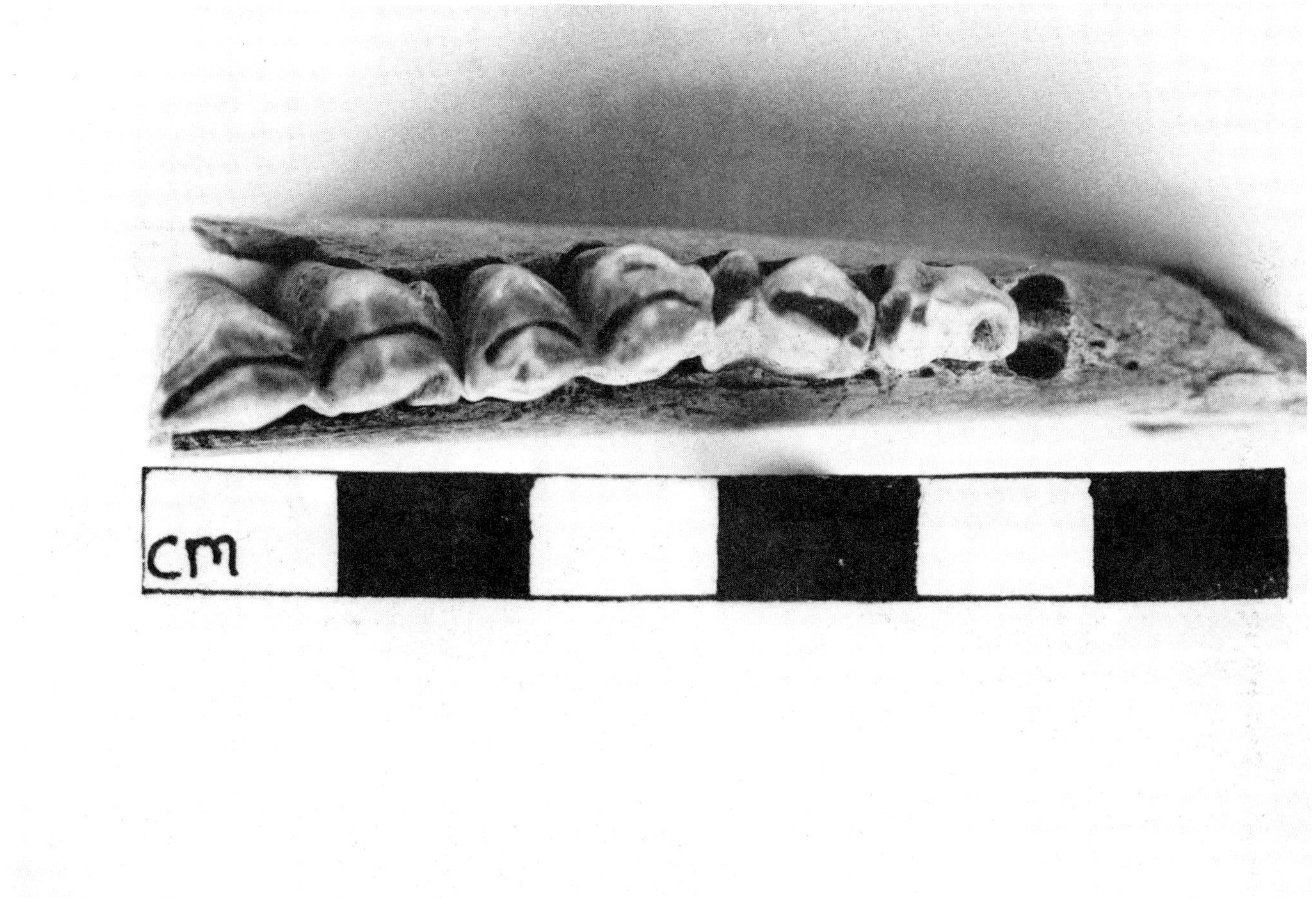

Plate 8. Tooth rotation: 90o rotation of second premolar.

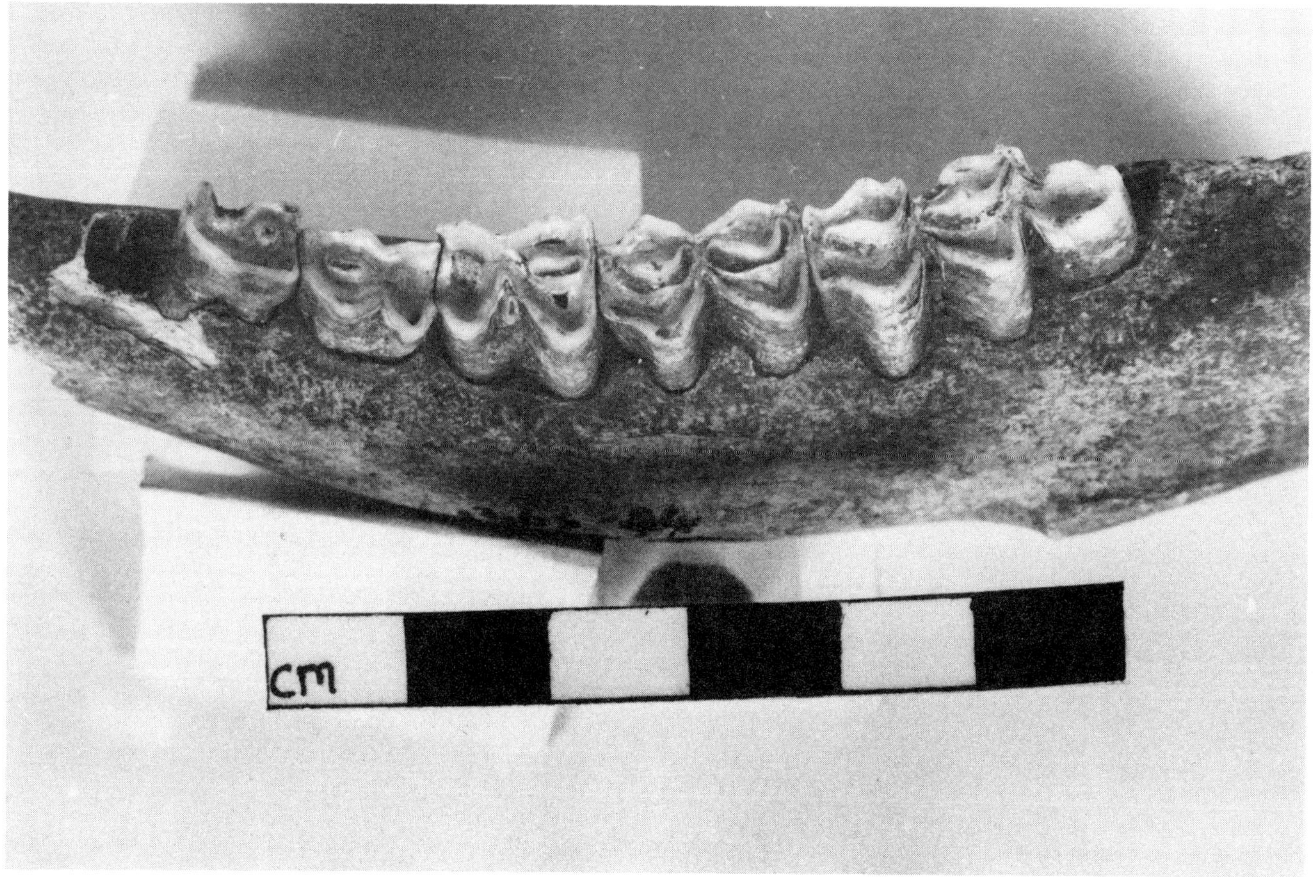

Plate 9. Pigmentation: note differential pigmentation of primary and secondary dentine. Pulp cavity exposure on first molar.

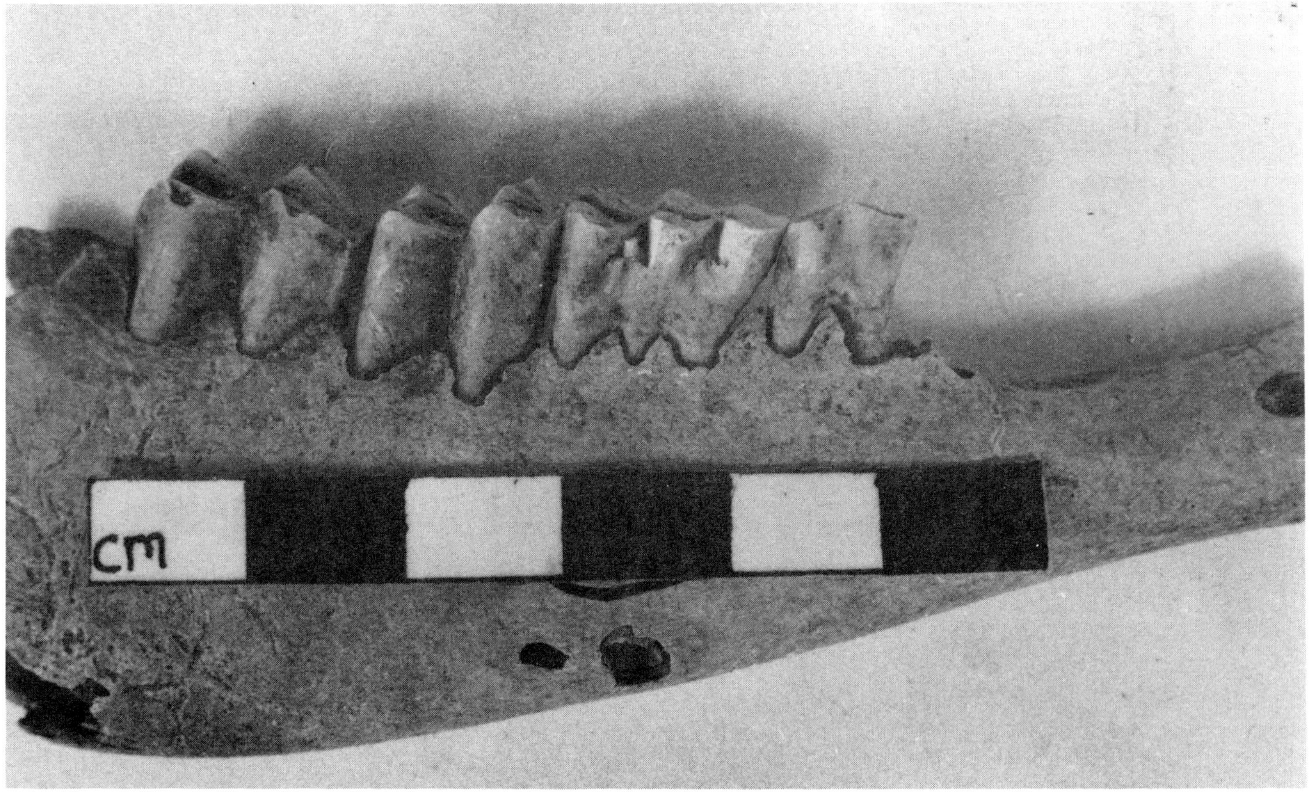

Plate 10. Disturbance and penetration of ventral margin by developing fourth premolar. 'Bovine' pillars.

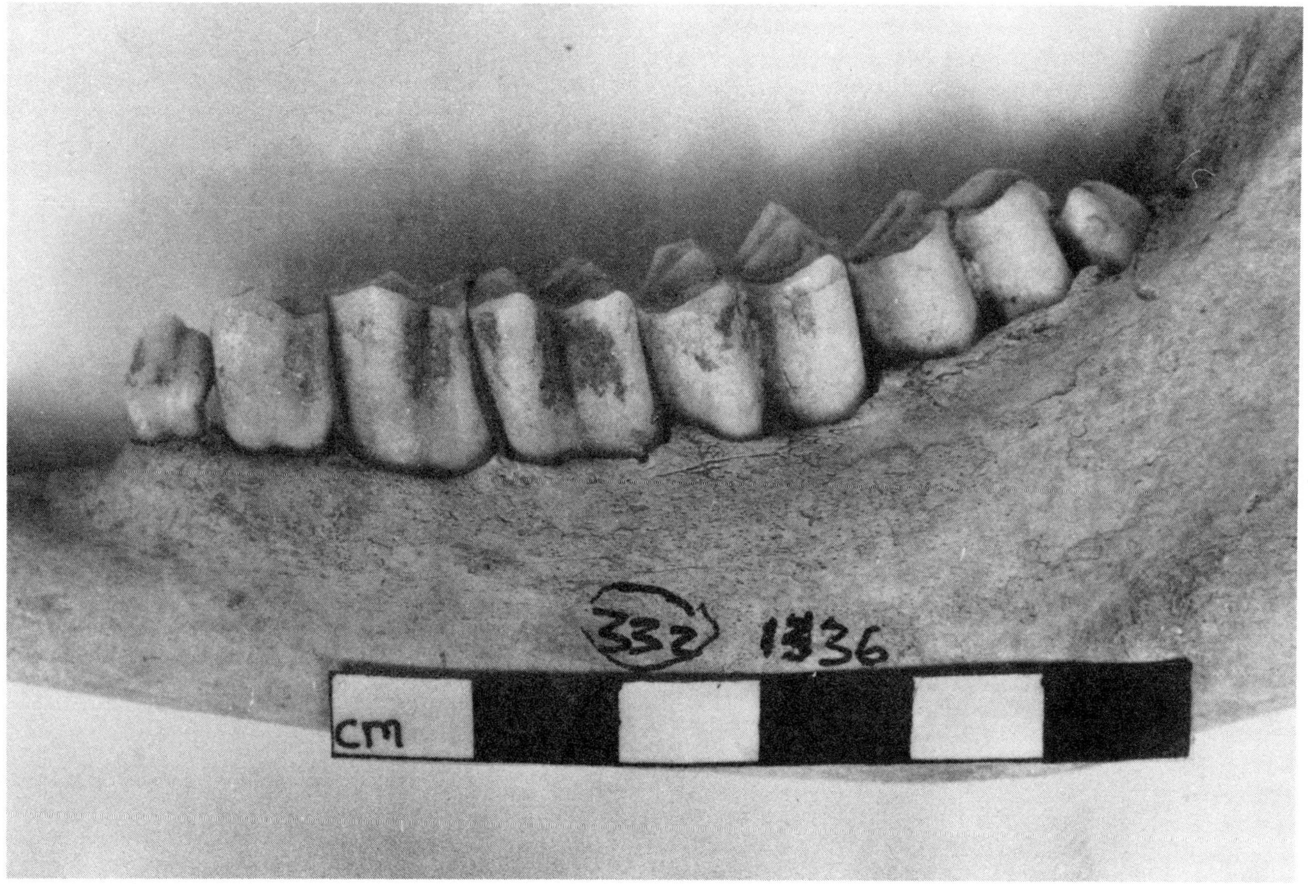

Plate 11. 'Platey' deposits of bone near alveolar margins.

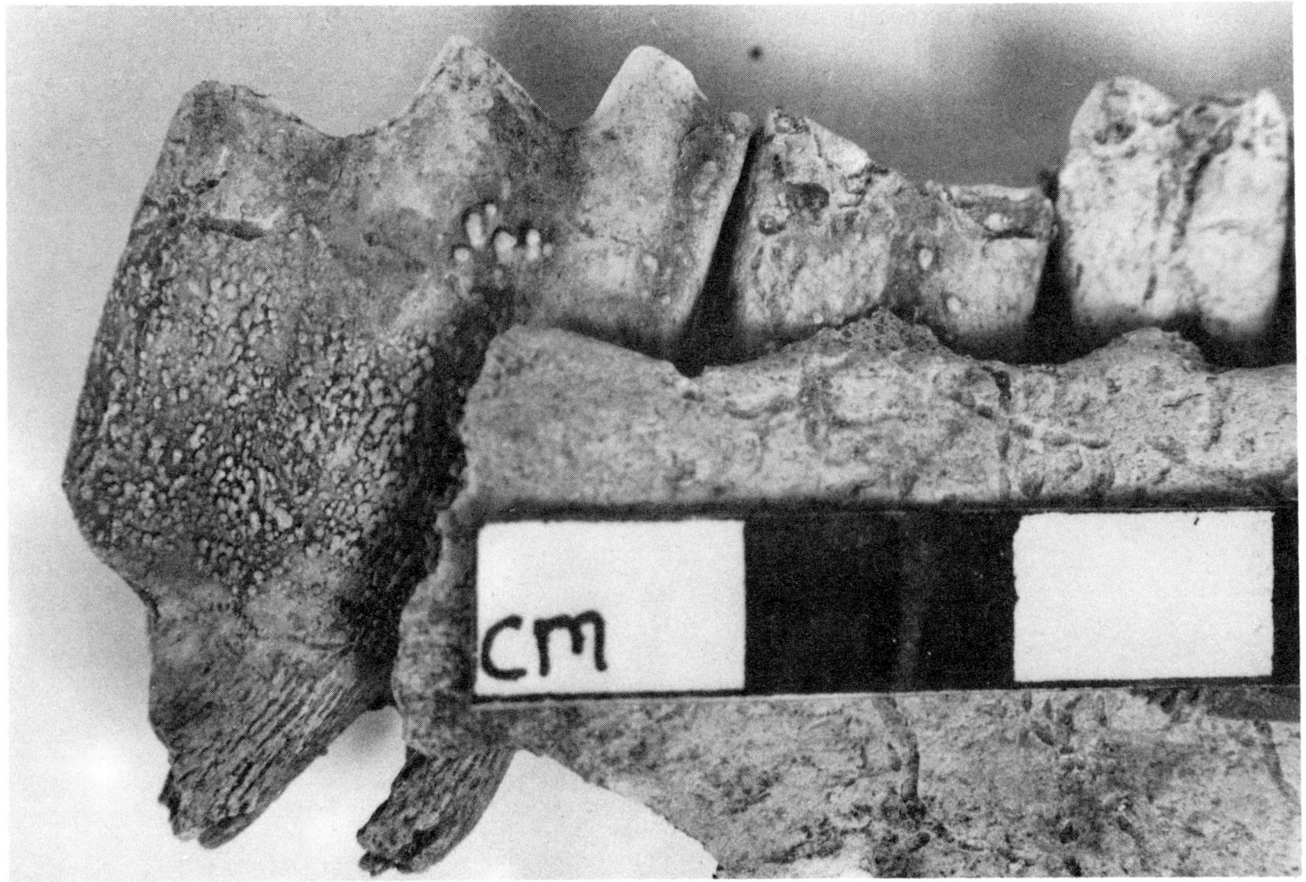

Plate 12. Developmental deposits on second and third molars.

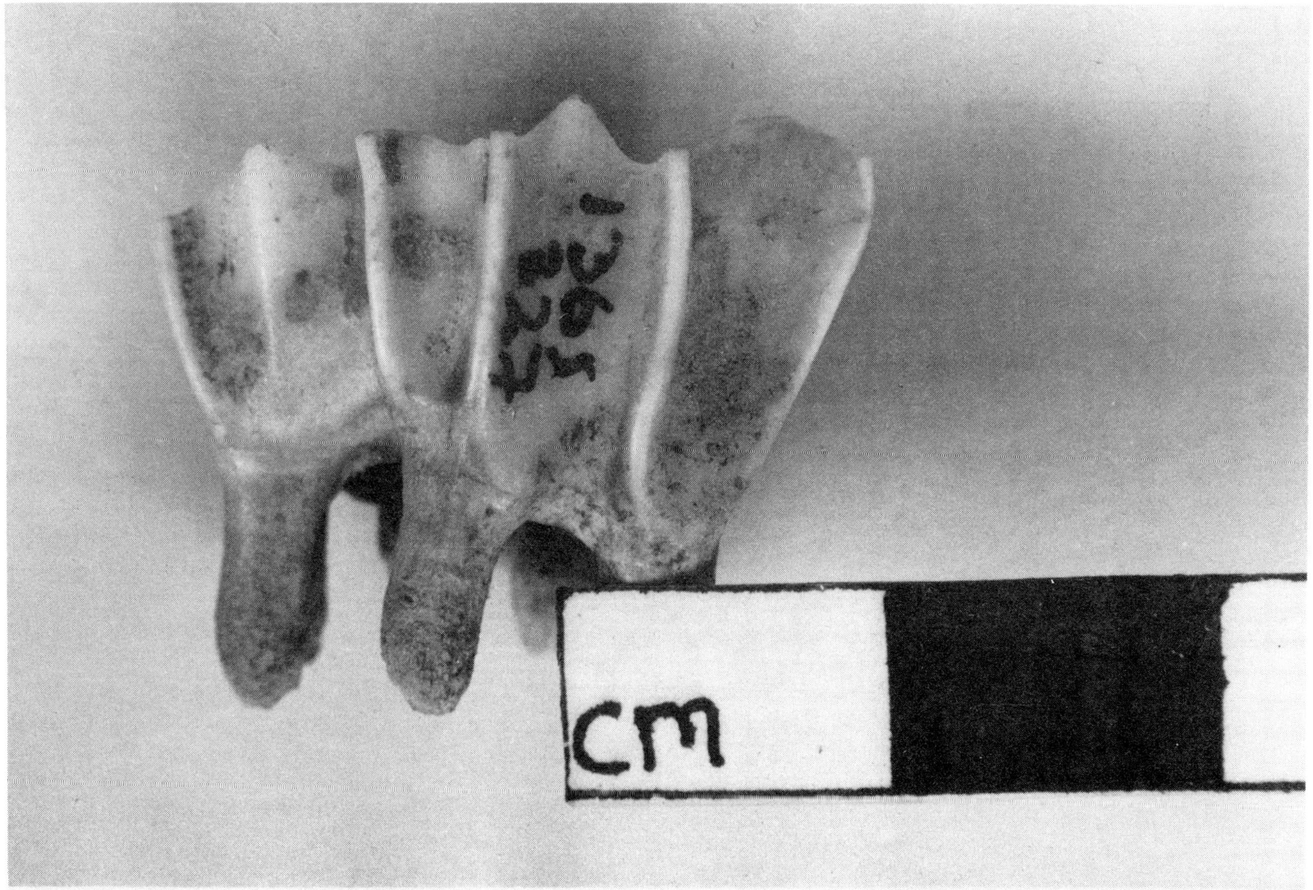

Plate 13. Compound deciduous upper premolars which have developed as one tooth.

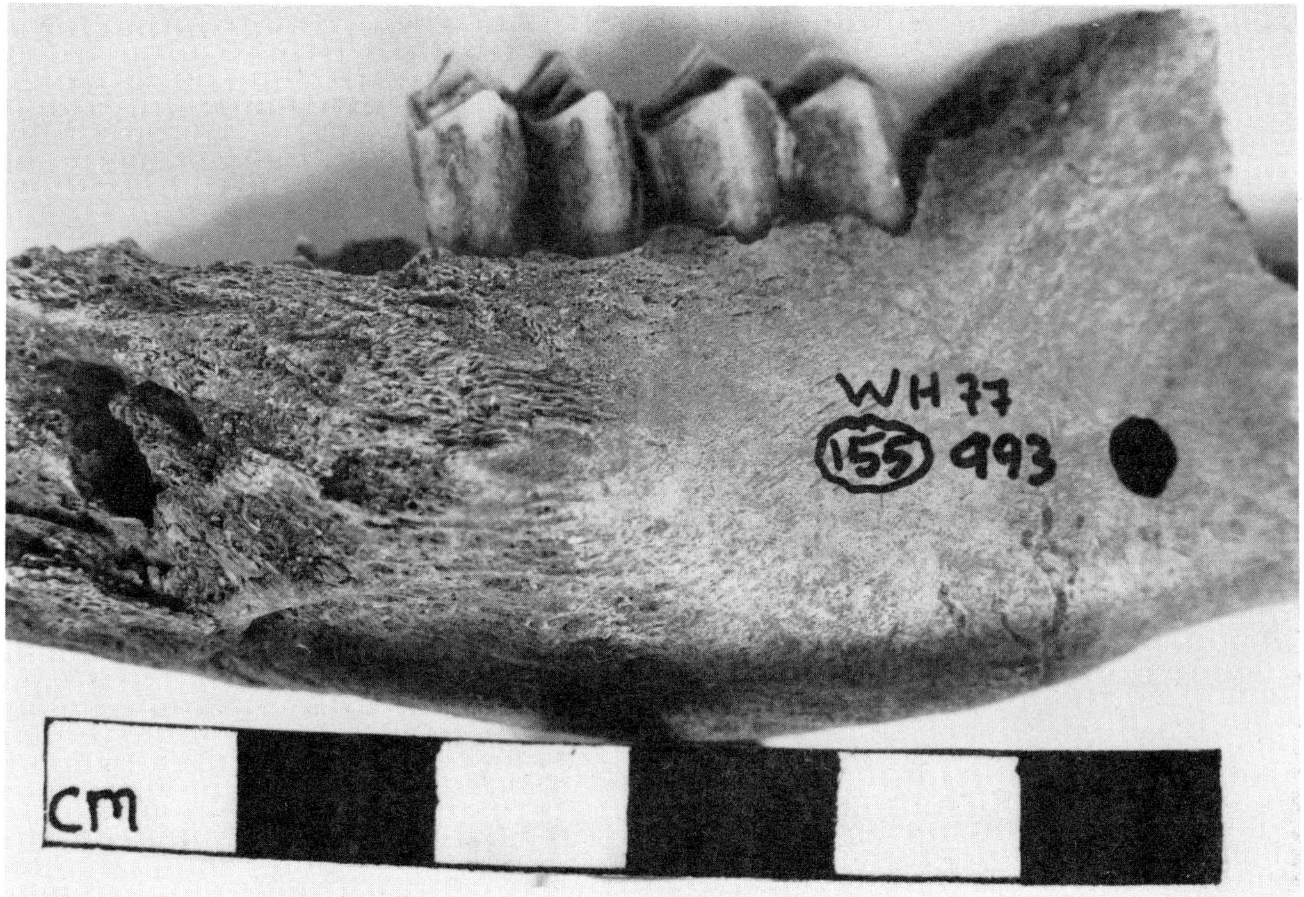

Plate 14. Severely infected mandible - ? Actinobacillosis.

Plate 15. 'Bovine' pillars. A- extra pillar on first molar,

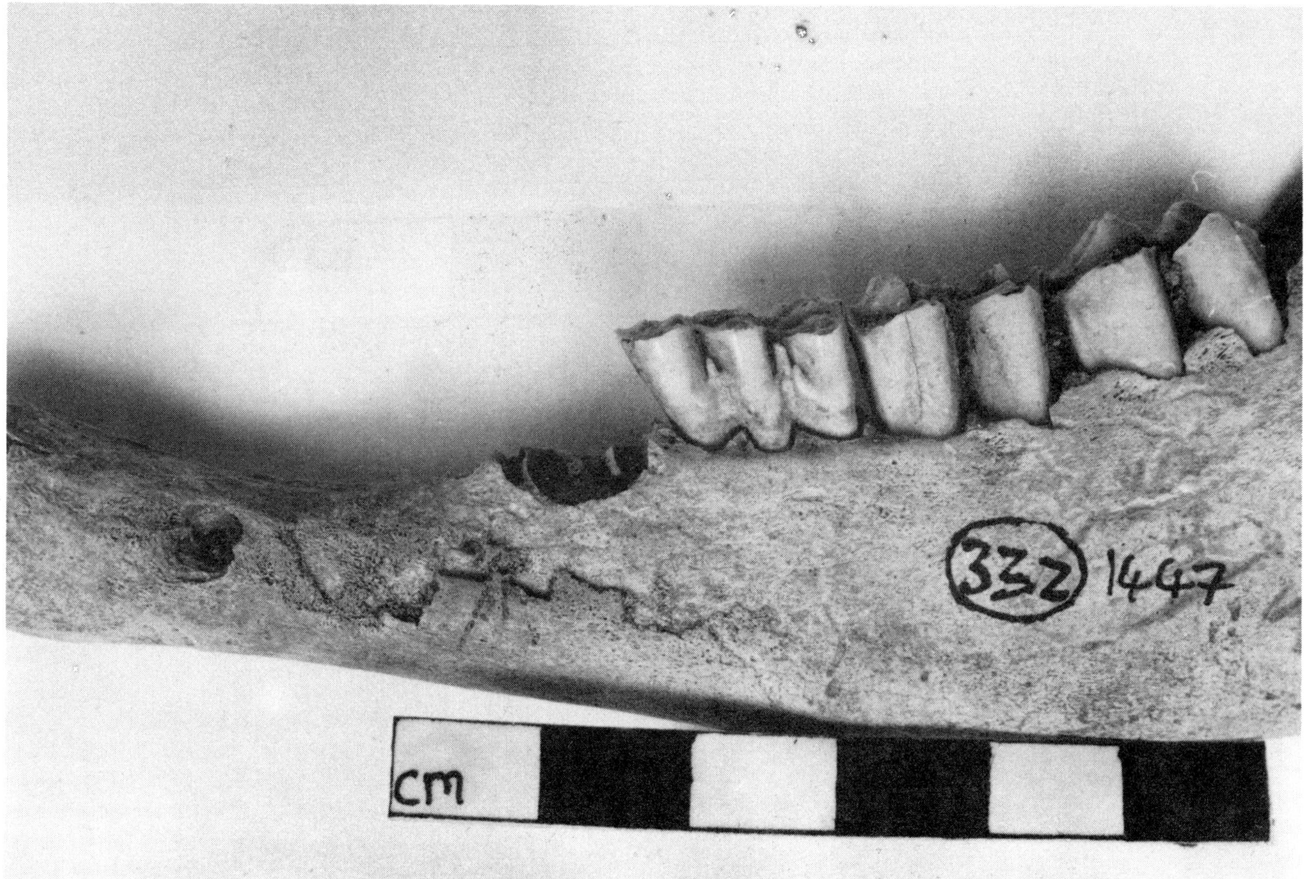

Plate 16. Extra bone growth on the diastema of a mandible: possibly due to muzzling.

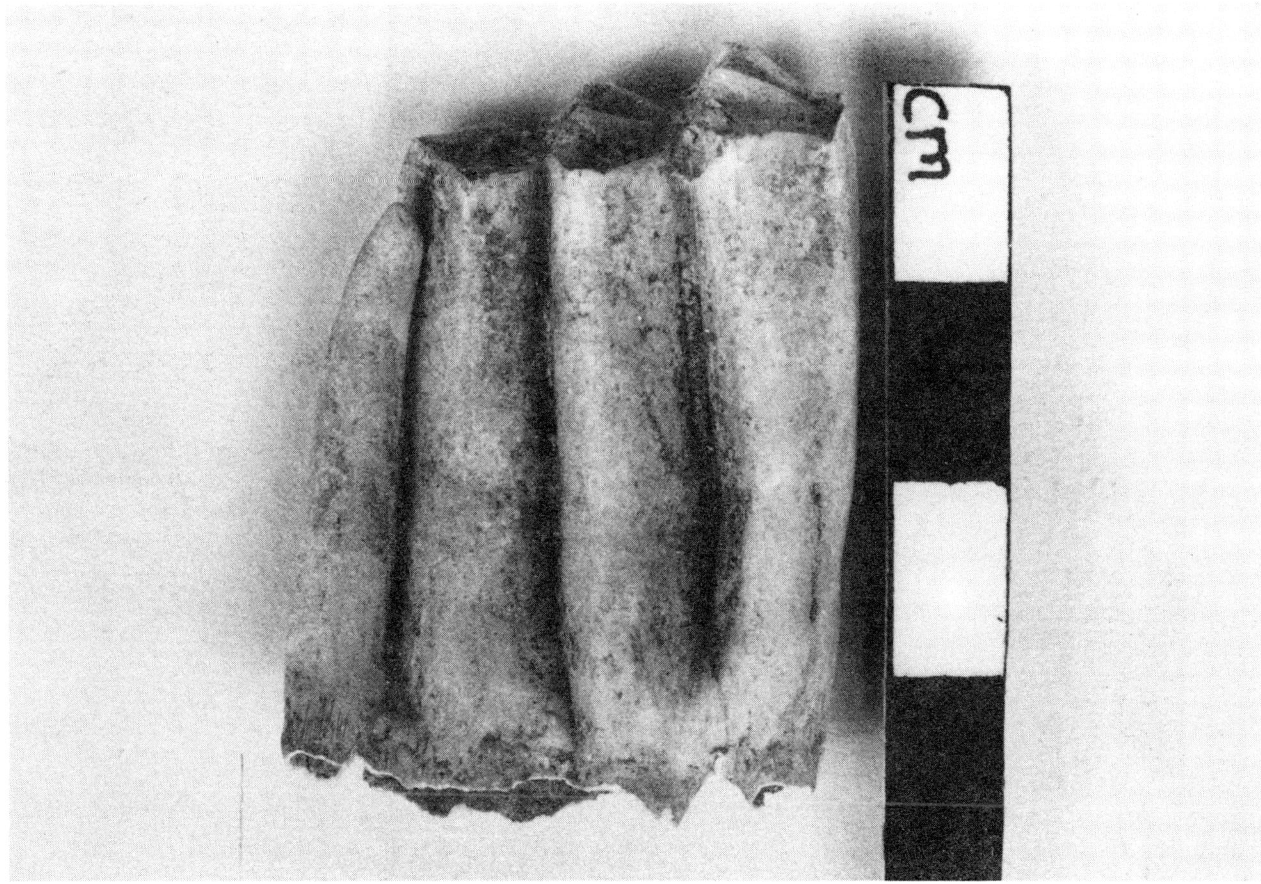

B- fourth pillar on third molar, typical of goat

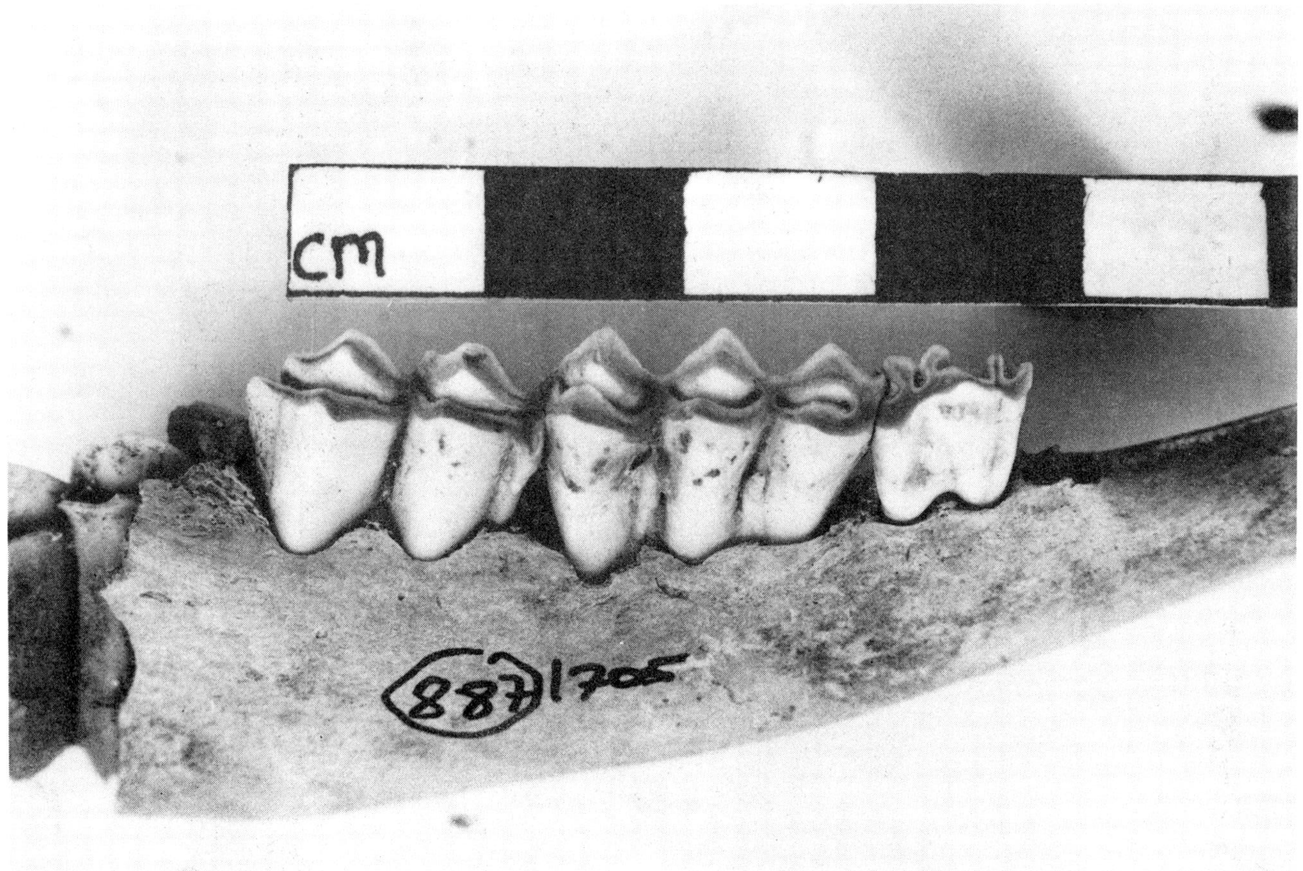

C- two pillars on deciduous fourth premolar: premolar is typical of sheep.

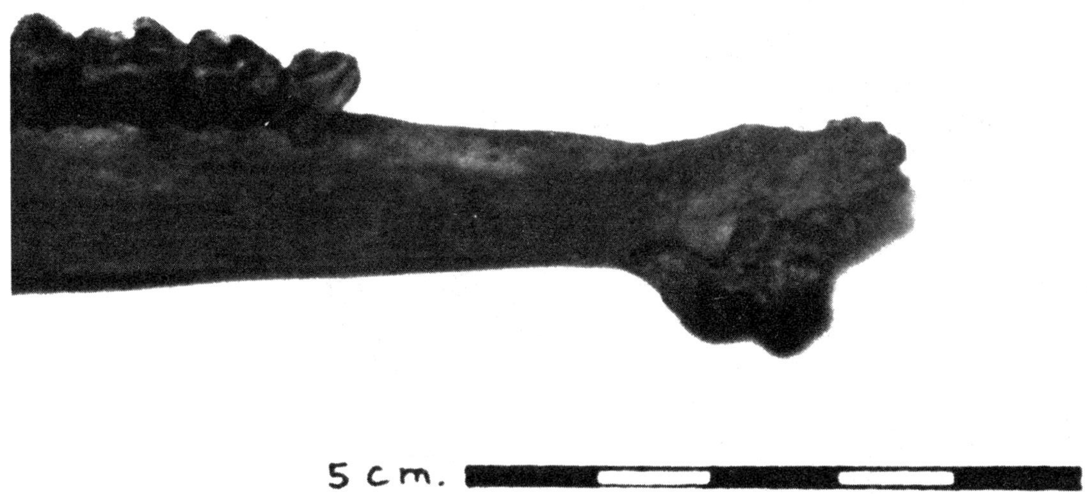

Plate 17. Abnormal extra bone development in mandibular symphysis - ? dentigerous cyst.

Plate 18. Severe malocclusion between fourth premolar and first molar.

ASSESSING THE ROLE OF PIGS FROM FAUNAL DEBRIS ON ARCHAEOLOGICAL SETTLEMENTS.

Jennie Coy, Faunal Remains Project, University of Southampton

ABSTRACT

In recent years both environmental and cultural explanations have been given to account for fluctuations in the relative numbers of pigs assessed for faunal material compared with the numbers of the other common domestic ungulates. This paper suggests that, while striving to develop the calculation of absolute values for pig presence, it should meanwhile be possible to make relative values more accurate and to define their method of calculation more precisely. Methodological safeguards must be combined with an assessment of settlement usage.

Results from some recent analyses conducted by the writer on Wessex material are discussed and are compared with earlier published results using an index 'PIG (A)' which represents the relative role of pig in the total ungulate count and 'PIG (B)' representing their role in those ungulate bones identified to species. Two additional useful indices which throw further light on possible biassing factors are an INDEX OF IDENTIFIABILITY for bones at a settlement and an index developed by King (1978) which gives the relative presence of jaw and limb bones.

INTRODUCTION

In recent years both environmental and cultural explanations have been given for fluctuations in the relative representation of pigs in faunal material compared with results for other common domestic mammals.

Grigson (1982) suggested relative increase in pig representation in the Late Neolithic in England compared with results for the Early Neolithic and linked this with the frontier aspect of pig-keeping in any expansion to a wider range of environments and especially to the pig's great benefits for clearance of regenerating woodland and areas of bracken (*Pteridium aquilinum*). The high frequency pig sites she discussed included Durrington Walls (SU 150453), Marden (SU 091583), and Mount Pleasant (SY 710899) all studies by Harcourt (1979).

Cunliffe (1978:182) suggested that the large variation in the relative number of pig bones on Iron Age sites may be a reflection of ecological variation, with woodlands being more suitable for pigs and open downland settlements therefore producing low values. Bradley (1978:37) also suggested that regional and environmental factors were important in any such analysis. King's excellent and provocative survey of Romano-British faunal material suggested that pig was commoner on 'Romanised settlements' than on native sites which might be caused by the former being near woodlands or by cultural factors associated with the Roman's high regard for pork. These new data presented here make a contribution to this debate.

For the Anglo-Saxon Period Clutton-Brock (1976:378) has shown that pigs are not so well represented on archaeological sites as would be expected from the documents and has suggested that carcases may have been boned and salted for bacon and their bones are therefore not found in kitchen waste.

Harcourt (1979) has put all this in the proper perspective by pointing out for his Late Neolithic material, cattle would have made the major meat contribution at Durrington Walls, Marden and Mount Pleasant despite the very high representation of pig fragments. (Harcourt 1979:222).

Of relevance to all such quantification of species representation, Maltby has pointed out that the results obtained may be highly biassed both by the methods of analysis used at all stages and by the contextual origins of the material (Maltby 1981:164). These two factors - methodology and intra-site variability - are discussed in detail below and the ideas are then applied to some recent results for the Wessex Iron Age.

THE QUANTIFICATION OF PIG REMAINS - METHODOLOGY

The value that we seek should really be an absolute one. For example, the number of fragments of pig, minimum number of individuals (MNI), weight of bone, or meat estimate measured against some unit of human occupation, such as 'deposit volume' or 'number of people years'. In exactly the same manner as Thomas (this volume) demonstrated for the mollusc *Pomatias elegans*, pig keeping may have been going on at the same level per human group throughout a period, although its relative value compared with the other major domesticates may vary.

Any calculation of 'deposit volume' would face the problem of soil compaction, mentioned by Gilbertson and Hawkins (this volume) in his discussion of bulk density and molluscan frequencies. Archaeozoologists could usefully carry out more work relating bone evidence with deposit volumes (perhaps using bone densities, e.g., number of fragments per cubic metre) and results would be expected to vary enormously between different types of feature. Maltby (per comm) has recently demonstrated the great differences between the content of the top halves of some Iron Age pits compared with the bottom halves. A thorough knowledge of the refuse disposal aspects of a settlement would therfore be essential for putting these variations on an intersite comparative basis just as it is for the current methodology discussed below.

Although absolute values may be the ultimate aim we are meanwhile reduced to using results for pig available now and comparing them with similar values for other species - that is, calculating a relative value. This is acceptable so long as we are only saying that the relative importance of pig declines or increases. There may have been exactly the same number of pigs per person. In addition it should also be remembered that pigs may not in fact compete with the grazing animals either for food or for people's attention. Three other factors should also be taken into account when discussing the relative value of the major species (i) the relative meat values, e.g., as discussed by Harcourt (1979:222); (ii) the life-span and prolificity of the species involved (Bourdillon and Coy 1980:91); and (iii) fluctuations which presumable occurred with the season. The last aspect is relatively unexplored in British Archaeology.

The use of percentages rather than the relative frequencies themselves is misleading and tempts archaeologists to make misleading interpretations from bone reports. Obviously original frequencies are what are subjected to statistical tests (like chi-squared) but percentages are what are usually published for intra- and inter-site comparisons and they are used in this paper in the hope that their limitations will be appreciated.

What we have actually quantified in the past has been numbers of fragments, MNI (calculated in various ways), or weight of bones of the various species. The suggestion of Fieller and Turner (1982) that the Petersen estimates used in biological capture-recapture methods could be used to give quite a different basis for such comparisons gives new possibilities. The results from these methods of representation vary widely and they can all be calculated in many different ways. Unless the way in which they were obtained is rigorously defined and consistent it is still not reliable to make inter-site comparisons, even when the same method of quantification appears to have been used. These methods of quantification will now be explored in turn.

Numbers of fragments

A crucial factor here is the 'level of identification', whether a fragment is identified to species or to a wider category. This has been discussed elsewhere (Coy 1980:42). On some sites there is very little input into the diet by horse or red deer (Cervus elaphus Linnaeus), so that workers assume that fragments of large vertebrae, for example, are those of cattle. Other workers may leave the majority (or even all) large ungulate ribs and vertebrae in the 'unidentifiable' category. Various ways round this have been introduced. Some workers use a method of only scoring bones with epiphyseal ends and therefore recognizable to species (Grant 1975: 379), Coy has adopted a recent policy of only contrasting results for 'small' and 'large ungulate' groupings rather than for individual species which avoids the problem that 'small ungulate' fragments may come from pig or sheep (Coy n.d. 1,2 and 3).

Therefore if a percentage is given for the representation of pig it should clearly state what it is a percentage of. I shall use in this paper two different percentage values for pig number of fragments - PIG (A) and PIG (B). These are defined in Table 1 using the codes in the Ancient Monuments Laboratory's system (Jones n.d.). A is the percentage of pig bones amongst the total number of fragments accepted as being from the major ungulates. There are many different factors which complicate the calculation of this including the accessibility of a good and relevant modern comparative collection, skill of the operative, and the extent to which the material bears modern breaks. The cut off for the scoring of fragments here is the point at which they can only just be identified as likely to come from the major ungulates. Those not identifiable to species are scored as SAR (small artiodactyl) or OXO (large ungulate). After that fragments are scored as UNM (unknown Mammal fragments) if it is just possible to say that they are likely to be mammalian. These and those from birds, fish, identifiable other mammals, and UNX (unknown vertebrate species) are excluded from the calculation of the PIG (A) index.

The results given in this paper assume that all modern breaks are discounted where possible. Obviously PIG (A) could also be calculated for cattle, sheep and goat, and for horse on settlements where this forms a part of the diet. The value of SAR and OXO is a useful indicator too of the level of identifiability of a collection and their values in fact control the Identifiability Index (Table 1).

Index B is the percentage of pig bones amongst the bones of the major domesticates that are identifiable to species (or to ovicaprid). In the sites discussed goat and horse have been ignored as they were not significant in the food supply.
So the B index totals for cattle, sheep and pigs, unlike the A totals, will add up to 100%. What is so confusing in the literature is that A indices are sometimes compared with B indices or with intermediate values. Often we do not know how the specific percentages were derived at all. Sometimes pig is quoted as a percentage of all the species including those for which there is no butchery evidence. This is permissible so long as other indices can be explored from the data provided, preferably the original frequencies.

Of comparative interest (Table 2) are the successive values derived from results for the Hamwic Pit (Colley, this volume) showing the effect of 'level of identification' on species percentage in an excavation where every bone up to 10mm in length was saved. These figures do not include the material retrieved by sieving the spoil, although on most sites most of the UNM and UNX material would only be retrieved by sieving.

Weights

Most of the above comments apply also to the derivation of indices from bone weights. Few workers in Britain use them but Bourdillon maintains that they have some comparative use in her continuing studies in Southampton. Their use was discussed for Hamwic for the Melbourne Street sites (Bourdillon and Coy 1980:83).

Minimum Number of Individuals

Grigson showed that the Late Neolithic increase for pig representation is so striking that it matters little whether numbers of fragments or minimum numbers of individuals (MNI) are used (Grigson 1982:306). Less striking changes might be obscured if only one method of representation were used. Numbers of fragments can severely under-represent pig compared with MNI in cases where there is a high frequency of mandibular remains. This point is made clearly in King's work (1978:211) which gives details for a number of settlements using an index which has been called the 'King Index' in Table 1. This is the number of jaw fragments as a percentage of the number of limb bones. Although King talks about the number of mandibular fragments this appears to include upper jaws (maxillae) as well.

On sites with a higher King Index for pig than for sheep and cattle, it is possible that minimum numbers would give a better indication of relative value of pig than numbers of fragments or weights. A number of other methods have been used in the past to throw up bias in anatomical representation and some of these have been used in association with minimum numbers (Binford and Bertram 1977:80 , Coy 1980:45, Reichstein and Tiessen 1974:21).

At Hamwic switching to MNI from 'numbers of fragments' doubles the apparent percentage of pig, whether it is counted for the site as a whole or cumulatively by feature (Table 3). For material studied from Bury Wood Camp, Wiltshire, where as at Hamwic fragment, weight and MNI methodologies were compared, the switch from overall to cumulative MNI raised the percentage of pig even more, from 10.5 to 17% (Coy unpublished manuscript referred to in Coy 1964:47). This material had a particularly high King Index and demonstrates that the calculation of that index is a useful indicator as it can explain in some cases why the different methods of calculating specific percentages give such different results.

Equally important when MNI is being used is some discussion of the method by which it was derived. One limitation of King's discussion (1978:208) is the use of differently derived MNIs but in such a general and large scale discussion this is not serious.

Capture-Recapture Methods

Fieller and Turner's use of capture-recapture as an answer to the problem of specific ratios represents a new departure which archaeozoologists may wish to apply (Fieller and Turner 1982). As they stress MNI is, after all, just the animals on the analyst's table. Use of the Petersen Index is, however, measuring something quite different so that this may indeed give quite new insight into archaeological material and act as a new comparative method. Winder, who has tried this technique on archaeological material is finding that proximal-distal matches may form a more practical alternative to left-right matches for long bones (Winder, personal communication).

THE QUANTIFICATION OF PIG REMAINS - INTRA-SITE VARIABILITY

There is a likelihood that bones of the three major ungulates, because of a number of factors including their likely differences in derivation and treatment and their very different bone shapes and sizes, are not deposited or preserved in the same way. Different context types might be expected to produce different results for pig and the analyses below were therefore very strongly tied to feature type. In fact, although there is a little evidence that poor preservation is linked with higher percentages for pig (Bourdillon and Coy 1980:104, Coy n.d.2) the pig values were remarkably consistent across the settlements in most cases. But feature types are indicated in Tables 4-6 for interest.

The indices already discussed were plotted for correlations and there was no direct obvious correlation between the percentage of pig bones and the degree of fragmentation.

RESULTS FOR SOME RECENT ANALYSES

Groundwell Farm, Blunsden St. Andrew, Wiltshire (SU 151891) showed Iron Age Banjo and Romano-British features of A.D. first and second century date and was excavated by Mr. C. Gingell for the Wessex Archaeological Committee. The bones came mostly from wall trenches of successive phases of huts. Much of the bone was very fragmentary, often as splinters, rather than cylinders, of long bones and may have been subjected to trampling and secondarily deposited in the wall trenches. This type of deposit is quite different from the available comparative material in Wessex of this period which tends to be from pits or ditches.

Table 4 shows that the PIG (B) index is extremely high, ranging from 27 to 49%. This is most unusual for an Iron Age site in Wessex and compares with values for PIG (B) of 8% for Iron Age phases of Winnall Down, Hampshire, (SU 498304), which had a similar level of identifiability, and 8% for Cowdery's Down Iron Age features, (SU 526369) with a very slightly higher level of identifiability (Maltby, personal communication). Unpublished results from the Micheldever Banjo, Hampshire (SU 658532) gave slightly higher values for PIG (B) of 15% for Iron Age features and 12% for Romano-British features. Bones from the Iron Age features were however largely from pits.

Cleavel Point, Ower, Dorset (SZ 003860) provided a variety of features from A.D. first to fourth century with the biggest samples from Phases 1 and 2 and the 3-4th century Phase, Phase 7. The excavations were carried out by Mr. P. Woodward for the Wessex Archaeological Committee.

Table 5 shows that samples were small and that values for PIG (B) were extremely variable and often very high. The King Index for this site was very high. These deposits were generally not good ones for preservation of animal bone and in some there is a high degree of fragmentation, erosion and dog-gnawing. The degree to which the King Index might indicate a considerable disappearance of pig fragments and the extent to which fragmentation and erosion might affect the different species differently are difficult to analyse, particularly for such a small sample. Surprisingly, the Identifiability Index for this site was much higher than for Wessex Iron Age sites normally.

Rope Lake Hole Kimmeridge, Dorset (SY 931776) is geographically and archaeologically very close to Eldon's Seat (Cunliffe and Phillipson 1968) and from what can be deduced from the account of faunal remains at the latter it has comparable relative results for pig. Table 6 shows an overall result for PIG (B) of 6% compared with an estimated result for Eldon's Seat of 5%. These results are more in line with the Wessex Iron Age results from earlier studies. Like Cleavel this settlement shows quite high Indices of Identifiability but it does not have particularly high King Indices.

The King Indices obviously vary madly for sites where pig is such a small proportion and, as King suggests, sites where fewer than ten individuals are represented cannot be expected to give reliable results (King 1978:210).

The phases of this settlement ranged from Early Iron Age through to Romano-British date.

CONCLUSIONS

Can we actually detect a shift in species ratios when it occurs? To work out a specific ratio there must be a careful combination of methodological safeguards ensuring continuity and comparability of methodology combined with an increased understanding of the complexity of site formation processes and seasonal effects. Once we have comparable specific ratios there may be secondary factors like meat value, life span and prolificity to be taken into account when comparing the roles of the different species.

The results obtained should also be viewed against the different husbandry practices and it should not be forgotten that the keeping of pigs may not compete with the keeping of grazing ungulates for resources of habitat, feed, or human labour. In many ways pigs, which can be dependent on waste products of human activity, might give a more reliable guide to human activity than cattle and sheep remains but we should need to calculate an absolute value not a relative one.

The differences seen in pig skeletal material with respect to fragmentation, distribution around a settlement, and the notably high King Index on some sites means that fragment numbers and MNIs must be examined carefully both with respect to fragmentation levels and differential anatomical survival. The results from the three Wessex settlements presented here show an interesting mixture of all these factors and, in addition, two of them have very high values for PIG (B) index which on the whole are consistent for all phases and feature types on the settlements. Groundwell has an unusually high number of pig bones for an Iron Age settlement but there is no ecological problem about pig keeping in such a rich area, near a river valley. Cleavel Point, Ower, is a completely different type of settlement but was at the time probably provided with a rich and varied hinterland and there may have been access to saltmarsh and to marine molluscs. There seem to have been good supplies of the latter but whether these supplemented pig as well as human diet is unknown. Dangers of the ecological explanation are brought home when viewing the area today as oaks are the only naturally seeding trees in any abundance but it is likely that this is the result of intensive cattle grazing in recent centuries as this is the only species which can survive this successfully (Oliver Rackham, personal communication). The only other coastal Wessex settlement of the period with similarly high pig values is Fishbourne (Grant 1971).

We are left then with a number of viable theories to explain these variations from settlement to settlement but it is certain that consistent methodology will be necessary if we are to learn more.

REFERENCES

Binford L.R. and Bertram J.B. (1977) Bone frequencies and attritional processes. In (L.R. Binford ed.) For Theory Building in Archaeology New York: Academic Press 77-153.

Bourillon J. and Coy J.P. (1980) The animal bones. In (P. Holdsworth ed.) Saxon Southampton: Excavations at Melbourne Street 1971-6. Council for British Archaeology Research Report 33 79-121.

Bradley R. (1978) The Prehistoric Settlement of Britain London: Routledge and Kegan Paul.

Clutton-Brock J. (1976) The animal resources. In (D.M. Wilson ed) The Archaeology of Anglo-Saxon England London: Methuen 373-392.

Coy J.P. (1969) The animal bones. In (D.Grant King) Bury Wood Camp. Wiltshire Archaeological and Natural History Magazine 64 47-48.

Coy J.P. (1980) The animal Bones. In (J. Haslam) Excavation of a mid-Saxon iron smelting site at Ramsbury, Wiltshire. Medieval Archaeology 24 41-51.

Coy J.P.(n.d.1) The animal bones. In (C. Gingell) Excavations at Groundwell Farm. To be published in Wiltshire Archaeological and Natural History Magazine.

Coy J.P. (n.d.2) The animal bones. In (P.J. Woodward) Excavation of a Late Iron Age trading settlement and Romano-British BB1 pottery production site at Ower, Dorset. To be published in the Dorset Natural History and Archaeological Society Monograph Series.

Coy J.P. (n.d.3) The animal bones. In (P.J. Woodward) Excavation of an Iron Age and Romano-British settlement at Rope Lake Hole, Corfe Castle, Dorset. To be published in the Dorset Natural History and Archaeological Society Monograph Series.

Cunliffe B. (1978) Iron Age Communities in Britain, London: Routledge and Kegan Paul, 2nd Edn.

Cunliffe B. and Phillipson W. (1968) Excavations at Eldon's Seat, Encombe, Dorset. Proceedings of the Prehistoric Society 34 191-237.

Grant A. (1971) The animal bones. In B.Cunliffe (ed) Excavations at Fishbourne 1961-1969 London: Society of Antiquaries Research Report 27, Vol.II. 377-383.

Grant A (1975) The animal Bones in B. Cunliffe (ed) Excavations at Porchester Castle Vol.I Roman London: Society of

Antiquaries Research Report 32, 378-408.

Grigson C. (1982) Porridge and pannage: pig husbandry in Neolithic England. In M. Bell and S. Limbrey(eds) Archaeological Aspects of Woodland Ecology Oxford: British Archaeological Reports International Series 146, 297-314.

Harcourt R.A. (1979) The animal bones. In (G.J. Wainwright) Mount Pleasant, Dorset: Excavations 1970-1971 London: Society of Antiquaries Research Report 37 214-223.

Jones R.T. (n.d.) Computer-based Osteometry. Ancient Monuments Laboratory Report No.3342

King A. (1978) A comparative survey of bone assemblages from Roman sites in Britain Institute of Archaeology Bulletin 15 207-232.

Maltby J.M. (1981) Iron Age, Romano-British and Anglo-Saxon animal husbandry - a review of the faunal evidence. In M. Jones and G. Dimbleby (eds) The Environment of Man: the Iron Age to the Anglo Saxon Period. Oxford: British Archaeological Reports. British Series 87 155-203.

Reichstein H. and Tiessen M (1974) Materialen zur Kenntnis der Haustiers Haithabus Berichte uber die Ausgrabungen in Haithabu 7 9-101.

Frequencies

P	number of pig fragments
J	number of pig jaw fragments
LB	number of pig long bone fragments (excluding fibula, metapodials, and phalanges)
A	total ungulate fragments
B	ungulate fragments identifiable to species

Indices Based on Above

index	how calculated	identifications
pig (A)	$\frac{P}{A} \times 100$	$\frac{\text{PIG}}{\text{PIG COW SHE S+G + SAR OXO}} \times 100$
pig (B)	$\frac{P}{B} \times 100$	$\frac{\text{PIG}}{\text{PIG COW SHE S+G}} \times 100$
identifiability	$\frac{B}{A} \times 100$	$\frac{\text{specific ungulate}}{\text{total ungulate}} \times 100$
King index	$\frac{J}{LB} \times 100$	$\frac{\text{pig jaws}}{\text{pig long bones}} \times 100$

Key to AML Codes

COW - cattle
SHE - sheep
S+G - sheep or goat
SAR - small artiodactyl (sheep, goat, pig, roe deer)
OXO - large ungulate (horse, cattle, red deer)

TABLE 1: Frequencies and Indices Used

SPECIES	% PIG
pig	
cow	
	13.4 pig (B)
sheep	
	7.0 pig (A)
goat	6.3
sh/go	5.9
SAR	
LAR	
UNM	
UN	
others	

TABLE 2: Percentage of Pig Fragments for Hamwic Pit Calculated for sucessive 'levels of identification'

(Bones identified to species only)

method	cattle	sheep	pig	total
no. fragments	23,896	14,606	6,953	45,455
%	53	32	15*	
MNI (overall)	211	265	192	668
%	32	40	29	
MNI (cumulative)	422	480	386	1,288
%	33	37	30	

* This corresponds to PIG (B). The PIG (A) value was 9%

TABLE 3: Methods used at Melbourne Street, Hamwic

phase	feature type	frequencies		indices %			
		A	B	pig(A)	pig(B)	ident	King
IA1	postholes	63	-	8	-	-	-
IA2	wall trench	2,189	837	12	31	41	81
IA3	wall trench +	413	128	10	32	31	40
IA4	wall trench	841	291	10	27	35	115
IA5	wall trench	3,251	1,363	18	42	42	72
IA6	hut entrance	517	225	12	28	43	92
IA	varied	2,497	886	11	31	35	81
RB	linear feature	108	43	19	49	40	-
TOTAL		10,050	3,815	13	35	38	74

Note: For meanings of all headings see Table 1. Blanks represent low frequencies unlikely to be reliable.

TABLE 4: Results for Groundwell Farm, Wiltshire (Coy n.d.1)

phase	feature type	frequencies		indices (%)			
		A	B	pig(A)	pig(B)	ident	King
1 + 2	mainly ditch	681	468	35	51	69	191
3	varied	239	131	18	34	55	200
4+5+6	varied	175	122	26	37	70	200
7	varied	561	436	14	18	78	41
TOTAL		1,658	1,159	25	35	70	141

Note For meanings of all headings see Table 1

TABLE 5: Results for Cleavel Point, Ower, Dorset (Coy n.d.2)

phase	feature type	frequencies		indices (%)			
		A	B	pig(A)	pig(B)	ident	King
I	collapsed stone huts	206	149	5	7	72	jaws only
II	shale and flint tips	253	163	4	7	64	100
IIIA	accumulation & occupation	430	257	5	8	60	117
IIIB	occupation	231	128	3	6	55	50
IVB	occupation	812	587	6	9	72	100
V	stone huts	517	333	3	4	64	-
VIB	pits into occupation	196	141	3	4	72	-
VIIA	varied	949	620	3	5	65	55
VIIC	disturbed occupation	1,355	719	4	7	53	-
TOTAL (inc small features not given)		5,160	3,229	4	6	62	60

Note For meanings of all headings see Table 1.

TABLE 6: Results for Rope Lake Hole, Dorset (Coy n.d. 3)

SOME EFFECTS OF PRESERVATIONAL BIASES ON INTERPRETATIONS OF ANIMAL BONES.

Sue Stallibrass
Dept. of Prehistory & Archaeology
The University of Sheffield
SHEFFIELD S10 2TN

ABSTRACT

A method for recording the preservational state of archaeological animal bones is described using material from two Romano-British sites in Castleford. Three variables are used - Texture, Minerals and Erosion. The variable of Texture seems to be the most useful as an indicator of differences in preservation conditions between context types. Combined with the other variables and, in some cases, average fragment sizes, these data are used to suggest different methods of bone disposal into the different contexts.

The relative numbers of fragments of Cattle, Sheep/Goat and Pig are given for each context group. Experiments by Nicki Cleminson of Southampton University have shown that Cattle bones tend to preserve better than Pig bones, and mature bones better than young ones. Using her results, together with the age structure of the Castleford material and the results from the analysis of preservation conditions, the relative numbers of Cattle, Sheep/Goat and Pig fragments are reanalysed taking preservational biases into account. This leads to a ranking of 'reliability' of the collections from different context groups, and a distinction is made between 'evidence of absence' and 'absence of evidence'.

Finally, these methods are used to make intersite comparisons, and an assessment is made of their usefulness in analyses of archaeological animal bones.

INTRODUCTION AND METHODS

Every stratified animal bone from the West Yorkshire Unit's excavations of Romano-British Castleford was recorded for several variables relating to its preservation, regardless of whether or not it could be identified to species level. The most important of these variables were: Texture, Minerals and Erosion. Each of these was, in fact, recorded for a variety of states, but this paper will be confined to the commonest and will only consider simple presence/absence data.

Texture was recorded as zero if the bone was simply clean and dry (ie was non-greasy). The commonest other state was 'brittle'. Brittle fragments are light and rather biscuity. They shatter easily. They have lost much of their organic content and are rather porous and, therefore, susceptible to leaching. I assume that a high proportion of brittle fragments indicates poor preservation.

Minerals: many of the bones are dusted with mineral stainings, mainly iron compounds, and some have encrustations of cemented grit and mineral deposits. The site at Castleford is on sands and clays near the river and has a high, sometimes fluctuating, water table. The worst affected bones tend to be from contexts with gleyed soils and, again, I assume that high proportions of fragments with Mineral deposits indicate poor preservation conditions, probably associated with water action.

Erosion: although the bones are not well preserved, very few of them are classified as 'eroded all over'. These few, Eroded, fragments have almost lost their structure. They have a powdery surface and are crazed throughout- they are liable to disintegrate when picked up.

The first site excavated at Castleford was part of the early fort, and had about 3,000 fragments from stratified Romano-British contexts. I have divided these contexts into seven groups - Pits, Ditches 1-5, Ditch 6, a turf Rampart, the Road surface, Linear features (gullies etc.), and a large Latrine. The occupation at this site was quite short (less than one century) and so the contexts have been treated as being broadly contemporaneous. A chi square analysis was calculated to test the null hypothesis that all attributes were equally distributed across the context types (see Figure 1).

In a chi square analysis all the variables have to be independent, whereas each fragment had been recorded three times over, therefore, they were separated out into Texture (ie brittle), Minerals (ie with minerals present) and Eroded (ie eroded all over), and then Texture with Minerals, Minerals with Erosion etc..

RESULTS AND DISCUSSION

In Figure 1, chi square values of 20 or more have been emphasised: encircled values eg 21 are Overrepresented (ie there are more bones in this category than would be expected if all attributes were equally distributed between context types), and values beneath a bar eg 20 are under-represented. The figure of 20 has been chosen as an arbitrary cut-off point to highlight the larger anomalies.

Figure 1 shows the following aspects:-

Pits: basically as predicted by the null hypothesis, except that there are fewer Mineral-affected fragments than expected. This is not surprising if the mineral deposits are, indeed, associated with the presence of water, since the Pits are less likely than some other features (eg the ditches), to have acted as water collectors.

Ditches: the bones from the ditches themselves are generally badly preserved. A total of seventy-three percent of the bones from Ditches 1-5 are brittle Textured, and an unexpectedly high number (56%) are both brittle Textured and affected by Minerals.

Ditch 6 has high scores for all three categories with brittle Texture, and for two each of the Eroded and Mineral categories. Both the Texture and Mineral aspects are thought to relate to the presence of water, and the ditches are thought to have acted as collectors of surface and ground waters, draining these down towards the river. Is the Eroded category related to water abrasion, or does it reflect residual material incorporated into the ditch?

Linears: no anomalies. These contexts were shallow, and so less likely to be affected by water than the ditches, since their deposits were probably always above the water table.

Road: high on brittle Texture. Probably due to fragments lying around on the Road surface exposed to sub-aerial weathering. Trampling may have further emphasised this aspect by comminuting the fragments into greater numbers. The average size of the Road fragments is, in fact, the smallest of all the context groups.

Rampart: this was built of clayey turves, and so it was expected that the bones would be well protected from leaching and water abrasion. As expected, there are few with Minerals, implying a relative absence of water compared with the other context types, and yet there are relatively high numbers of Eroded and brittle Textured, brittle Textured, and Eroded fragments. These are probably earlier material incorporated into the turf Rampart when it was constructed. Pieces lying around on the ground surface or in the turf would probably have become leached and brittle like the ones on the Road surface. It is important to be able to distinguish residual material- although the ditches and Rampart were constructed at the same time, their contents relate to two different periods - one, the Rampart, relating to a pre-defensive stage, and the other - the accompanying ditch fill, to activities post-dating their construction. On some sites, especially prehistoric ones where pottery is rare or poorly preserved, residual bones may be able to give important clues about earlier phases of the site. Certainly, their presence should be noted before any attempt is made at interpretation.

Latrine: three categories are under-repesented ie the bones are better preserved than expected. The Latrine is low on brittle Texture, brittle Texture and Erosion, and on brittle Texture, Minerals and Erosion. This is probably the best preserved collection from the site, and the fragments also tend to be larger than average. Most of the material seems to have been discarded directly into the context and not left around to be weathered or trampled.

In this chi square analysis, very low numbers were expected in all context types for two of the three categories to include Eroded fragments (vis. Erosion, and Minerals plus Erosion). Because of this, no precise statistics have been given here, although there are clearly very significant differences between the states of preservation of the bones from different context groups. The analysis has been included here, mainly to show the anomalous numbers of Eroded fragments in the Rampart and Ditch 6 collections.

Figure 2 repeats the chi square analysis but ignores the variable of Erosion, thereby avoiding the statistical problem of low values for expected numbers. Otherwise, it shows exactly the same pattern of differences as Figure 1, and it gives a chi square value of 765 with 27 degrees of freedom, $p<0.001$.

One of the reasons for grouping the contexts was to see if there was any lateral variation in the species represented. Figure 3 shows the relative proportions of the three major domestic species - Cattle, Sheep/Goat and Pig, using numbers of fragments. Although all seven groups are low on Pig (all are less than 10%), the relative proportions for Cattle and Sheep/Goat vary considerably- Cattle from 64% to 87%, Sheep/Goat from 12% to 32%. But are these differences valid, or are they artefacts of the preservation conditions? Are some species more susceptible to destruction, especially by leaching, than others?

I have not yet had time to do my own experiments testing this in the lab., but a few years ago Nicki Cleminson at Southampton did her undergraduate dissertation comparing the relative survival of young and mature bones of Cattle and Pig in controlled acid solutions (Cleminson, 1979). She found clear differences showing that Cattle survived better than Pig, and that mature bones survived better than young ones. Unfortunately, she did not test any Sheep bones.

At Castleford, most of the Cattle bones were mature and, therefore, should have preserved well, whereas Pig bones were nearly all immature, and so were disadvantaged on both factors. Many of the Castleford Sheep/Goat bones were also immature, and so were disadvantaged on at least one factor.

Looking again at Figure 3, showing the percentages of the species present in the different contexts, the earlier discussion of preservation conditions can be used to evaluate the reliability of the data.

On the grounds of brittle Texture and presence of Minerals, the Ditches had the worst preservation conditions; and Ditch 6 contained almost no Pig bones and very few of Sheep/Goat. Ditches 1-5 also had high numbers of brittle Textured and Mineral-affected bones, but far fewer Eroded fragments; they had a higher proportion of Pig bones than Ditch 6, but were still low on Sheep/Goat.

The Road had high numbers of brittle-Textured bones too, but far fewer with Mineral deposits and almost no Eroded fragments. Despite the difference in the number of Mineral-affected fragments, the species proportions in the Road and Ditches 1-5 are very similar. On the basis of the variables considered in this study, there are two possible explanations of this phenomenon: (1) the proportion of Mineral-affected bones has little relevance to the differential survival of species. The assumption here is that the species compositions of the two collections were originally similar and that, having been subjected to similar preservation conditions (reflected by the numbers of brittle-Textured and Eroded fragments), these species proportions have remained similar despite differences in the relative numbers of fragments affected by Minerals; or (2) the species compositions of the two collections were originally dissimilar but that, whilst the variables of brittle Texture and Erosion have been held constant, the bias caused by differential preservation (reflected by Mineral-affected bones) exactly counteracts the original difference in species composition. On the grounds of parsimony, the former hypothesis seems more likely than the latter.

The Latrine had the best preservation conditions of all the contexts. This may be reflected by its having (marginally) the most Pig (9%). But it still has only 20% Sheep/Goat. These ratios I would accept with reasonable confidence as a representative sample of the original proportions of fragments in the deposit (allowing for the lack of sieving), since the bones appear to have been deposited directly into the context, and to have suffered relatively little subsequent destruction.

Linear features were average on all aspects of preservation and yet had more Sheep/Goat than the Latrine, which I accept as a 'genuine' difference (ie one not explicable by preservation conditions). It should be noted that the age structures of all three species are similar throughout the site, therefore, differences in species proportions cannot be explained by different ratios of mature to immature bones.

The Rampart had dreadful preservation- lots of Eroded fragments compared with the other contexts, and has very little Pig, as expected. But it also has the highest percentage of Sheep/Goat, whereas the preservation conditions would suggest that Cattle would be overrepresented. However, any interpretations of the Rampart collection are complicated by the presence of residual material. That is, although the relatively high proportion of Sheep/Goat fragments may even be an under-statement of the anomalous position of this group, it is unclear whether the anomaly reflects simply lateral, or lateral and temporal variation.

In the case of Ditch 6, it was unclear whether the relatively quite high numbers of Eroded fragments denoted residual material or not. This was because the general level of preservation was so poor (reflected by the numbers of brittle Textured and Mineral-affected fragments), that it seemed possible that fragments might have become Eroded whilst being tumbled along by water in the gritty, rubbly matrix. For this reason, the Ditch 6 collection was not treated as a temporally mixed deposit.

The Pits, the context group with second best preservation, also have more Sheep/Goat than the Latrine, so this difference again appears to be 'genuine'. Whether or not a 10% difference in the proportion of Sheep/Goat is significant in archaeological terms is a matter for personal interpretation, but the disparity between 20% Sheep/Goat in the Latrine and 30% Sheep/Goat in the Pits cannot be explained away by the preservation factors considered in this paper.

BONE PRESERVATION AT CASTLEFORD

So, summing up the reliability of Figure 3, I would suggest that the Latrine has the most reliable set of data, and that the Pits and Linear features are quite good. Differences between these three context types, therefore, are 'genuine' and available for interpretations. The fact that the Rampart and the Pits, despite being disadvantaged by preservation conditions have higher percentages of Sheep/Goat than does the Latrine is, I think, significant, and may be used in discussions of the material. However, the lower numbers of Sheep/Goat in the disadvantaged ditch and Road collections might be due to bad preservation and, therefore, reflect absence of evidence rather than evidence of absence. It should also be remembered that all context groups are likely to be low on Pig due to the poor preservation qualities of young bones of this species. Generally, what variation there is in the percentages of Pig fragments at Castleford Site 1 can be explained by the measurements of preservation conditions given here eg: the highest percentage of Pig is in the context with the best preservation conditions, the Latrine (9%), whilst the lowest percentages are in contexts with the worst conditions - Ditch 6 and the Rampart (1% each). Within this study, therefore, spurious conclusions referring to these contextual differences can be avoided. But comparisons with other sites are more difficult, and may be invalidated if preservation conditions are not taken into consideration.

INTER-SITE COMPARISONS

To show how this approach can be applied to inter-site comparisons, I now want to consider briefly another site at Castleford that I have recorded in the same way.

Castleford Site 9 is a bathhouse area close to Site 1, but was separated from it by a few Victorian terrace houses with cellars, which precluded contiguous excavations. The soils are similar, but situated somewhat closer to the river. The contexts are mainly slightly later in date than those of Site 1, and have been grouped into four types - Baths, a Road, Miscellaneous (pits, post-holes etc.) and a Ditch.

Perhaps because of their proximity to the river, the Site 9 bones were generally worse preserved than those at Castleford Site 1.

Figure 4 shows the percentages of fragments affected by Minerals in each context group at the two sites. Although the Site 1 groups varied considerably, all seven are less affected than any of the four Site 9 groups. The averages for the two sites show that about one third of the fragments at Site 1 were affected, compared with about 90% at Site 9.

Figure 5 shows that there were similar differences in the proportions of fragments with poor texture. At Castleford Site 9 many of the bones were not just brittle Textured, but had become permineralised. Although the brittle fragments often had mineral staining or deposits on their surfaces, the permineralised fragments showed staining throughout their cross-sections, most easily seen in edges freshly exposed by new chips or breaks. Because of the extra susceptibility to breakage of these 'semi-fossilised' fragments, I consider the permineralised fragments to indicate even worse preservation than the brittle Textured fragments. No permineralised bones were found at Site 1 and yet Site 9 had very high proportions in three of the four context groups (the exception is the group from the Baths themselves).

In Figure 5, the solid black bars represent the percentages of brittle Textured fragments at the two sites, and the striped sections show the permineralised fragments at Site 9.

Despite the low numbers of affected fragments in the Site 9 Bath group, the overall averages show clear differences between the two sites, just as the Minerals did: at Site 1, about one third of the fragments have a brittle Texture whereas at Site 9, not only are about two-thirds affected, but these are more likely to be permineralised rather than just brittle.

If we look at the proportions of species found at these two sites, then, we would expect, on preservational grounds, to find that Site 9 is disadvantaged against the recovery of Pig and Sheep/Goat bones (N.B. age structures at the two sites are very similar). But, if the data from the two sites are superimposed, this does not appear to be the case - see Figure 6.

Here, the circles represent the four context groups from Site 9, and all are clearly higher in their proportions of Pig than any of the Site 1 groups, and all are above average in their percentage of Sheep/Goat. In other words, despite the poorer preservation conditions, more Pig and Sheep/Goat were found at the bathhouse site of Castleford Site 9 than at the defensive site of Castleford Site 1, and I think that this is a 'genuine' difference which can be used as a basis for discussion.

CONCLUSIONS

I hope, in this paper, to have shown the importance of taking preservation conditions into consideration before attempting any interpretations of archaeological animal bones. I hope, too, that it has shown not only the negative side - in which some collections may be proven to be untrustworthy for certain lines of analysis, but also a more positive aspect, by which we can gain insights into the taphonomic history of the site. This positive aspect can reveal not only commonsense laws (eg that bones on a road surface tend to get leached and broken), but more interesting facts such as the probable presence of a pre-Rampart phase to Site 1, and the use of the Latrine as a vast dustbin for quick-disposal refuse (where was this refuse coming from; and why did it require quick disposal compared with the rubbish in the Pits, for instance?).

Within a site, the methods described here seem to work quite well, although the variable of 'Mineral-affected' may prove to be of little importance compared with the Texture and Erosion variables (see the discussion of the Road fragments). Log linear modelling of the data by John Biggins of the Department of Probability and Statistics of Sheffield University, showed that the variables of Minerals and Erosion were both dependent on Texture, although independent of each other. Further work will look at the relationship between Texture and the relative numbers of different bone elements within species.

But recording Texture has a big drawback, since it is a subjective classification. Without a set of standard 'weights and measures' to refer to, intersite comparisons are very difficult unless done by the same person (and even then, I know that my own cut-off points between categories may waver from time to time). Perhaps an agreement to use a standard number of categories ranging from, say, (1) well preserved, to (5) almost totally disintegrated, might make different people's assessments broadly comparable.

At the moment, other criteria for measuring preservation do not seem to be any more satisfactory: for instance, the ratio of loose teeth to other fragments has plenty of its own biases (eg degree of comminution of 'other fragments', sieving versus nonsieving, lateral variation in disposal of head or post-cranial material etc etc). However, it would be regrettable if studies of this potentially valuable aspect of the data were abandoned because there is no obvious, absolute measure. The author would appreciate any comments or suggestions on this matter.

ACKNOWLEDGEMENTS

I should like to thank Nick Fieller for advice regarding the chi square analyses, and John Biggins for conducting the log linear analysis. Both these gentlemen work in the Dept. of Probability and Statistics, Sheffield University.

REFERENCE

Cleminson, Nicki (1979) The preservation of long bones - an experimental approach Unpub. undergrad. diss., Department of Archaeology, University of Southampton.

Fig. 1: Chi square forCastleford Site 1 fragments using three variables:

Texture, Minerals and Erosion

	T	M	E	TM	TE	ME	TME	na
Pits	0	$\overline{21}$	0	12	8	0	1	59
Ditches1-5	8	2	0	(124)	2	1	14	245
Ditch 6	$\overline{20}$	12	0	(23)	(52)	16	(44)	203
Linears	4	10	·5	0	0	0	3	29·5
Road	(35)	14	8	3	5	1	7	73
Rampart	(58)	$\overline{20}$	(25)	12	(34)	1	0	153
Latrine	$\overline{20}$	3	5	10	$\overline{35}$	1·5	$\overline{21}$	133·5

T Texture

M Minerals

E Erosion

na not affected

$\overline{20}$ underrepresented

(20) overrepresented

Fig. 2: Numbers of fragments and chi square values for Castleford Site 1 using two variables: Texture and Minerals

	Fragments					Chi square				
	T	M	TM	na	tot.	T	M	TM	na	tot.
Pits	71	30	35	211	347	1	$\overline{20}$	9	(17)	47
Ditches 1-5	64	32	138	14	248	7	5	(229)	$\overline{88}$	329
Ditch 6	60	109	116	97	382	1	(18)	(42)	$\overline{35}$	96
Linears	52	65	38	67	222	3	13	0	12	28
Road	88	30	29	148	295	(21)	12	8	1	42
Rampart	81	8	9	63	161	(87)	$\overline{17}$	12	2	118
Latrine	201	364	192	938	1695	$\overline{39}$	5	$\overline{28}$	(33)	105
tots.	617	638	557	1538	3350	159	90	328	188	765

Texture: brittle

Minerals: present

na: not affected

Fig. 3: Percentages of fragments of Cattle, Sheep/Goat and Pig

at Castleford Site 1

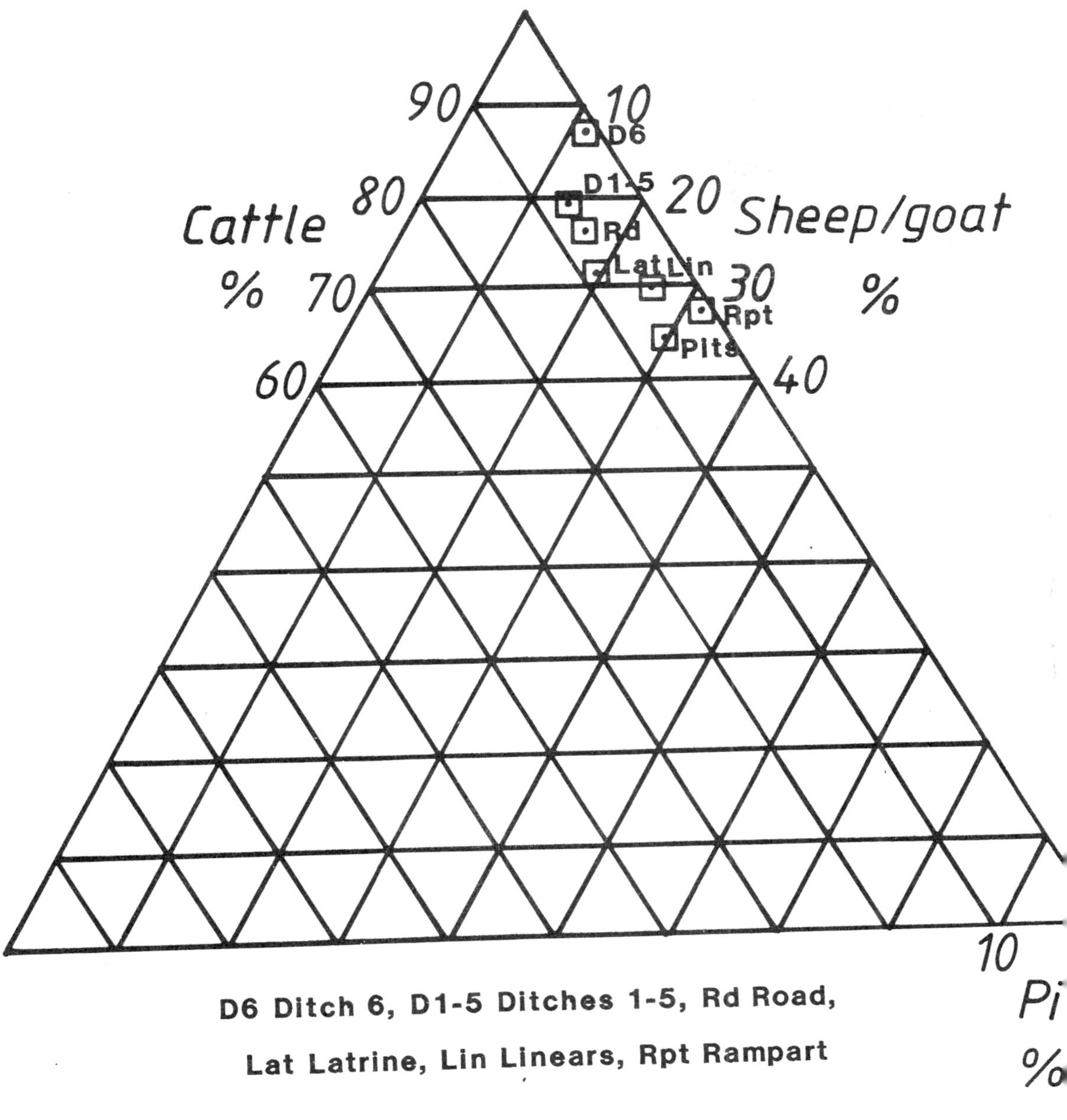

D6 Ditch 6, D1-5 Ditches 1-5, Rd Road,

Lat Latrine, Lin Linears, Rpt Rampart

Fig.4: Percentages of fragments affected by minerals

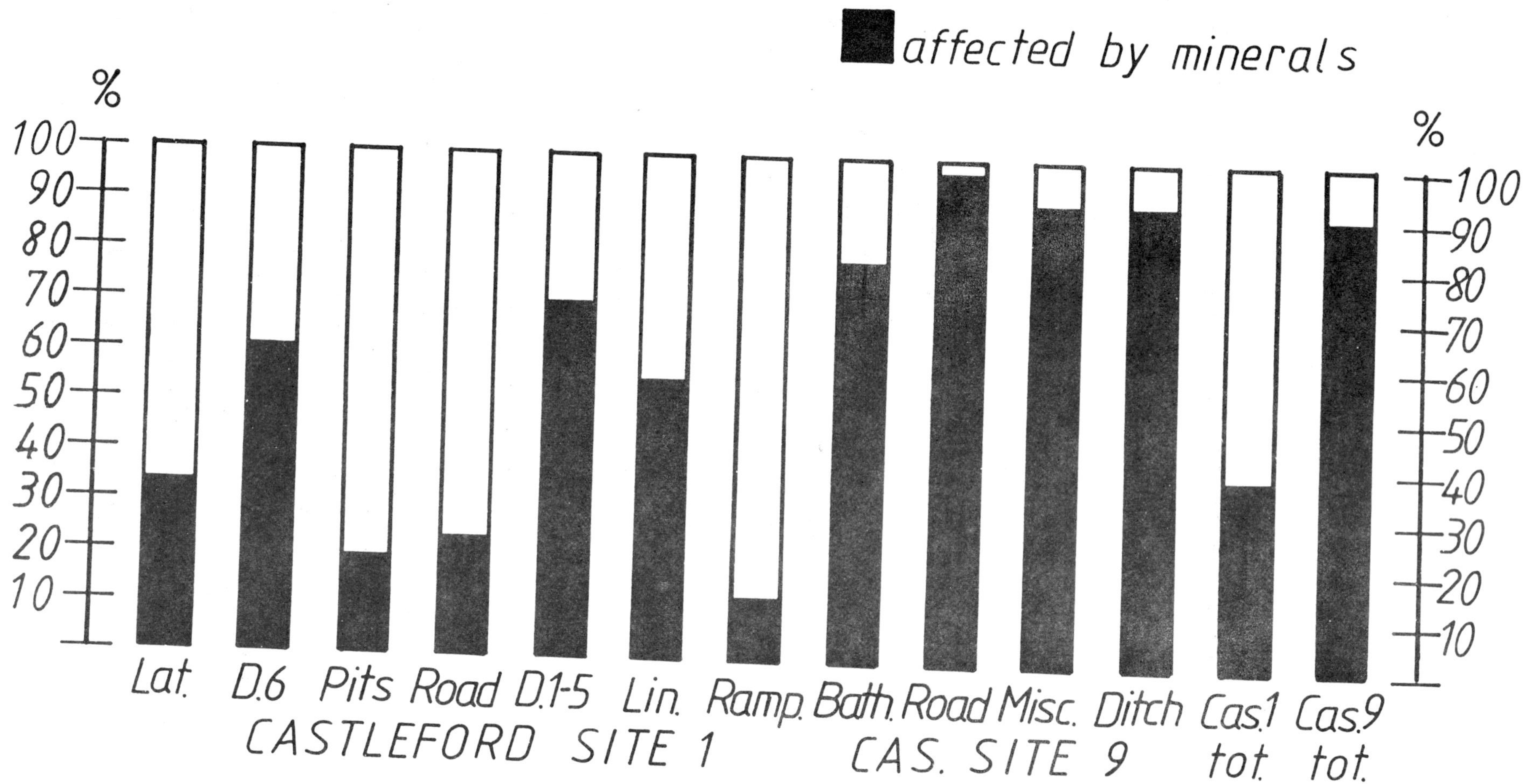

Fig.5: Percentages of fragments affected by texture

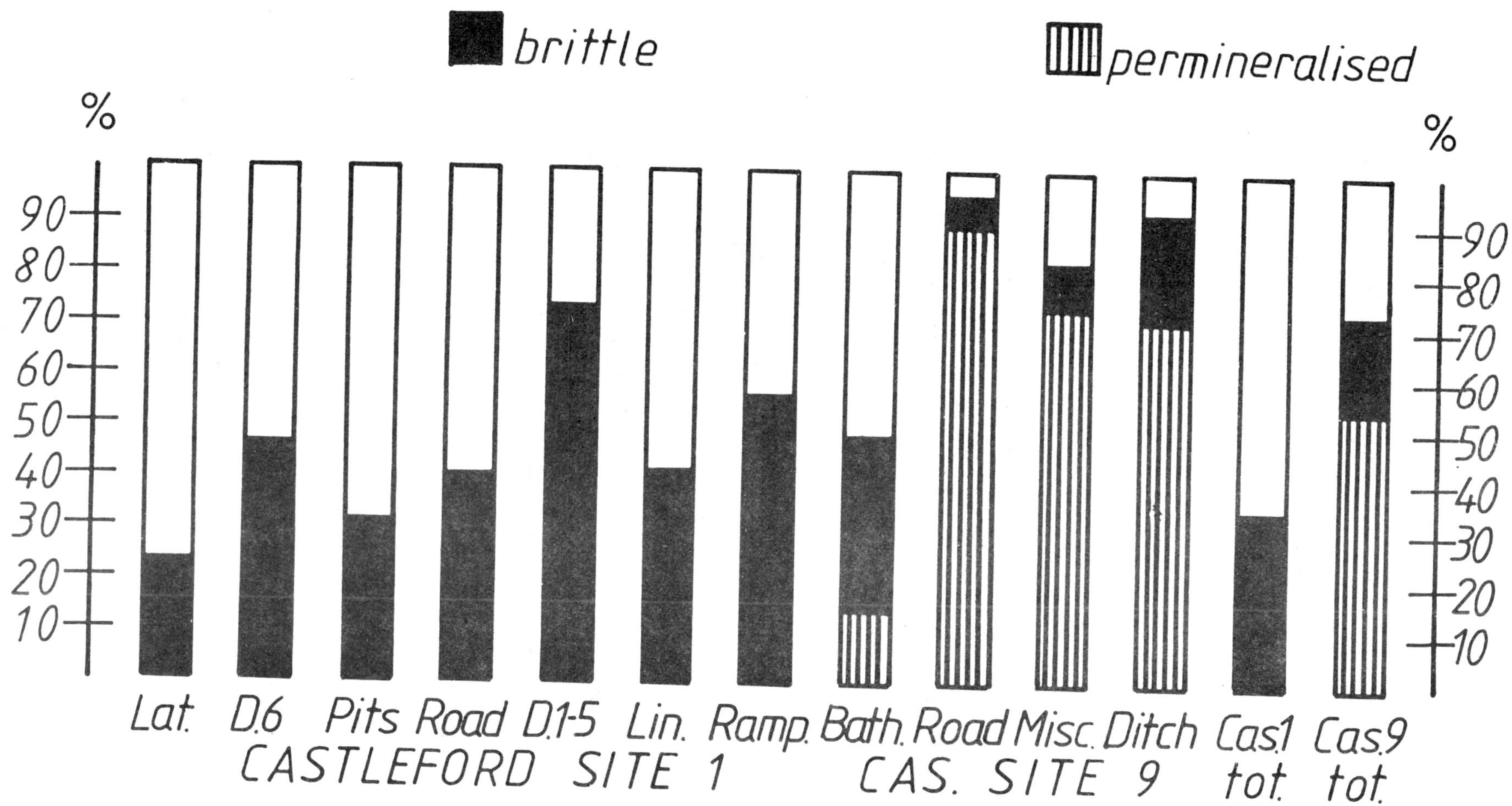

Fig. 6: Percentages of fragments of Cattle, Sheep/Goat and Pig

at Castleford Site 1 and Castleford Site 9

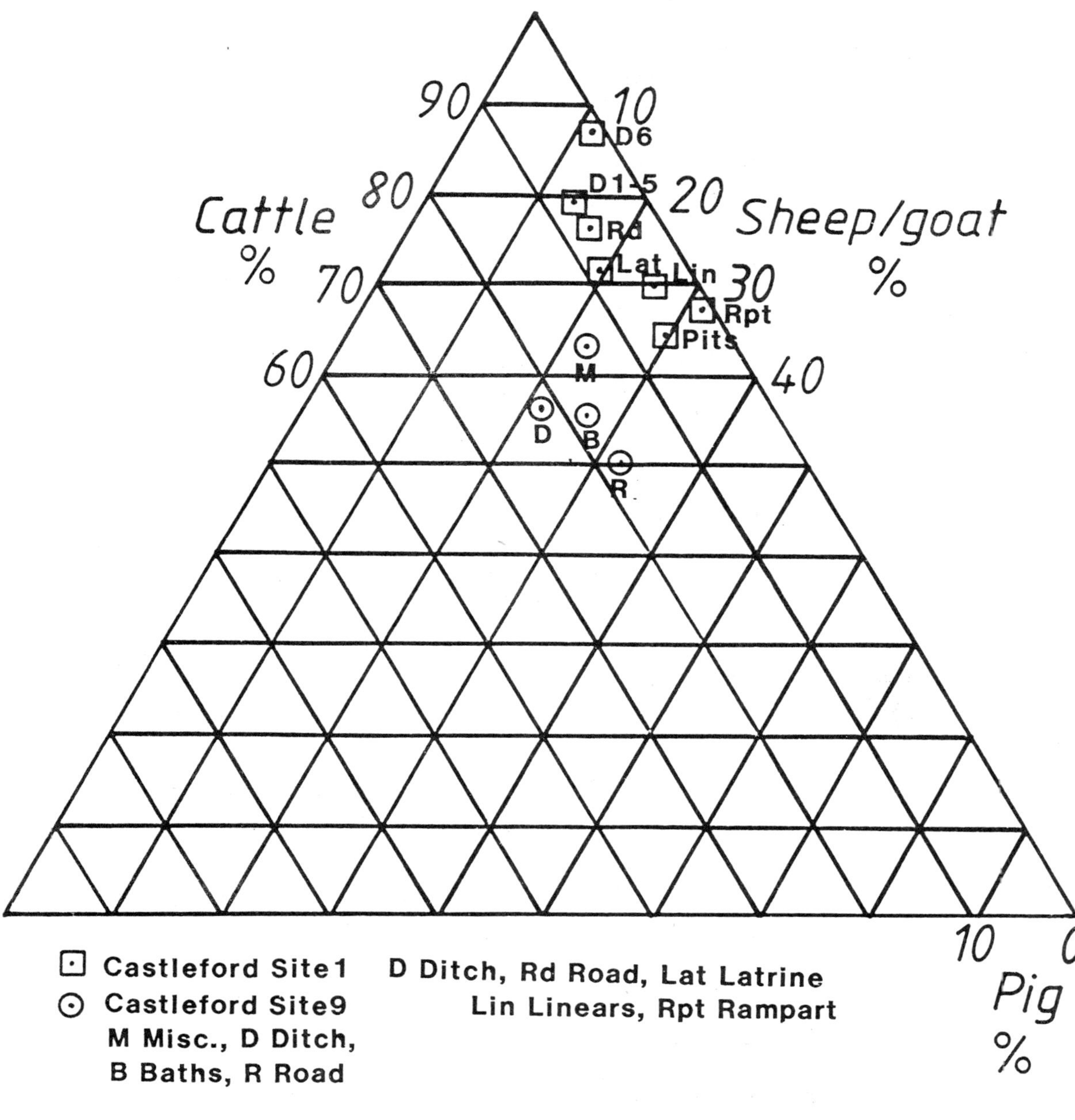

⊡ Castleford Site 1
⊙ Castleford Site 9
M Misc., D Ditch,
B Baths, R Road

D Ditch, Rd Road, Lat Latrine
Lin Linears, Rpt Rampart

THE ENVIRONMENTAL IMPLICATIONS OF CHANGEOVER RATES IN VERTEBRATE FAUNAS.

R.D.S. Jenkinson
Creswell Crags Visitor and Cave Research Centre
Welbeck
Worksop
Notts S80 3LH

ABSTRACT

Change within vertebrate communities can be predicted as a complex series of processes affecting vertebrate population structure in response to a wide range of stimuli. Modern studies demonstrate that the amplitude of change for vertebrate communities varies within biome fluctuations associated with decreasing latitude. This observation allows differentiation of biome extremes on the basis of their vertebrate faunas. In this study vertebrate faunal changeover rates are considered for the Late Pleistocene site of Pin Hole Cave, Creswell Crags, in an attempt to differentiate arctic and woodland faunal communities.

INTRODUCTION

Vertebrate fossils recovered from archaeological sites, potentially offer a wealth of environmental information.

For many species environmental information can be deduced from anatomical characteristics, this in turn can be supported by studies of associated species.

These types of studies combined with those of fossil invertebrates, floras, and sedimentary contexts allow a general reconstruction of prehistoric environments. The application of such techniques to vertebrate faunas from Quaternary sites allows palaeontologists to consider particular faunas or parts of faunas as indicative of warm or cold stage conditions. Much of our vertebrate faunal history for Britain is constructed in this way (Stuart 1983).

There are two major problems with this approach; first too much stress is placed upon postulated habitat-type fossils. For example the presence of woolly rhinoceros (Coelodonta antiquitatus), mammoth (Mammuthus primigenius), and reindeer (Rangifer tarandus) is sometimes regarded as sufficient to justify claims of a cold stage fauna being present. In these situations little significance has often been placed upon less indicative species or upon relative frequency of individual species within a fauna.

A second major problem is caused by our poor understanding of the palaeo-zoogeography of many species. This situation presents particular difficulties when changes take place within the make-up of a vertebrate fauna and where individual species' distribution can be complex. In such situations, the components of a fauna may vary on a localised basis.

This study sets out to develop methods to overcome these problems by

(1) considering change within whole vertebrate faunas; and

(2) in relation to the important Palaeolithic cave and fissure sequences at Creswell Crags (Jenkinson 1983; Jenkinson, Bramwell, Briggs, Gilbertson, Griffin, Stebbings, Watts and Wilkinson 1982, Jenkinson and Gilbertson 1984a, 1984b in press).

VERTEBRATE POPULATION CHANGE

Change within past vertebrate communities can be postulated as a complex series of processes affecting vertebrate population structure in response to a wide range of stimuli. Short term fluctuation, often cyclic in nature, is well known from modern ecological studies, and can be identified to causal factors which vary from cycle to cycle, between species and habitat (Bliss 1977, Hawkioja and Hakala 1975). The amplitude of change is also known to vary: the number of species in a region with a favourable climate is thought to be very much higher than in extreme (i.e. polar) regions. In favourable climates the ecological factors governing the distribution and frequency of species take the form of inter species competition. In regions with unfavourable climatic conditions, abiotic factors are thought to play a decisive role (Remmert 1980).

The detection from palaeontological data of such cyclic fluctuations in the ecosystem is difficult, given the longevity of individuals of many vertebrate species. Nevertheless, the existence of such fluctuations, particularly in arctic vertebrate faunas, does offer a means of evaluating synecological change in fossil vertebrate faunas.

Further potential is offered through the study of palaeo-biomes. The importance of amplitude in biome changes with decreasing latitude away from the polar mass, is a widely recognised factor of ecological importance. On the basis of this observation, Remmert (1980) differentiated two types of arctic biome extremes: a dry/continental arctic characterised by severe winters and short, warm summers, and a second wet/oceanic arctic characterised by mild winters and cool summers, and where water is an not a limiting factor.

Evidence of the character of past landscapes is available from pollen studies. In general terms the floristic changes associated with the last cold stage have been demonstrated by Godwin (1975).

These accounts, however, deal with general developments and can rarely account for the complexity of local biome variation.

A third factor relevant to these data is that of vertebrate population stability and constancy. In general terms modern ecological studies show that ecosystems display varying sensitivity to change. The rate of change can be considered in terms of constancy or stability of population. In a recent review of this problem, Remmert (1980) has put forward two observations. Firstly, that in some ecosystems, particularly arctic terrestrial systems, the member species fluctuate considerably, a situation which contrasts with small fluctuations characteristic of other ecosystems, particularly forests. Secondly, inconstant systems are resilient to ecocatastrophes and are capable of returning to their original state.

The large number of species characteristic of constant systems is recognised as a mechanism ensuring high turnover under very constant environmental conditions, a situation commonly destroyed by ecocatastrophes. Inconstant systems are conversely characterised by their resilience and regeneration (Ives and Barry 1974, Tierzan 1978).

These observations from modern faunal studies highlight the potential of considering vertebrate population turnover rates based on autotrophic species to provide details of a restricted section of the ecosystem. It is likely that heterotrophic (including vertebrates) activity in any biome is a more complex problem. Study of vertebrate faunas in this sense may therefore add details of local biome type and change where other research methods are inadequate.

INDEX OF CHANGE FOR FOSSIL FAUNAS

A key factor allowing this analysis of vertebrate fauna is the design of a uniform means of calculating vertebrate faunal turnover which is applicable to fossil vertebrate faunas. This can be achieved by evaluating the differences in species and their frequency in the faunas recovered from arbitrary levels on archaeological and palaeontological sites. For such palaeontological collections the species present in any one level or horizon can be described in two ways: originating species, i.e. those present in a particular level but not present in the underlying level. A second description is that of extinction where a species present in one level is absent in the overlying level. Neither of these descriptions necessarily records the first appearance or last presence of any of the species being considered. A relative quantity measurement can be calculated by quantifying this index on the basis of numbers of species and minimum numbers of individuals for each level being considered.

This calculation can be expressed thus:

$$\frac{TS}{TI} \times \frac{So/Io}{Se/Ie} \%$$

Where S = species, I = minimum numbers of individuals, o = originating species, e = extinct species, TS = total species present in the unit and TI = total individuals present in the unit.

Expression of change in the make-up of a vertebrate fauna present in a series of archaeological levels in these terms provides a means of considering the temporal constancy or stability of a given fauna, at specific sites where the data are sufficiently well recorded. These observations would allow a consideration of vertebrate faunal response to environmental stimuli (e.g. climatic change) detected within the fossil record.

CAVE STUDY : PIN HOLE CAVE

This study aims to test the potential of this method of elucidating faunal changeover rates and consequent implications for environmental change upon the vertebrate faunas from the Late Pleistocene site of Pin Hole Cave, Creswell Crags.

This site has been excavated by Mello (1875) and Armstrong (1928). The archaeological and palaeontological evidence available from this site has been reviewed subject to many studies (Jackson 1936, Kitching 1963, Campbell 1977 and Collcutt 1975).

Many of these studies have suffered from difficulties in reconstructing the spatial distribution of vertebrate remains from the cave. Recent re-analysis of the excavators' records clarifies this issue and allows for the separation of vertebrate remains according to twenty-one stratigraphic layers each of 0.30 metre thickness (Jenkinson 1983). From a total of twenty-two thousand, five hundred and sixteen bone fragments, identified to one hundred and sixty-six species, it has been possible to separate fifteen thousand, three hundred and fifteen vertebrate bones into twenty separate horizons (Jenkinson 1983).

The species present within the cave are listed in Table 1. These are shown against minimum number of individual estimations within each stratigraphic layer. Many of these species are common in Late Pleistocene sediments particularly cave hyaena (Crocuta crocuta), wolf (Canis lupus), brown bear (Ursus arctos), mammoth (Mammuthus primigenius), woolly rhinoceros (Coelondonta antiquatis), giant deer (Megaloceros giganteus) and reindeer (Rangifer tarandus). An unusual aspect of the cave is the preservation of large quantities of remains of smaller vertebrates : the evidence for birds, fish and amphibians is of particular note.

Figure 1 illustrates the quantity of vertebrate bone, species and minimum number of individuals for each stratigraphic layer of the cave. It can be seen from this figure that considerable minor fluctuations occur within the vertebrate community whilst there is an overall trend toward a large community in the highest stratigraphic levels of the cave. A higher proportion of bone remains of a greater number of species are present within the upper levels.

It is clear from the range of species present within the cave that the vertebrate community is predominantly a cold stage one.

A number of the species present are to be found only in Late Pleistocene cold stage contexts (Stuart 1977).

It is equally clear, from the presence of partially or wholly aquatic species, tree dwellers and burrowers that a wide range of habitats are indicated. Figure 2 shows the percentage frequency, based upon minimum numbers of individuals, of species grouped according to their habitat preference. This figure illustrates the potential diversity of ecosystems present in and around the cave.

The short-term fluctuations revealed by this type of analysis can be used to identify the presence of causal factors changing through time which affect species and habitat.

ENVIRONMENTAL IMPLICATION OF VERTEBRATE FAUNAL TURNOVER RATES

The variations in vertebrate faunal turnover rates demonstrated from the Pin Hole Cave sequence have a number of implications for elucidating palaeoenvironments.

High levels of turnover rate (inconstancy) may well indicate unfavourable conditions for the vertebrate community. In this instance these may be equated with arctic conditions. The vertebrate community in such circumstances are more likely to be controlled by abiotic factors to which there may be dramatic response particularly in terms of large and cyclic variations.

Turnover rates calculated for the vertebrate communities within this cave are shown in Figure 3. Assessment of the vertebrate community in these terms expressed the degree of constancy within the community for each stratigraphic level in respect to the others. A key issue for understanding the development of the vertebrate community is the relative degree of constancy displayed by the community in respect to changing stimuli. As can be seen in Figure 3 there is considerable variation through the cave sequence in constancy of the vertebrate fauna.

Stratigraphic levels 13-16 and to a less marked extent levels 8-9 and 1-2 show a high degree of turnover. Intervening levels display more constancy in the vertebrate community.

The sequence of change is of considerable interest: the high turnover at the base of the cave is followed by constancy (levels 10-12) - high turnover (levels 8-9) - constancy (levels 5-7) - a slight rise in turnover rate (level 4) - a marked decrease to constancy(level 3) - an equally rapid rise to high turnover (level

2) and a subsequent decrease in turnover rates.

An ecosystem of this type is both prone to, and would also display a low resilience to, ecocatastrophes. The anticipation that such vertebrate communities are composed of low numbers of species is borne out by the data from Pin Hole where particularly high turnovers in levels 13-16 have an average of 64 species present, levels 8-9 an average of 22 species.

Low levels of turnover rate (constancy) within the vertebrate community are likely to indicate more favourable conditions for vertebrates. Such a community is more likely controlled more by ecological than abiotic factors, particularly in the form of inter-species competition. The latter observation is clearly shown from the nature of the vertebrate community in levels of constancy. An example demonstrating this point is the large number of smaller carnivores and raptorial birds present in stratigraphic level 3 and which must have been in competition for similar prey species.

Clearly a number of competition avoidance mechanisms must have been present with this section of the vertebrate community. Similarly, a fluctuation within the community is less marked. An ecosystem of this type would be resilient to ecocatastrophes and would be characterised by the ability for regeneration.

The sequence of change within the vertebrate community from Pin Hole Cave, raises a number of issues concerning environmental change during the Late Pleistocene of eastern Britain. Analysis of habitat types likely to be preferred by particular species opens the possibility of comparison between vertebrate faunal evidence and the vegetational change known from pollen studies.

The latter suggests a cold stage vegetation for the Devensian dominated by open, herbaceous and aquatic plants, broken during the interstadials by the occasional development of boreal forest (Godwin 1975).

Habitat types as indicated by vertebrate species from Pin Hole agree well in general terms with the developments suggested by pollen and plant macro fossils studies. This analysis shows the virtually continuous nearby presence of aquatic habitats. Arctic habitats are also present throughout the whole sequence but are particularly important in stratigraphic levels 10 and 5. Open ground or montane habitats, present in the lower part of the sequence, decline with the development of woodland preferring species within the upper half of the sequence.

These data suggest a sequence of environmental change represented within the cave which in the lowest stratigraphic levels 13-17 developed from arctic environments, to arctic/montane (levels 9-12). Levels 8 and 9 can be interpreted as a dramatic return to unfavourable arctic conditions followed in successive levels by the development of montane to woodland environments, a development interrupted in stratigraphic levels 4 and 2 by the return of unfavourable (arctic) conditions.

CONCLUSIONS

The methodology outlined in this study for analysis of vertebrate communities for archaeological, especially Palaeolithic cave contexts using turnover estimations has the considerable potential of differentiating long-term trends from short-term fluctuation. The existence of constraints acting upon the community associated with the two extremes of turnover: abiotic factors in inconstant communities and ecological factors in constant communities can be anticipated and tested upon palaeontological data.

The variation in resilience to environmental change implicit in the variations in faunal constancy is an important concept and provides a means of examining dramatic changes known in Late Devensian vertebrate faunas.

REFERENCES

Armstrong, A.L. (1928) Excavations in the Pin Hole Cave, Creswell Crags, Derbyshire, Proceedings of the Prehistoric Society 6, 330-4.

Bliss, L.C. (ed) (1977) Truelove Lowland, Devon. Island Canada - A High Arctic Ecosystem. Edmonton: University of Alberta Press.

Campbell, J.B. (1977) The Upper Palaeolithic of Britain. Oxford: Clarendon Press, 2 vols.

Collcutt, S.N. (1975) The Stratigraphy of Creswell Crags. unpublished M.A. thesis. University of Edinburgh, Department of Archaeology.

Godwin, H. (1975) History of the British Flora. London: Cambridge University Press.

Haukioja, E. and Hakala, T. (1975) Herbivore cycles and periodic outbreaks, formulation of a general hypothesis. Report of the Kevo Subarctic Research Studies 12, 1-9.

Ives, J.D. and Barry, R.G. (1974) Arctic and Alpine Environments. London: Methuen.

Jackson, J.W. (1946) The Upper Pleistocene fauna and its relation to the Ice Age, Proceedings of the Liverpool Geological Society 19, 165-82.

Jenkinson, R.D.S. (1983) The Archaeological Site of Creswell Crags. unpublished PhD thesis. Department of Prehistory and Archaeology, University of Sheffield.

Jenkinson, R.D.S., Bramwell, D., Briggs, D.J., Gilbertson, D.D.,

Griffin, C.M., Stebbings, R.E., Watts, C.M., Wilkinson, M. (1982) Death of a Wolf. Nottingham. Nottinghamshire County Council/Creswell Crags Visitor Centre, Report No. 3, 23.

Jenkinson, R.D.S. and Gilbertson, D.D., (1984a) In the Shadow of Extinction: A Quaternary Geology and Palaeoecology of the Lake Fissures and Smaller Caves at Creswell Crags. Sheffield: University of Sheffield, Monographs in Prehistory and Archaeology, 122pp.

Jenkinson, R.D.S. and Gilbertson, D.D. (1984b in press) A Quaternary Geology and Palaeoecology of Robin Hood's Cave, Creswell Crags. Sheffield: University of Sheffield Monographs in Prehistory and Archaeology.

Kitchen, J.W. (1963) Bone, Tooth and Horn Tools of Palaeolithic Man; an Account of the Osteodontokeratic Discoveries in the Pin Hole Cave, Derbyshire. Manchester: Manchester University Press.

Mello, J.M. (1875) On some bone caves in Creswell Crags, Quarterly Journal of the Geological Society of London 31, 679-91.

Remmert, H. (1980) Arctic Animal Ecology. Berlin: Springer-Verlag.

Stuart, A.J. (1977) British Quaternary Vertebrates. In British Quaternary Studies: Recent Advances (ed) F.W. Shotton, 69-81. Oxford: Clarendon Press.

Stuart, A.J. (1983) Pleistocene Vertebrates in the British Isles. London: Longmans.

Tierzen, L.L. (1978) Vegetation and productive ecology of an Alaskan Arctic Tundra, Ecology Studies 29, 686.

1. Table showing the numbers of individuals of vertebrate species from each stratigraphic level of Pin Hole Cave. The original raw data is published in Jenkinson (1983).

SPECIES	STRATIGRAPHIC LAYER 0	1	2	3	4	5	6	7	8	9	10	11	12	13	14	15	16	17	18	19	20
INSECTIVORA:																					
Erinaceus europaeus (Hedgehog)	3	-	-	-	1	-	-	-	-	-	-	-	-	-	-	-	-	-	-	-	-
Sorex minutus (Pygmy shrew)	-	-	-	2	-	-	-	-	-	-	-	-	-	-	-	-	-	-	-	-	-
Sorex araneus (Common shrew)	-	-	-	1	-	-	1	-	-	-	-	-	-	-	-	-	-	-	-	-	-
Talpa europaea (Mole)	-	1	-	1	-	-	1	-	-	-	-	-	-	-	-	-	-	-	-	-	-
CHIROPTERA:																					
Rhinolophus hipposideros (Lesser horseshoe bat)	11	11	14	14	4	4	2	-	-	-	-	-	-	-	-	-	-	-	-	-	-
Myotis daubentoni (Water bat)	-	2	-	1	-	-	-	-	-	-	-	-	-	-	-	-	-	-	-	-	-
Myotis mystacinus (Whiskered bat)	4	-	2	3	-	3	-	-	-	-	-	-	-	-	-	-	-	-	-	-	-
Myotis nattereri (Natterers bat)	-	3	2	5	2	-	-	-	-	-	-	-	-	-	-	-	-	-	-	-	-
Plecotus auritus (Long eared bat)	1	2	-	1	1	-	-	-	-	-	-	-	-	-	-	-	-	-	-	-	-
Barbastella barbastella (Barbastelle bat)	2	1	-	-	-	-	-	-	-	-	-	-	-	-	-	-	-	-	-	-	-
Pipistrellus pipistrellus (Common bat)	-	1	-	-	-	-	-	-	-	-	-	-	-	-	-	-	-	-	-	-	-
Nyctalus leisleri (Leisler's bat)	-	1	-	1	-	-	-	-	-	-	-	-	-	-	-	-	-	-	-	-	-
CARNIVORA:																					
Crocuta crocuta (Cave hyaena)	-	2	4	3	3	4	3	6	5	4	4	11	5	1	1	1	1	-	1	-	-
Felis sylvestris (Wild cat)	1	1	1	1	1	-	-	1	-	-	-	-	-	-	-	-	-	-	-	-	-
Felis leo (Cave lion)	-	1	1	1	1	-	-	1	1	3	2	1	1	1	-	-	1	-	-	-	-
Mustela putorius (Polecat)	-	1	-	1	1	-	1	1	2	1	-	1	-	-	-	-	-	-	-	-	-
Mustela erminea (Stoat)	-	1	1	2	-	1	-	1	-	-	-	1	1	-	-	-	-	-	-	-	-

SPECIES	0	1	2	3	4	5	6	7	8	9	10	11	12	13	14	15	16	17	18	19	20
Mustela nivalis (Weasel)	1	-	-	-	-	1	-	-	-	-	-	-	-	-	-	-	-	-	-	-	-
Meles meles (Badger)	1	2	2	1	1	1	-	1	1	-	-	3	-	-	-	-	-	-	-	-	-
Canis lupus (Wolf)	1	1	1	3	3	2	3	3	2	1	1	2	2	1	1	1	-	-	-	-	-
Vulpes vulpes (Red fox)	1	1	3	5	18	3	-	4	4	5	2	3	2	1	-	1	-	-	-	-	-
Ursus arctos (Brown bear)	1	2	2	3	3	3	2	2	2	1	1	4	1	1	-	1	-	-	-	-	-
<u>PROBOSCIDEA</u>:																					
Mammuthus primigenius (Woolly mammoth)	-	-	2	1	1	-	1	4	1	2	1	2	3	1	-	1	-	-	-	-	-
<u>PERISSODACTYLA</u>:																					
Coelodonta antiquitatis (Wolly rhinoceros)	2	1	2	3	2	3	4	2	3	4	2	5	2	1	1	1	-	-	-	-	-
Equus sp. (Horse)	3	2	2	2	2	2	5	4	3	3	2	10	6	3	1	1	-	-	1	-	-
<u>ARTIODACTYLA</u>:																					
Sus scrofa (Wild pig)	-	2	1	-	-	-	1	-	-	-	-	1	-	-	-	1	-	-	-	-	-
Megaloceros giganteus (Giant deer)	-	-	1	-	1	-	1	1	1	1	1	1	1	-	-	-	-	-	-	-	-
Rangifer tarandus (Reindeer)	2	5	14	17	12	17	9	7	7	6	4	13	4	1	1	1	1	-	-	1	-
Bos	1	1	1	1	1	1	2	1	1	1	1	2	1	1	-	1	1	-	-	-	-
<u>RODENTIA</u>:																					
Sciurus vulgaris (Red Squirrel)	-	1	-	-	-	-	-	-	-	-	-	-	-	-	-	-	-	-	-	-	-
Muscardinus avellanarius (Dormouse)	-	1	-	-	1	-	-	-	-	-	-	-	-	-	-	-	-	-	-	-	-

SPECIES	0	1	2	3	4	5	6	7	8	9	10	11	12	13	14	15	16	17	18	19	20
Clethrionomys glareolus (Bank vole)	4	4	-	3	-	1	1	-	-	-	-	-	-	-	-	-	-	-	-	-	-
Arvicola terrestris (Water vole)	1	-	1	-	1	-	1	2	-	2	-	1	1	-	-	-	-	-	-	-	-
Microtus arvalis (Common vole)	-	7	-	-	1	-	-	-	-	-	-	2	-	-	2	-	-	-	-	-	-
Microtus agrestis (Field vole)	-	1	-	-	-	-	-	-	-	-	-	-	-	-	-	-	-	-	-	-	-
Microtus oeconomus (Root vole)	-	2	-	2	2	2	-	-	-	-	-	5	1	-	-	-	-	-	-	-	-
Microtus gregalis (Gregarious vole)	-	1	1	1	2	-	-	1	-	3	-	3	4	1	-	-	-	-	-	-	-
Lemmus lemmus (Norway lemming)	-	4	6	5	3	22	21	7	2	7	37	36	39	1	-	-	-	-	-	-	-
Dicrostonyx torquatus (Arctic lemming)	7	29	18	38	6	36	23	5	2	10	33	24	19	1	1	-	-	-	-	-	-
Apodemus flavicollis (Yellow-necked field mouse)	-	3	4	1	1	-	-	-	1	-	-	2	1	-	-	-	-	-	-	-	-
Mus musculus (House mouse)	-	-	-	-	-	-	-	-	-	-	-	-	-	-	-	-	-	-	-	-	-
Rattus rattus	-	1	-	-	-	-	1	-	-	-	-	1	-	-	-	-	-	-	-	-	-
LAGOMORPHA:																					
Lepus timidus (Mountain hare)	1	2	4	3	2	3	2	1	1	1	-	1	1	-	-	-	-	-	-	-	-
GAVIIFORMES:																					
Gavia arctica (Black-throated diver)	-	-	-	-	-	-	-	-	-	-	-	-	1	-	-	-	-	-	-	-	-
CICONIIFORMES:																					
Ardea cinerea (Grey heron)	1	1	1	1	1	-	-	1	-	-	-	-	1	-	-	-	-	-	-	-	-
Ciconia ciconia (White stork)	-	1	-	-	-	-	-	-	-	1	-	-	-	-	-	-	-	-	-	-	-

SPECIES	0	1	2	3	4	5	6	7	8	9	10	11	12	13	14	15	16	17	18	19	20
<u>ANSERIFORMES</u>:																					
Cygnus columbianus (Bewick's swan)	-	-	-	-	1	-	-	-	-	-	-	-	-	-	-	-	-	-	-	-	-
Branta bernicula (Brent goose)	-	-	-	1	-	-	-	-	1	1	-	-	1	-	-	-	-	-	-	-	-
Branta leucopsis (Barnacle goose)	-	-	1	-	-	1	-	-	-	-	-	-	-	-	-	-	-	-	-	-	-
Anser anser (Greylag goose)	-	-	1	1	-	-	-	-	-	1	-	-	-	-	-	-	-	-	-	-	-
Anser albifrons (White fronted goose)	-	-	-	1	-	-	-	-	-	1	-	-	-	-	-	-	-	-	-	-	-
Anser brachyrhynchus (Pink footed goose)	-	-	1	-	1	1	-	-	1	-	1	1	-	1	-	-	-	-	-	-	-
Anas platyrhynchas (Mallard)	-	-	-	1	1	-	-	-	-	1	-	-	-	-	1	-	-	-	-	-	-
Anas penelope (Wigeon)	-	-	-	1	-	-	-	-	-	-	-	-	-	-	-	-	-	-	-	-	-
Anas crecca (Teal)	1	-	-	1	-	-	-	-	-	-	-	-	1	-	-	-	-	-	-	-	-
Anas querquedula (Garganey)	-	-	-	-	-	-	1	-	-	-	-	-	-	-	-	-	-	-	-	-	-
Tadorna ferruginea (Ruddy shelduck)	-	-	-	-	1	-	-	-	1	-	-	-	-	-	-	-	-	-	-	-	-
Aythya fuligula (Tufted duck)	-	-	-	-	-	-	1	-	-	-	-	-	-	-	-	-	-	-	-	-	-
Melanitta nigra (Common scoter)	-	1	-	-	1	-	-	-	-	1	-	1	1	-	-	-	-	-	-	-	-
Mergus merganser (Goosander)	-	-	2	-	-	1	-	-	-	-	-	1	-	-	1	-	-	-	-	-	-
<u>FALCONIFORMES</u>:																					
Aquila chrysaetus (Golden eagle)	-	-	1	-	-	-	-	-	-	-	-	-	-	-	-	-	-	-	-	-	-
Buteo lagopus (Rough-legged buzzard)	-	-	-	-	-	-	-	-	-	-	-	1	-	-	-	-	-	-	-	-	-
Accipiter gentilis (Goshawk)	-	-	-	-	-	-	-	-	-	-	-	1	-	-	-	-	-	-	-	-	-
Pandion haliaetus (Osprey)	-	-	-	1	-	-	-	-	-	-	-	-	-	-	-	-	-	-	-	-	-
Falco tinnunculus (Kestrel)	1	3	1	1	-	2	-	1	1	-	1	-	1	-	-	-	-	-	-	-	-

SPECIES	0	1	2	3	4	5	6	7	8	9	10	11	12	13	14	15	16	17	18	19	20
Falco columbarius (Merlin)	-	-	-	1	-	-	-	-	-	-	-	-	-	-	-	-	-	-	-	-	-
Stercorarius sp. (Skua)	-	-	-	-	1	-	-	-	-	-	-	-	-	-	-	-	-	-	-	-	-
Larus canus (Common gull)	-	-	-	-	-	-	1	-	-	-	-	-	-	-	-	-	-	-	-	-	-
Cepphus grylle (Black guillemot)	1	-	-	-	-	-	-	-	-	-	-	-	-	-	-	-	-	-	-	-	-
Fratercula arctica (Puffin)	-	-	1	-	-	-	-	-	-	-	-	-	-	-	-	-	-	-	-	-	-
<u>COLUMBIFORMES</u>:																					
Columba palumbus (Woodpigeon)	-	-	-	-	-	-	-	-	-	1	-	-	-	-	-	-	-	-	-	-	-
Columba oenas (Stock dove)	-	-	-	1	1	-	-	-	-	-	-	-	-	-	-	-	-	-	-	-	-
<u>STRIGIFORMES</u>:																					
Aegolius funereus (Tengmalm's owl)	-	-	1	-	-	-	-	-	-	-	-	-	-	-	-	-	-	-	-	-	-
Asio flammeus (Short eared owl)	-	-	1	-	-	2	-	-	-	-	-	-	1	-	-	-	-	-	-	-	-
Strix aluco (Tawny owl)	-	1	-	-	-	-	-	-	-	-	-	-	-	-	-	-	-	-	-	-	-
Tyto alba (Barn owl)	-	1	-	1	1	-	-	-	-	-	-	-	-	-	-	-	-	-	-	-	-
Survia ulula (Hawk owl)	-	-	-	1	-	-	-	-	-	-	-	-	-	-	-	-	-	-	-	-	-
<u>APODIFORMES</u>:																					
Apus melba (Alpine swift)	-	-	-	-	-	-	-	1	-	-	-	1	-	-	-	-	-	-	-	-	-
<u>CORACIIFORMES</u>:																					
Alcedo atthis (Kingfisher)	-	-	-	-	-	-	1	-	-	-	-	-	-	-	-	-	-	-	-	-	-

SPECIES	0	1	2	3	4	5	6	7	8	9	10	11	12	13	14	15	16	17	18	19	20
PICIFORMES:																					
Dendrocopos minor (Lesser spotted woodpecker)	-	-	-	1	-	-	-	-	-	-	-	-	-	-	-	-	-	-	-	-	-
PASSERIFORMES:																					
Alauda arvensis (Skylark)	1	1	1	2	1	2	-	1	-	-	1	-	1	-	-	-	-	-	-	-	-
Lullula arborea (Woodlark)	-	-	-	-	-	-	-	-	-	-	-	1	-	-	-	-	-	-	-	-	-
Ptyonoprogne rupestris (Crag martin)	-	-	-	-	1	-	-	-	-	-	-	-	-	-	-	-	-	-	-	-	-
Hirundo rustica (Swallow)	-	1	-	1	2	2	-	-	-	-	-	-	-	-	-	-	-	-	-	-	-
Anthus pratensis (Meadow pipit)	-	-	-	-	1	-	-	-	-	-	-	-	-	-	-	-	-	-	-	-	-
Sturnus vulgaris (Starling)	3	1	1	1	-	1	-	1	-	-	-	-	-	-	-	-	-	-	-	-	-
Bombycilla garrulus (Waxwing)	-	-	-	-	-	-	-	-	-	-	-	-	1	-	-	-	-	-	-	-	-
Garrulus glandarius (Jay)	-	1	1	-	-	-	-	-	-	-	-	-	-	-	-	-	-	-	-	-	-
Pica pica (Magpie)	-	1	-	1	1	-	-	-	-	-	-	-	1	-	-	-	-	-	-	-	-
Nucifraga caryocatactes (Nutcracker)	-	-	1	-	-	-	-	-	-	-	-	-	-	-	-	-	-	-	-	-	-
Corvus monedula (Jackdaw)	8	12	4	12	3	3	1	1	1	-	-	-	-	-	-	-	-	-	-	-	-
Corvus frugilegus (Rook)	-	-	1	-	-	-	-	-	-	-	-	-	-	-	-	-	-	-	-	-	-
Corvus corone (Crow)	-	-	-	-	-	-	-	-	1	1	-	-	-	-	-	-	-	-	-	-	-
Corvus corax (Raven)	-	-	-	1	-	-	-	1	-	1	1	1	1	-	1	-	-	-	-	-	-
Cinclus cinclus (Dipper)	-	1	-	-	-	-	-	-	-	-	-	-	-	-	-	-	-	-	-	-	-
Troglodytes troglodytes (Wren)	-	-	-	-	1	-	-	-	-	-	-	-	-	-	-	-	-	-	-	-	-
Prunella modularis (Dunnock)	1	-	-	-	1	-	-	-	-	-	-	-	-	-	-	-	-	-	-	-	-
Sylvia atricapilla (Blackcap)	-	1	-	-	-	-	-	-	-	-	-	-	-	-	-	-	-	-	-	-	-

SPECIES	0	1	2	3	4	5	6	7	8	9	10	11	12	13	14	15	16	17	18	19	20
Oenanthe oenanthe (Wheatear)	-	-	3	1	2	-	-	-	-	-	-	-	-	-	-	-	-	-	-	-	-
Phoenicurus phoenicurus (Redstart)	2	-	1	1	-	1	1	-	-	-	-	-	-	-	-	-	-	-	-	-	-
Erithacus rubecula (Robin)	1	2	1	-	-	1	-	-	-	-	-	-	-	-	-	-	-	-	-	-	-
Turdus torquatus (Ring ouzel)	3	4	-	1	-	-	1	-	-	-	-	-	-	-	-	-	-	-	-	-	-
Turdus merula (Blackbird)	2	2	1	3	3	-	1	2	-	-	-	-	-	-	-	-	-	-	-	-	-
Turdus iliacus (Redwing)	1	1	3	1	1	2	-	-	-	1	-	-	-	1	-	-	-	-	-	-	-
Turdus philomelos (Song thrush)	9	2	7	3	2	3	4	-	-	-	-	-	1	-	-	-	-	-	-	-	-
Turdus viscivorus (Mistle thrush)	-	1	3	1	1	-	1	2	-	-	-	-	-	-	-	-	-	-	-	-	-
Turdus pilaris (Fieldfare)	3	2	3	-	1	-	-	-	-	-	-	-	-	-	-	-	-	-	-	-	-
Parus major (Great tit)	-	1	-	-	1	-	-	-	-	-	-	-	-	-	-	-	-	-	-	-	-
Aegithalos caudatus (Long tailed tit)	-	-	-	-	1	-	-	-	-	-	-	-	-	-	-	-	-	-	-	-	-
Sitta europaea (Nuthatch)	-	1	1	-	1	-	-	-	-	-	-	-	-	-	-	-	-	-	-	-	-
Passer domesticus (House sparrow)	-	1	-	2	-	-	-	-	-	-	1	-	-	-	-	-	-	-	-	-	-
Passer montanus (Tree sparrow)	1	-	-	-	-	-	-	-	-	-	-	-	-	-	-	-	-	-	-	-	-
Fringella coelebs (Chaffinch)	-	-	1	-	-	-	-	-	-	-	-	-	-	-	-	-	-	-	-	-	-
Fringella montifringilla (Brambling)	1	1	-	1	-	-	-	-	-	-	-	-	-	-	-	-	-	-	-	-	-
Pyrrhula pyrrhula (Bullfinch)	-	1	1	-	-	-	-	-	-	-	-	-	-	-	-	-	-	-	-	-	-
Coccothraustes coccothraustes (Hawfinch)	-	1	-	-	1	-	1	-	-	-	-	-	-	-	-	-	-	-	-	-	-
Carduelis chloris (Greenfinch)	-	-	1	-	1	-	-	-	-	-	-	-	-	-	-	-	-	-	-	-	-
Carduelis cannabina (Linnet)	-	-	-	-	-	-	-	1	-	-	-	-	-	-	-	-	-	-	-	-	-
Loxia curvirostra (Crossbill)	-	-	-	1	1	-	-	-	-	-	-	-	-	-	-	-	-	-	-	-	-
Pheuticus ludovicianus (Rose breasted grosbeak)	-	-	-	1	1	-	-	-	-	-	-	-	-	-	-	-	-	-	-	-	-

SPECIES	0	1	2	3	4	5	6	7	8	9	10	11	12	13	14	15	16	17	18	19	20
Miliaria calandra (Corn bunting)	-	1	-	1	-	-	-	-	-	-	-	-	-	-	-	-	-	-	-	-	-
Emberiza schoeniclus (Reed bunting)	1	-	-	-	-	-	-	-	-	-	-	-	-	-	-	-	-	-	-	-	-
Plectrophenax nivalis (Snow bunting)	2	-	-	-	-	1	1	-	-	-	-	-	-	-	-	-	-	-	-	-	-
AMPHIBIA:																					
Rana sp. (Frog/toad)	47	53	21	104	98	4	3	7	-	-	1	3	4	3	-	-	-	-	-	-	-
PISCES:																					
Salmo sp. (Salmon)	-	-	-	-	1	1	-	-	-	-	-	-	-	-	-	-	-	-	-	-	-
Thymallus thymallus (Grayling)	-	-	-	1	2	-	1	1	1	1	6	1	1	-	-	-	-	-	-	-	-
Esox lucius (Pike)	-	-	1	-	1	1	1	-	2	3	1	5	3	-	-	-	-	-	-	-	-
Leuciscus cephalus (Chub)	-	-	-	-	1	-	-	-	-	-	-	-	-	-	-	-	-	-	-	-	-
Cyprinid sp. (Carp)	-	-	1	-	-	-	-	-	1	-	-	1	-	-	-	-	-	-	-	-	-
Perca sp. (Perch)	-	-	-	1	-	-	-	1	-	-	-	-	-	-	-	-	-	-	-	-	-
Platichthys flesus (Flounder)	-	-	1	-	-	-	-	-	-	-	-	-	-	-	-	-	-	-	-	-	-

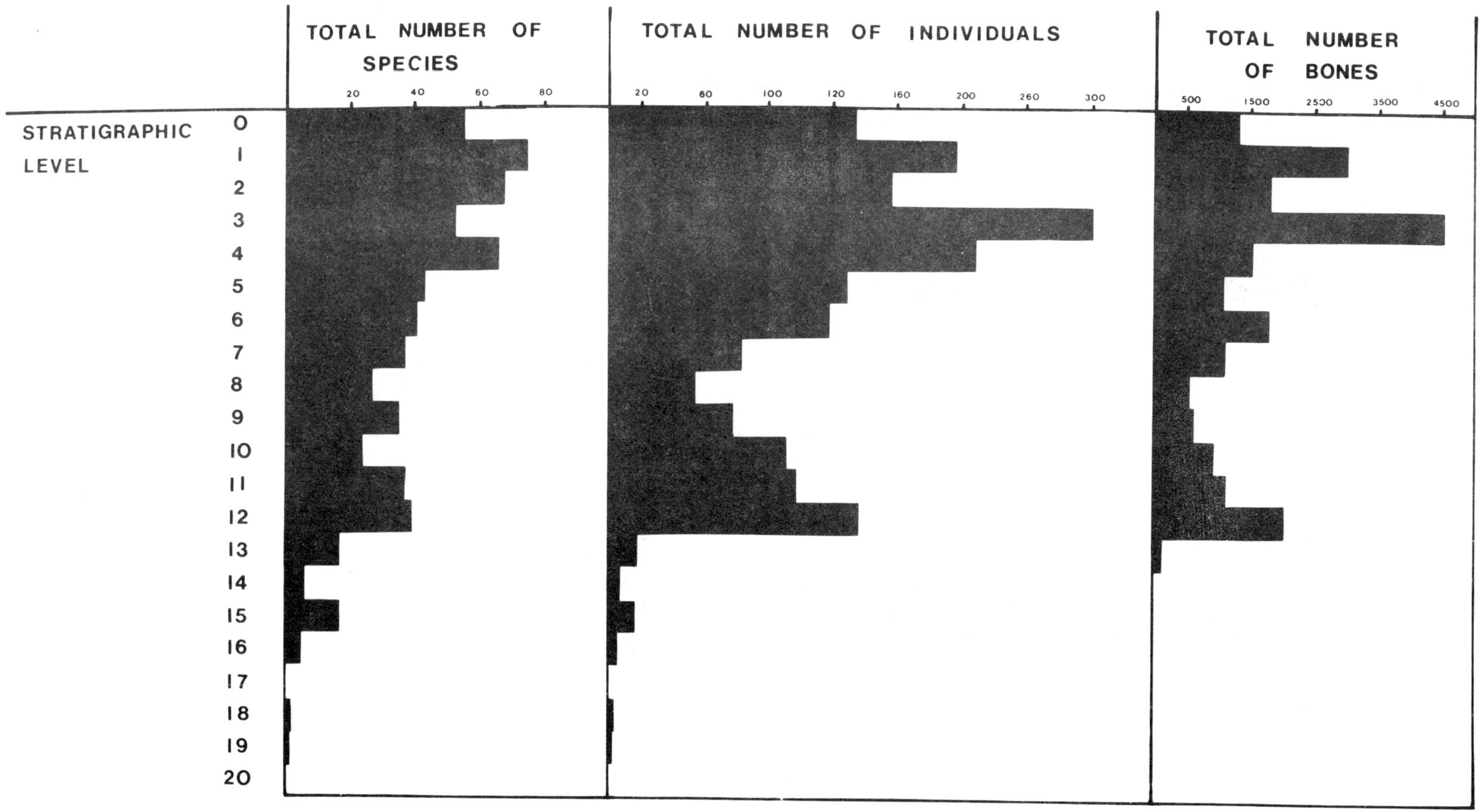

1. Graph showing the percentage frequency variation of species, minimum numbers of individuals and bone fragments from each stratigraphic level of Pin Hole Cave.

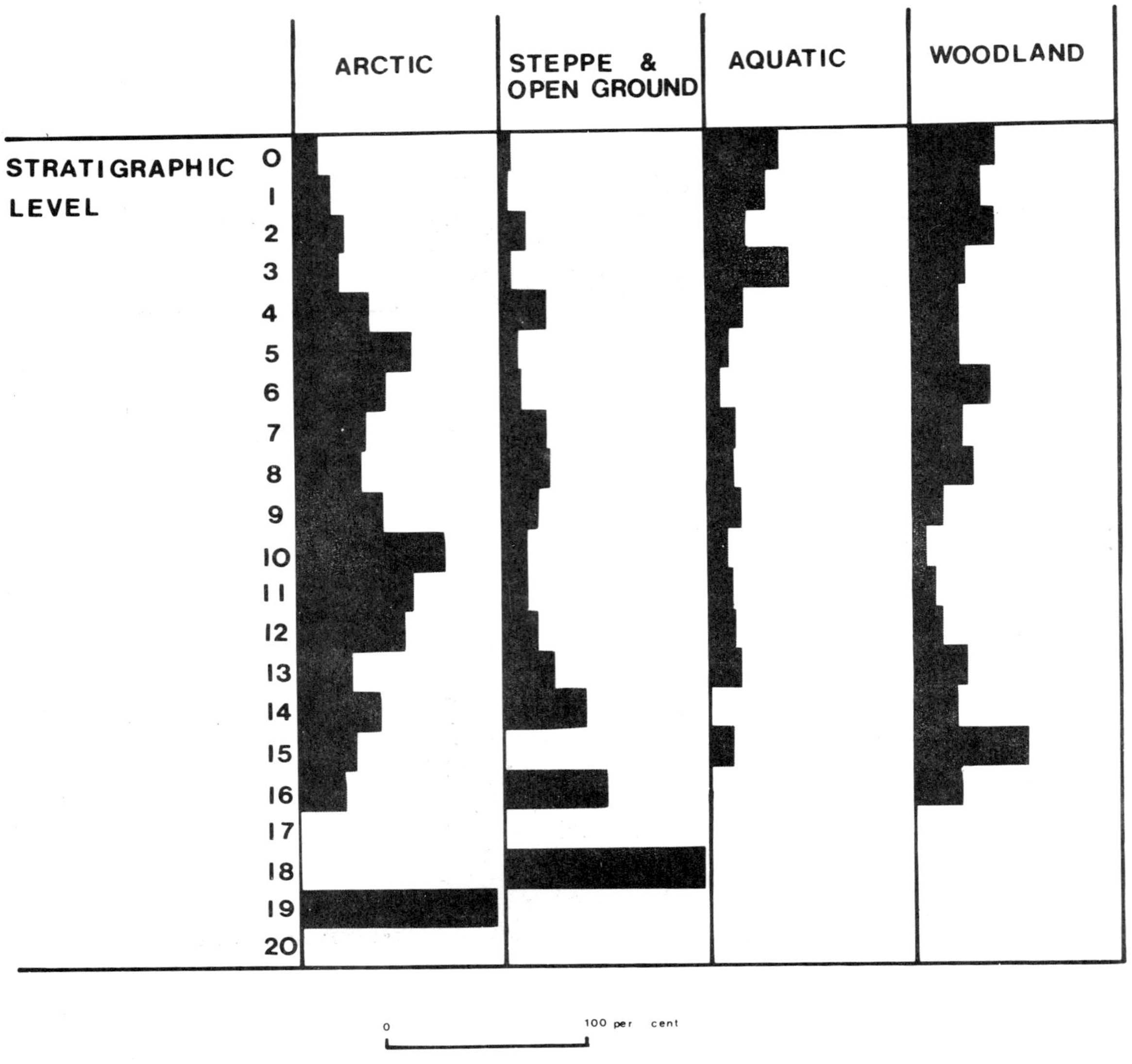

2. Graph showing the percentage frequency, based upon minimum numbers of individuals, of vertebrate species from Pin Hole Cave grouped according to generalised habitat type.

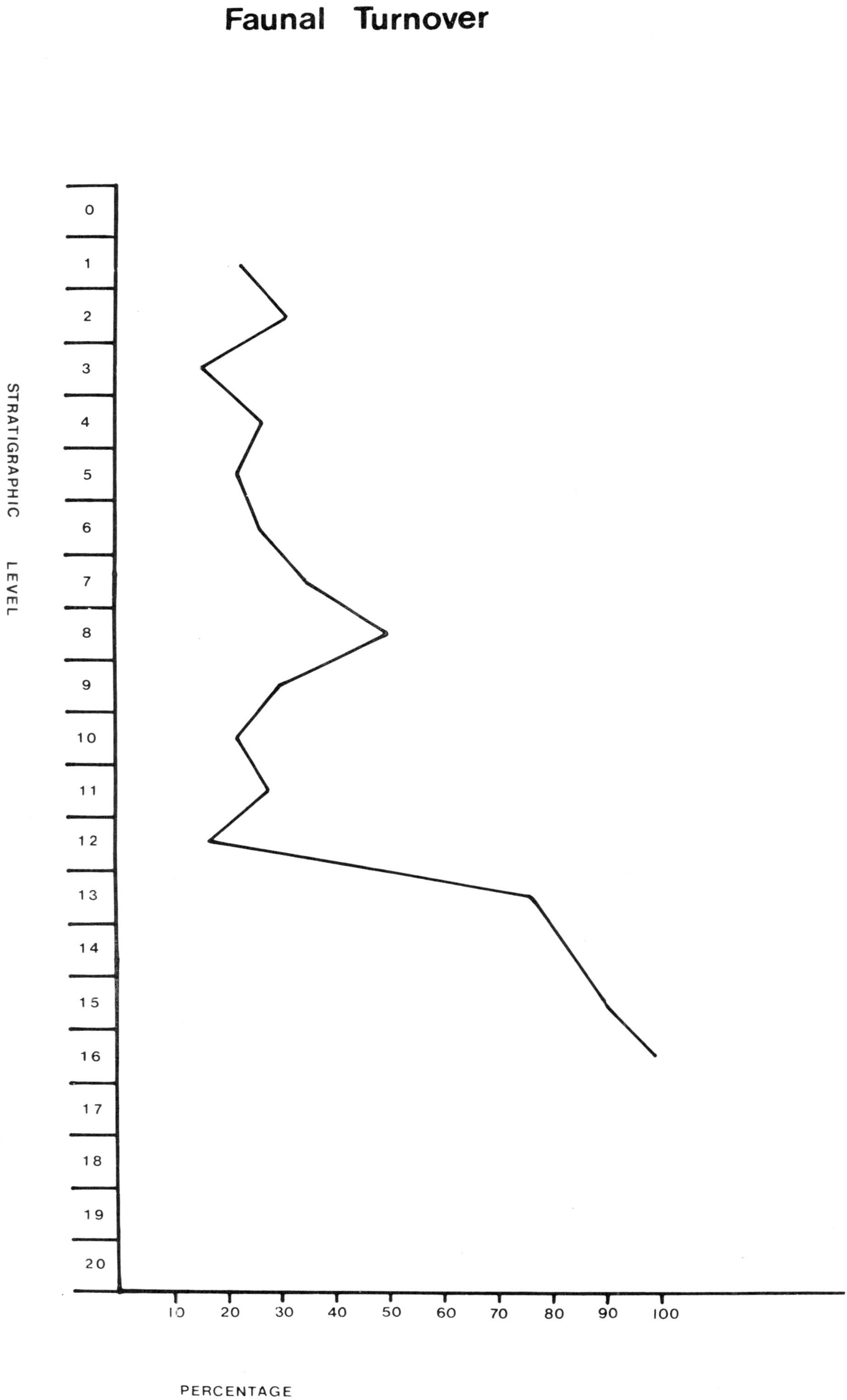

3. The rate of vertebrate faunal changeover from Pin Hole Cave.

DEGRADED BONES, FEATURE TYPE AND SPATIAL PATTERNING ON AN IRON AGE OCCUPATION SITE IN OXFORDSHIRE, ENGLAND.

Bob Wilson
University Museum
Parks Road
Oxford OX1 3PW

ABSTRACT

Highly leached and fragmented domestic mammal bones from an Iron Age occupation site on the gravels of the Upper Thames Valley are described. General indices of bone degradation are used to compare groups of bones from different types of archaeological feature and within a concentric sample pattern based on the centre of the site. Bones from deeper and peripheral features are better preserved than those from the commonly shallow central features. Although skeletal elements and epiphyses of sheep and pig were especially vulnerable to degradation, the concentric representation of these species relative to the more robust and larger boned cattle and horse is greater at the centre of the site and appears opposite to that expected to be produced by degradation. Discussion elaborates where cultural or ecological explanations are not applicable to some sets of data from bones but are required of others. Implications for excavation strategies are noted.

INTRODUCTION

This paper describes methods for investigating the taphonomy of bones at archaeological sites. Useful, if general, relationships between differentially degraded bones, feature type, and spatial distribution are demonstrable for a collection of highly degraded Iron Age bones excavated from a small enclosed settlement on the flood plain of the River Windrush in the Upper Thames Valley.

Mingie's Ditch, Hardwick with Yelford, Oxfordshire (SP391 059)

The site comprised a circular double ditched enclosure containing the foundations of five or more buildings, other structures, and the end of a trackway (Figure 1). Excavation was relatively complete and included widespread and relatively superficial occupation layers of alluvium and turf (Allen and Robinson 1978 and 1979).

THE BONES AND THEIR DEGRADED STATE

Cultural Typicality

As cultural debris the bones are similar to those from other Iron Age settlements on the river gravels of Oxfordshire. Bones from small sized sheep and cattle are abundant and the patterns of slaughtering indicated by the mandibles and teeth are paralleled elsewhere in the region. Collection of some relevant cultural data, for example, butchery marks, was however much restricted by the severity of bone degradation.

Fragmentation

The collected bones are highly fragmented and 84% were not identified to species.

External and Internal State

Fragments from the upper levels at the site are extensively eroded, leached white, soft textured and rounded. Some are encrusted with iron oxide or hydroxide. A proportion show internal deposition of this mineral and a blacker one, perhaps of manganese. The better preserved bones are stained light to dark brown.

Species and Skeletal Element Representation

Table 1 includes general information about the abundance of species among the bones collected by normal excavation. The numbers of fragments identified per individual animal of a species are similar to figures for a range of sites of different periods (Wilson 1978a:113 and in prep., Table 7). Unlike the other sites, however, loose teeth were easily the elements most commonly identified, viz. 54% of sheep bones and 40% of cattle. For sheep the next most common elements were tibia 12%, mandible 8%, and radius 7% (in prep., Table 3). Due to their disintegration, most cranial elements, vertebrae, ribs and small bones are scarcely represented in the collection.

Epiphysial Fusion Data

Table 1 also shows that relatively few epiphyses were recorded, particularly of sheep, pigs and cattle. The survival rate per individual is very low. Compared to other Iron Age sites with similar kill-off patterns provided by mandibles, there is an appreciable drop in the representation and survival of unfused epiphyses as indicated by the percentages of fused epiphyses.

Usually the unfused epiphyses of sheep and pig vanished without trace.

Degradation of Bones

It is clear that the assemblage has been highly degraded by a variety of cultural and natural processes, most obviously by leaching. The disintegration of bone was increased by breakages from excavation.

Sheep bones were degraded most, pig, cattle and horse in lessening amounts. Degradation appears inversely related to the general robustness of bone in terms of: fragment size, especially thickness; bone density, and the extent of mineralization at the age stages at which animals were killed. Some 40-50% of the sheep killed were lambs or hoggets but, apart from the teeth, their bones and particularly their epiphyses survived rarely.

Degradation of Bones According to Type of Feature

Figure 2 shows relationships between three different quantifications of the degraded bone debris according to feature type and depth at Mingie's Ditch, with additional information from Ashville Trading Estate, Abingdon (Wilson 1978a).

Feature Depth

For lack of closer definition the depth of feature types is depicted as the range of depth of all features of that type. Deposits of Iron Age bones ere found at all depths between the deepest limit and the ground surface although the most abundant bone-bearing layers need not have occurred close to either extreme (e.g. due to the presence of sterile primary fills and post-occupational deposition, particularly in the enclosure ditches).

Bone degradation as a percentage of four skeletal elements in the total of sheep bones

From the information above it is reasonable to propose that the higher the total percentage of loose teeth, mandible, radius and tibia, the more degraded is a group of bones. In the report on bones (in prep.), the major index of degradation used is the sum of the percentages of cattle and sheep teeth in their respective totals. This index gave similar results but is not an independent measure of the simpler one preferred in Figure 2.

Both percentage indices are higher where archaeological deposits were shallow and lower where features were deep (Figures 2a and 2b).

The trend is made more explicit with the information from Ashville, where bones were preserved well although other site conditions, such as the river terrace level, differ from that of Mingie's Ditch.

Percentage of Identified Bones

Difficulties with this measure will be discussed later but there is an interesting decrease in the already low rate of identification as degradation by other measures increases (Figure 2c). These results tend to confirm that as bones disintegrate they become less identifiable. The percentage identified at Ashville was 38%.

Rank order of sieved bone sample according to their state of preservation

Ten samples of sieved bones (2mm mesh) were examined and their state of preservation was ranked before there was any thought that this information might be correlated satisfactorily with the indices of degradation for normally collected bones. Preservation, however, was ranked qualitatively because of suspected problems of accounting for recent breakages and because of doubts about the value of time spent in a systematic assessment of this material.

The least well preserved fragments were considered to be heavily eroded, rounded, small, and perhaps few per bucketful of soil. Encrustation by iron oxides or hydroxides was associated with fragments in the five least well preserved samples. Survival of burnt bones appeared better than of the unburnt.

In Figure 2d the samples of sieved bones are categorised according to the type of feature and the depth below ground level that the original soil samples were taken from. The location of soil samples must be allowed some imprecision since a bucketful of soil is not all excavated from exactly the same level. Sample number 7 had no proper data on depth. Nevertheless the information on sample depths was supplied 'blind' according to a list of sample numbers which was randomly different to the ranked order of preservation.

The figure shows there is a rough correspondence between the rank order of preservation in bone samples and the rank order of bone degradation in feature types. Rank order of preservation corresponds better to the depth of soil at which samples were taken. For example, the best preserved groups of bones (Samples 1 and 2) were excavated from a pit at a deeper level than the groups from the trackway ditches (Samples 3 and 4).

Possibly the combination of both factors of feature type and depth gives the best correspondence to the rank order of degradation indicated by the sheep bones. Samples occurring at a similar soil level appear better preserved or less degraded on the left rather than the right hand side of the scheme. Thus the results from sieved soil samples appear to be a good and independent, if somewhat imprecise, validation of other indices of bone degradation.

Feature Type Variations

Figure 2 shows that bones from layers, postholes and post occupation deposits are degraded most. Those from the ditches and the 'pits' (bath-shaped features not interpreted as storage pits) were better preserved. The degree of degradation corresponds roughly to the depth of deposits in the features and suggests that degradation is substantially linked to processes effective just below the ground surface and, by implication, upon it.

Qualifications to the use of the indices of degradation

The measures and the relationships between them make sense particularly for later stages of the disintegration of bones. The percentage indices however need not represent equal units of degradation in their scales nor need these percentages result from identical processes, excluding methodological variation, at any single stage. At Mingie's Ditch, degradation, particularly of bone surfaces, proceeded so far as to make virtually unmeasurable the extent to which butchery and scavenging contributed to breakdown of bones.

Some results could be interpreted differently. The varied percentages of identification and the element representation of sheep might be taken as evidence of differential recovery. In part this rationale has been applied, perhaps mistakenly, to medieval urban evidence in Oxford (Wilson 1980). At Mingie's Ditch it would imply that less material was recovered from the ditches and pits than from layers and postholes, but reasons for this are not easy to substantiate.

It is possible that modern excavation was less careful for the former types of large and deep feature than for the latter shallower and smaller ones. Possibly teeth and small bones were difficult to see amongst water-logged silts and peats in the main enclosure ditches but this does not explain the variation observed among other features.

As will be seen, at least one additional and probably more important factor involved is the distribution of different sized fragments around the site. Thus at least two factors affect the percentage of unidentified bones so this index should be regarded cautiously and will not be used further here.

DEGRADATION AND RELIABILITY OF DATA TO IDENTIFY CULTURAL AND CONTEMPORARY ECOLOGICAL PROCESSES

Age Data

Where the degree of degradation is fully reported, cultural phenomena are unlikely to be boldly or unconditionally misinterpreted afterwards, particularly where the evidence, for example of slaughtering patterns, is documented by methods which provide contrasting results. At this site certainly epiphysial fusion data from moderate to heavily degraded sheep and pig bones should not be used to discuss patterns of animal husbandry. Even the data of tooth eruption and wear in sheep at Mingie's Ditch is only just adequate to determine whether their slaughtering pattern was similar to those at other Iron Age sites where the mandibles are relatively complete.

Fragment numbers and species abundance

Equally to the considerations above, however, the evidence of degradation may lead to the total rejection of data which would serve a general comparative purpose. Fragment number counts are regularly criticised in the literature and must be suspect here. If the bones of medium sized mammals, and juvenile and immature individuals are thought to have been destroyed, then sheep and smaller species ought to be under-represented by fragment number counts.

The implication must be true of small mammals and birds (Wilson, in prep.). Yet sufficient remained of the sheep to demonstrate, through greater effort than usual with recording teeth (Note b, Table 1), that the overall abundance of species did not differ greatly between methods of estimation by minimum numbers of individuals and by fragment number counts. Further, better preserved local Iron Age site collections do not support the inference by showing substantially greater proportions of sheep among the common species (Wilson 1978a:136). Other evidence was mentioned earlier to indicate that, despite the disappearance of some elements, the number of fragments per individual was not markedly reduced compared to other sites.

The evidence suggests that, as less robust fragments, in and on the ground, were broken up by scavengers and trampling and disintegrated and dissolved by leaching, the more resistant or deeply buried elements were rendered fragile and broke under continued processes of degradation including modern excavation, thus producing further fragments, especially loose teeth, which numerically replaced those already lost to fragment number counts of species.

Surprisingly at this site, therefore, fragment numbers of sheep and other domestic animal bones appear to give broadly useful species data despite severe and differential degradation which substantially altered the frequencies and percentages of the skeletal elements present.

SPATIAL DISTRIBUTION OF BONES

The previous discussion assists the isolation of types of data which allow the similarity of occupations at differently preserved sites to be compared approximately. A more difficult problem arises where poorly preserved site material yields unusual or even uniquely systematic configurations of data. This point is illustrated by Table 2 where selected elements of the main species are assigned within a concentric grouping pattern based on the site grid system and the centre of the site, as depicted in Figure 1. The details of the method (Wilson, in prep.) are not necessary here except to note that the selection of skeletal elements by this method tended to exclude the less well preserved bones.

In general, Table 2 confirms an earlier sampling exercise which showed that the majority of bones occurred in the central occupation area (Wilson 1978b). More relevantly, the percentages of sheep bones are very high for the central areas and lower near the periphery. While the bones of the larger animals are common centrally, their percentage proportions rise markedly on the outside of the area of occupation. Other site detritus, for example burnt bones and pottery, have similar distributions to bones of sheep, while that of large bone fragments, perhaps not surprisingly, resembles those of cattle and horse.

CONCENTRIC PATTERN OF BONE DEGRADATION

Interpretations of the concentric or radial distributions of site material are complicated and will not be entered into here except to ask whether these interesting distributional trends are distorted or have been produced by the differential degradation of bones.

Obtaining an answer to these problems brought difficulties since:

a) direct measures of preservation were not undertaken when normally collected bones were recorded. It was doubted that these would yield more than the obvious state of the bones, and, in any case, this aspect of reporting was not rated as highly as other priorities.

b) plotting of data in spatial sampling schemes used information from identified material which was least vulnerable to degradation so as to minimise the effect of this factor. Data plotting was time consuming and carried out for the most necessary comparisons, that is, those leading to inferences about Iron Age activities which could be tested at better preserved sites.

c) the necessity of degradation data in concentric groupings became imperative only as the scope of analysis was extended beyond its initial findings.

At first the only data available on degradation in concentric form were the frequencies of all loose or attached teeth among the selected elements of each species. These figures did not allow for differential degradation of maxillae or mandibles although differential degradation of the other moderately preserved elements might have indicated a trend. The concentric distribution of these data for cattle and sheep was irregular and probably caused by the clusters of teeth in more complete jaws and was most noticeable for teeth in single crushed skulls of cattle in two peripheral features. With allowance for this bias, the evidence suggested that better preserved bones occurred in the peripheral features: without allowance, no trend and relative evenness of preservation across the site would be inferred.

This inconclusiveness was unsatisfactory although the first inference of differential preservation is supported by observations about the general state of bones (Wilson in prep.). Eventually, the best established index of bone degradation from the elements of sheep was selected and data was replotted and counted according to a slightly modified concentric sampling scheme for the area inside the circular enclosure ditches. The modification is related to an extension of the concentric sampling scheme to centre on specific site structures. Differences with the previous sampling scheme are minor.

Table 3 shows that there is a smaller percentage of the four most common elements in the totals of sheep bones for each sample group outwards from the centre of the site from 84% to 64%. Unfortunately, the sample sizes of the outermost groups are small, but this evidence shows a trend toward the better preservation of bone in the peripheral features.

SIGNIFICANCE OF FRAGMENT FREQUENCIES IN THE INDEX OF BONE DEGRADATION PROVIDED BY SHEEP BONES

Although scepticism was felt about general statistical testing of fragment numbers obtained from badly fragmented debris, it appeared acceptable to use data in order to assess the variability of bone degradation.

In Table 4 the results show a significant difference in the representation of skeletal elements of sheep recovered from the central and peripheral areas of the occupation debris. In Table 5 even more significant differences are shown for sheep bones from different types of feature. The values of chi-squared are calculated from the original 'raw' data.

Thus the evidence of major differences in the type of deposit, and the degree to which bones in them are degraded, is reinforced by statistical testing. The variability of such differential degradation is averaged and reduced where results from different feature types are added together. Summation of feature data within the concentric sample pattern, however, does not entirely cancel the difference in bone degradation between inner and outer areas of the site.

FEATURE TYPE, DEPTH AND SPATIAL PATTERN

The conclusion that bone preservation in the concentric groupings is better for peripheral features is now seen to be predictable by an examination of the arrangement of features on the site plan and of evidence that degradation was significantly less for bones in the enclosure and trackway ditches. Conversely, greater degradation was most characteristic of the large group of bones from the layers surrounding the site centre. In general, features near the centre of the site were shallow - there are exceptions - and those toward the periphery were deepest, see Figures 1 and 2.

The logic of identifying the location of feature type with concentric patterns of data implies also a differentiation of species representation where feature types tend to fall into either peripheral or central categories, for example, contrast between superficial occupation layers and enclosure ditches. Evidence of such differences is documented already (Wilson in prep., Table 11).

DEGRADATION, DEPTH AND SOIL

Central deposits of bone were thus exposed most to processes of degradation during and after the occupation, and before and after burial. Trampling of bones directly or indirectly within superficial occupation layers could occur at all these stages. Although leaching by rain or floodwaters is regarded as a major cumulative and long term effect, the rate of leaching after burial must have been slight for any bone to have survived in the superficial layers.

Leaching was associated with the uppermost levels of brown silts or clays in the central area, was greatest in the top 0.2m or so of soil, and of little effect by 0.6m. Deposition of iron oxides or hydroxides on bones of the sieved samples occurred between 0.2m and 0.6m below ground level but this process may also have occurred above the upper level given. The water-table perhaps fluctuated about these levels of iron-panning effects. Internal mineralization of leached fragments from upper levels suggests that these fragments were reservoirs of alkalinity in a neutral soil or one occasionally made weakly acidic - precipitation of iron hydroxide and manganese oxide occurs in oxidising and alkaline conditions. Bone destruction by plant rootlets also occurred here but this was probably a subsidiary effect.

Better preserved material was incorporated in deeper and increasingly anaerobic clays and silts of grey or bluish colour, and dark coloured peats. At all levels better and worse preserved material occurred. Preservation in peripheral features was assisted by later and deeper deposits of alluvium which capped layers of occupational debris.

EXPLANATIONS OF CONCENTRIC DISTRIBUTIONS OF SPECIES DATA

It is fortunate that the remains of sheep were abundant in the central area where degradation appears greatest. Low percentages of these might have led to the supposition that poor preservation was the cause. Instead, the lowest percentages of sheep and pig were found where preservation was better and cannot therefore be ascribed to the disintegration and disappearance of bones of these species from peripheral areas.

This conclusion is supported by a previous argument that species information is broadly useful in assessing site phenomena from degraded material because the eventual breakdown of relatively resistant elements tends to replace the numbers of fragile ones which disintegrate and disappear at an earlier stage of degradation. It was suggested that bone fragmentation of this kind was not excessive so that sheep are unlikely to be much over represented by fragment number counts for the central area. A small percentage increase of sheep might be argued from the apparent increase in unidentified material with degradation but this information has been seen to be somewhat ambiguous.

It is therefore concluded that the concentric pattern of species data is mainly due to Iron Age human activity or contemporary scavenging by animals. A range of such explanations will be discussed in subsequent publications.

STRATEGIES OF FUTURE EXCAVATION AND INTERPRETATION

The described variation of bone degradation and culturally relevant information shows how excavation and sampling priorities and strategies could influence the collection and analysis of bones.

a) Top soil stripping (as usual with plough soil) and excavation of deep features will avoid the poorest preserved bones and collect the better preserved. At Mingie's Ditch, the removal of all soil to the level of the gravel, would have missed the largest group of fragments to be found in any feature type and would be biased against the recovery of sheep and pig bones.

b) An excavation focus on the central occupation area would collect greater numbers of sheep and pig bones and species percentages would be favoured against cattle and horse. Examination of the excavated and unexcavated areas of the site shown in Figure 1 suggests that some of this bias occurred at Mingie's Ditch. The density of bones in peripheral features was low but the area of unexpected enclosure ditches is obviously considerable. Part of the remaining inner enclosure ditch was excavated subsequently and produced eight bones of cattle and horse.

c) A reversed bias of species percentages would occur with the excavation or sectioning of peripheral features, or of the enclosure and outer trackway ditches which show on the cropmark photo of Mingie's Ditch. This approach, more common in the past, would provide the most biased results by indicating that 50 - 80% of the bones identifiable from the site are comprised of cattle and horse.

Nevertheless, the trends obtained for this site, where taken as a general interpretative model, also indicate how strategies and analyses of limited excavations can anticipate and allow for such problems. Suitably balanced investigations of whole sites are required in order to confirm the trends argued for at Mingie's Ditch.

CONCLUSIONS

Obtaining quantitative measures of the highly degraded bones from Mingie's Ditch have been limited by the complexity and obscurity of taphonomic factors originally involved, by uncertainty from the priorities of recording site data, and by difficulties in the control of methodological variables. Nevertheless, crude but useful measures of the degree of degradation have contributed to the following conclusions.

1. Degradation of bones was caused chiefly by leaching and earth pressure following or contemporary with destruction by trampling, scavenging, and butchery.

2. The degree of degradation is primarily related to the depth of bone bearing deposits below ground level - an 'obvious' finding but one poorly documented in the literature.

3. The type of feature that bones were found in is also important. Variability of bone degradation between deposits in different feature types was very great.

4. Degradation of bones in deposits was greater at the centre of the site and less at the periphery but the difference was less than among bones from features of different type and depth.

5. Even crude indices of bone degradation indicate that guidelines can be devised to limit the use of particular sets of data in overall site interpretations of variously degraded collections of bones.

6. Bone degradation differentially destroys skeletal elements and their epiphyses but the overall effect on comparative species

data appears less than expected because the fragmentation of relatively robust elements tends to replace numerically the loss of the most fragile elements to counts of fragment number.

7. The concentric pattern of varied representation of species in inner and outer areas of this site does not appear to be explained by differential degradation.

8. The concentric pattern of species data is also related to feature type where any one of them tends to occur in either central or peripheral areas of the site.

9. Irregular differences between the type and the amount of features excavated, and irregular differences in the extent of excavation in central and peripheral areas of sites, may produce difficulties of interpreting all of the bone collected.

10. Sampling, excavation, and interpretative strategies which allow for the kind of intrasite variation described should produce more reliable reports on bones.

ACKNOWLEDGEMENTS

Tim Allen and Mark Robinson provided site information and many useful comments on the script. Assistance with several points of information was given by Brian Atkins and Richard McPhail. Help from the referees in drawing attention to the problems of the text is appreciated. Roger D'Arcy typed the manuscript and Wendy Page helped to prepare the figures. Organisational and financial support in preparing this paper was generously provided by the Oxford Archaeological Unit, the Committee for DoE funded environmental archaeology in the South Midlands, and the Ancient Monuments Laboratory, London.

REFERENCES

Allen, T. and Robinson, M. (1979) Hardwick with Yelford, Mingie's Ditch. CBA Group 9 Newsletter 9, 115-117.

Robinson, M. and Allen, T. (1978) Hardwick with Yelford, Mingie's Ditch. CBA Group 9 Newsletter 8, 112-114.

Siegel, S. (1956) Non parametric statistics for the behavioural Sciences. (International Student Edition).

Wilson, R. (1978a) The animal bones. In: The Excavation of an Iron Age Settlement at Ashville Trading Estate, Abingdon, Oxfordshire 1974-76. (Ed) M. Parrington. CBA Research Report 28, 110-115.

Wilson, R. (1978b) Sampling bone densities at Mingie's Ditch. In: Sampling in Contemporary British Archaeology (Eds) J.F. Cherry, C. Gamble, and S. Shennan. BAR British Series 50, 355-361.

Wilson, R. (1980) Animal bone and shell. In: A Beaker burial and Medieval tenements in the Hamel, Oxford. (ed) N. Palmer. Oxoniensia xlv 198 and Fiche E06-E09.

Wilson, R. (in press) Faunal remains: animal bones and shells. In: Archaeology at Barton Court Farm, Abingdon. D. Miles. CBA Research Report.

Wilson, R. (in prep.) Report on the bones from Mingie's Ditch, Hardwick with Yelford, Oxon.

Table 1: General information on the non-sieved bones collected from the site at Mingie's Ditch

	Total number	Species representation sheep/goat[d]	cattle	pig	horse	do
Identified bones[a]	1746	52%	30%	6%	12%	0.3%
Min. No. Individuals[b]	98	60%	23%	8%	6%	3.0%
Fragments/individual		16	22	c15	34	nc
No. of epiphyses	46.0	65.0	7.0	57.0	nc	
Epiphyses/individual	0.8	3.0	1.1	9.5	nc	
Percentage of epiphyses fused	80.0	92.0	(71)	97.0	nc	
% at other sites[c]	53-54	74-79	25-27	85-100	nc	

[a] Fragment numbers counted from all identifiable elements except ribs and sternal segments. Freshly broken fragments which fitted together to give an identifiable element were counted as a single fragment.

[b] MNI determined from teeth and mandibles. Particularly for sheep 'mandibles' were frequently recorded as single (especially p4 and M3) or associated teeth of a tooth row - the method relied on detailed locations of find spots. these results provided the apparently high MNI from such fragmentary debris.

[c] Ashville, Abingdon (Wilson 1978a, Table 18) and combined data from Burton Court Farm, Farmoor and Appleford (Wilson in press, Table 27).

[d] Goat is probably present in small numbers.

Table 2: Percentage of species bones outward from the centre of the site (concentric sample pattern)

Sample numbers	Mean distance (m) from centre of site	Fragment number	Percentage sheep/goat	pig	cattle	h
1	7.2	698	67	4	23	
2	12.4	363	59	10	22	
3	16.0	278	50	9	33	
4	18.9	158	15	9	48	
5-6	22.6	52	25	2	31	
7-10	28.6	75	25	5	47	
11-16	36.4	58	29	2	55	
17-25	45.7	19	47	5	26	

Table 3: Percentages of the four most common elements in the total of sheep bones (index of degradation) for concentric sample areas.

Me fr ce	Sa nu	Fr nu	Pe of fo
7.2	1	465	84
12.	2	198	83
16.	3	135	79
18.	4	28	79
23.	5-7	22	64

[a] Tooth, mandible, radius and tibia.

Table 4: Chi squared values of testing component frequencies of degraded sheep bones given in Table 3 (2 x 2 contingency tables, where $p \geq 0.05$, $X^2 \geq 3.84$; 1df: Siegel 1956:107. Raw data in Wilson, in prep.).

Group Sample Numbers	Central Area 1	2	Peripheral 3	4	
2	0.04				83%
3	1.56	0.64			79%
4	0.29	0.13	0.03		79%
5-7	5.06[a]	3.85[a]	1.81	0.72	64%
	4%	83%	79%	79%	Percentages given by degradation index (Reference only).

[a] Significantly different results.

Table 5: Chi squared values of testing component frequences of degraded sheep bones from different types of feature as depicted in Figure 2 (2 x 2 contingency tables, where $p \geq 0.05$, $\chi^2 \geq 3.84$; 1df: Siegel 1956:107. Raw data in Wilson, in prep., Table 13).

	ED	TD	P	HPH	HG	
Enclosure ditches (ED)						
Trackway ditches (TD)	3.96[a]					72%
Pits (P)	4.03[a]	0.00[+]				73%
House post-holes (HPH)	7.34[a]	1.30	0.94			82%
House gullies (HG)	13.42[a]	4.25[a]	0.22	0.06		86%
Layers (L)	27.52[a]	19.54[a]	10.22[a]	0.81	0.53	88%
	50%	72%	73%	82%	86%	[b]

[a] Significantly different results.

[b] Percentage given by degradation index (reference only).

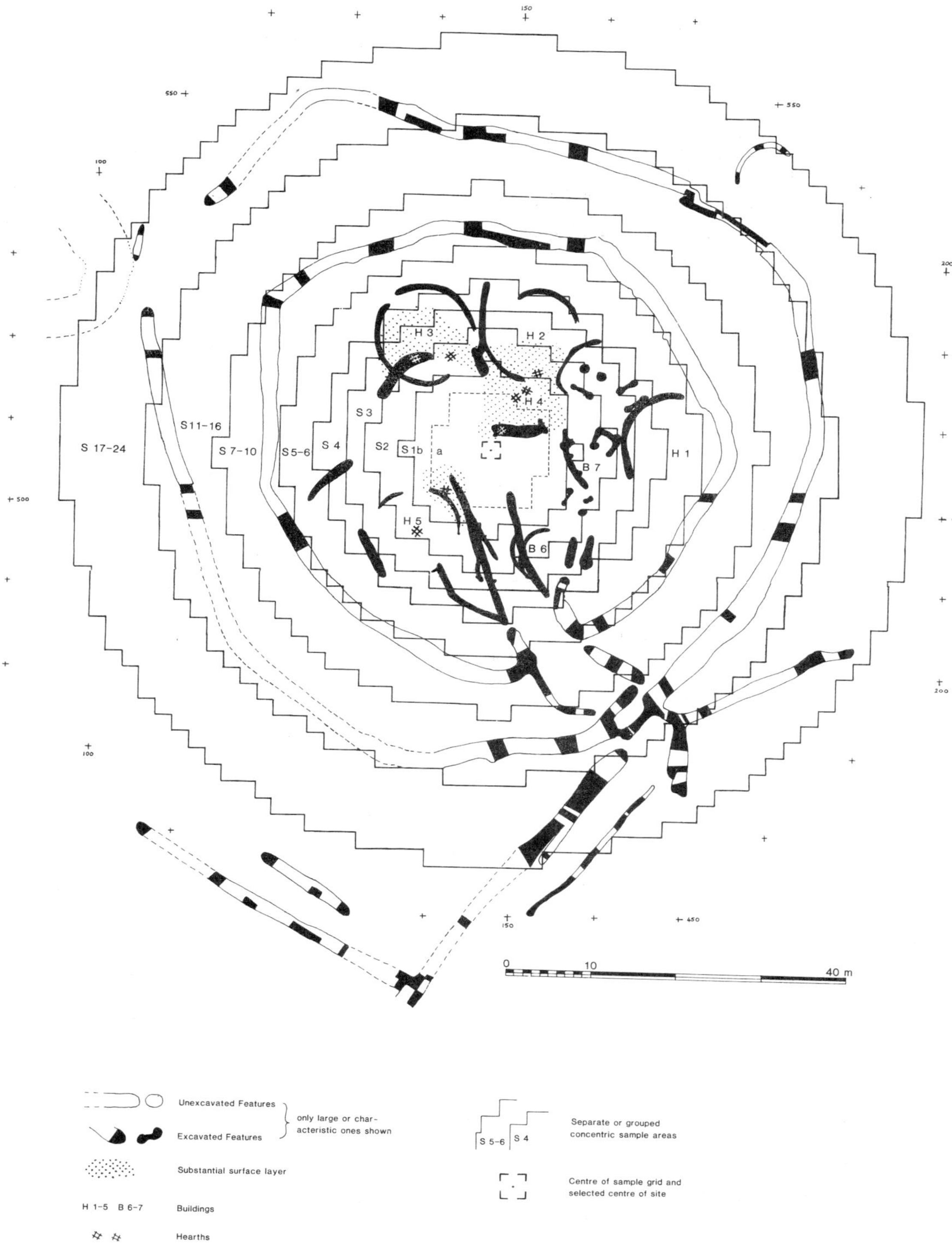

Figure 1: Plan of the Iron Age settlement at Mingie's Ditch, Hardwick, Oxon, showing major features, excavated areas, and the boundaries of a concentric sampling pattern for grouping bones.

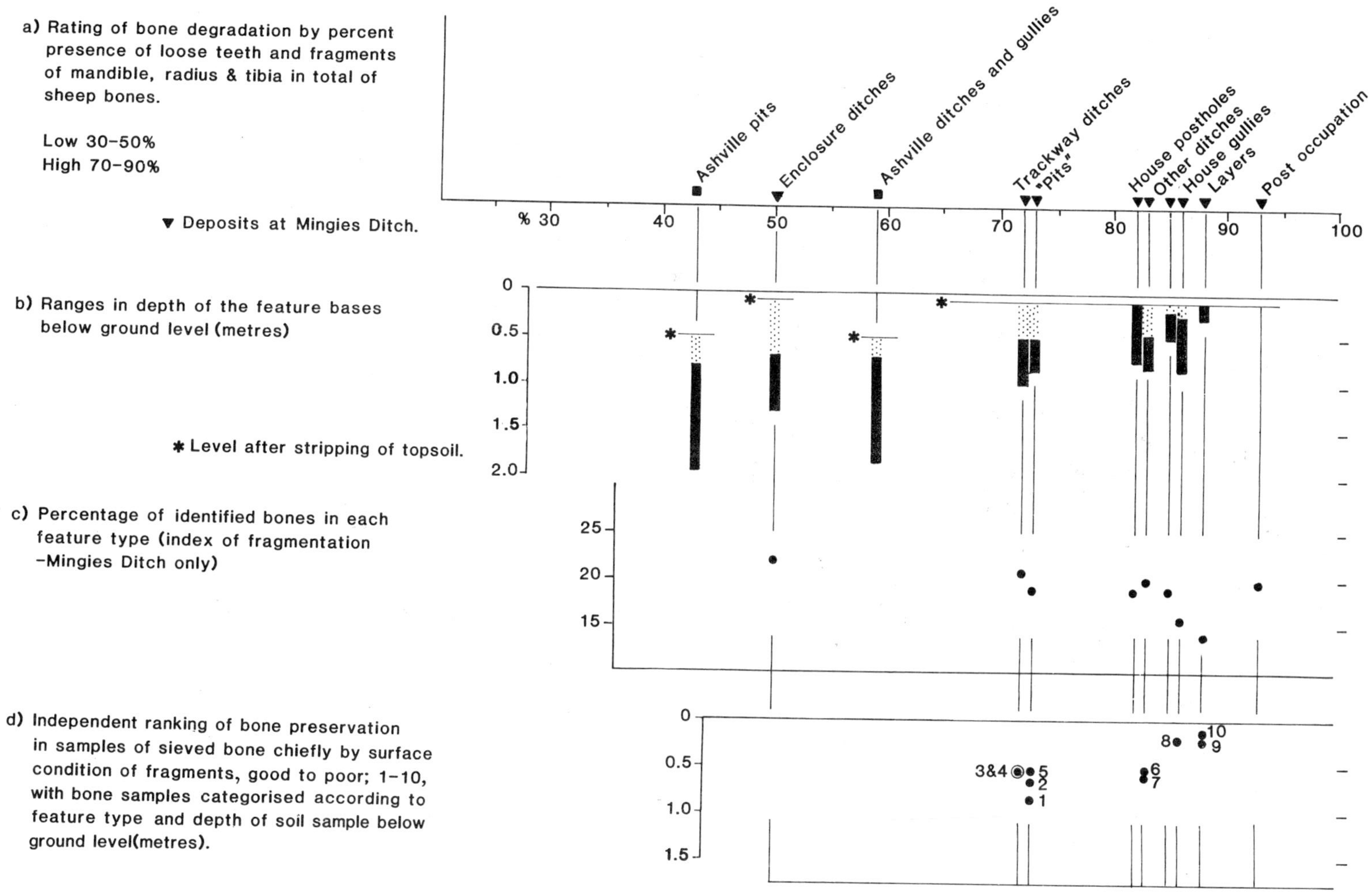

Figure 2: Comparison of feature types at Mingie's Ditch and Ashville, Abingdon, according to feature depth and three different measures of bone degradation.

INSECTS

OUTDOORS - INDOORS? THE OUTDOOR COMPONENT OF ARCHAEOLOGICAL INSECT ASSEMBLAGES.

Harry Kenward
Environmental Archaeology Unit, University of York, YO1 5DD, England.

ABSTRACT

Insect corpses deposited in modern buildings have been examined. The death assemblages may include many `outdoor' forms, while the number and proportion of insects associated with buildings may vary from near zero to very large. On the basis of preliminary observations three factors appear important: the presence of birds' nests; distance from openings; and airflow pattern. Deposition of flying insects in the `dead space' between narrow openings may lead to the incorporation of a large background fauna much like that found in modern roof gutters. Interpretation of archaeological assemblages from buildings or putative buildings requires re-thinking as a result of these observations.

INTRODUCTION

An `outdoor component' in insect death assemblages may be defined as comprising those species unable to find habitats in a building of any kind (Kenward 1978, 14). This concept of an outdoor component has been found to be very useful in the interpretation of urban archaeological insect assemblages. It should, however, be remembered that most species cannot be categorised as literally `indoor' or `outdoor' insects; their ability to breed within a typical building will in most cases have varied with the period. Roman stone buildings may have supported only half-a-dozen or so species, but Viking-age earth-floored wattle ones several hundred. For the same reason, species excluded from the `outdoor' category do not invariably live indoors; indeed, very few insects are restricted to buildings in Britain.

For most purposes the number of certain-plus-probable outdoor specimens, abbreviated 'OB', is the most useful statistic. It is believed that two things are being measured by estimates of OB: (1) the importance of `outdoor' habitats at or circumjacent to the site of deposition; (2) how much background fauna (Kenward 1975, 88) was incorporated. The latter factor is important in attempts to reconstruct ecological conditions in restricted areas, as is usually the case with occupation sites. Obviously, it must be shown that the insects used for the reconstruction are autochthones (ie species breeding on the spot) and did not arrive in flight or in the droppings of predators; both mechanisms have been shown to be able to contribute a considerable allochthonous component to modern deposits (Kenward 1976).

The proportion of background fauna is seen as a measure of several factors: (1) the rate of deposition and total time taken for the deposit to form; (2) the abundance of autochthones; and (3) the degree of protection from flying insects and such sources of corpses as bird droppings. This paper is concerned primarily with the last factor.

It has been argued that background fauna may be measured by the combination of OB and an index of diversity (Kenward 1978, 16-26), and that these two parameters can be used to determine whether deposits formed within buildings, as in the case of the Lloyds Bank site, 6-8 Pavement, York (loc. cit., 35-6). At this site four narrow trenches were dug within modern cellars, cutting a succession of essentially horizontal layers of richly organic deposits. In one trench (II) most layers were sampled for insect analysis. Fisher's Alpha (an index of diversity appropriate to random collections, Fisher, Corbet and Williams 1943, May 1975) and OB were measured and two groups of assemblages, the first of high Alpha and OB and the second of low Alpha and OB, were detected. These, it was originally postulated, might represent deposits of outdoor and indoor origin respectively. In the final environmental report (Hall, Kenward, Williams and Greig 1983, 196-201) this conclusion was revised in the light of further analyses. The samples certainly fall into two major groups (loc. cit., fig. 44), but even the higher group, best termed intermediate, is of much lower Alpha and OB than many other samples from the Lloyds Bank site and, more especially, from other sites. The two groups from Trench II most probably represent two kinds of indoor deposit, a view which accords better with the archaeological interpretation offered by Addyman and Hall (forthcoming) and with observations of Viking-age buildings at the 16-22 Coppergate, York, site.

MODERN INDOOR DEPOSITS

In order to provide a better basis for archaeological interpretation, a study of modern indoor deposits has been undertaken. The first sample was taken from a small modern thatched barn (Kenward, Hall and Jones forthcoming). The assemblage of beetles and bugs was found to have very high values of both Alpha and OB, containing many species in small numbers. This assemblage is in some ways more reminiscent of roof-gutter assemblages (regarded as essentially background fauna, Kenward 1976) than of those from deposits inside ancient buildings. One reason may be the relatively small number of autochthones in the modern barn assemblage (although the five most abundant species probably bred in its roof); another may be the fact that it has two openings (see below). Further samples from within modern buildings have subsequently been examined by the author and by Fionnagh Stone (Kenward and Stone, forthcoming). The material has been collected from unoccupied roof spaces; from an occupied roof space; from a cellar; and from under ground-floor floorboards. One-kilogramme samples were processed in each case. Although few samples could be examined, the results give a great deal of enlightening and cautionary information.

SAMPLES FROM UNOCCUPIED ROOF SPACES

Two samples of dust from between the rafters in the roofs of two late-Victorian terraced two-storied houses in south east urban York have been examined. Both roofs were unlined and contained dry birds' nests. The first of these (Belle Vue Street, MOD FAS 1/1) gave a large and diverse assemblage of beetles and bugs (minumum number of individuals (MNI) = 222; 99 species; Alpha = 69, SE = 8). The most abundant species is Anotylus tetracarinatus (Block) (22%), never likely to have bred in the roof although it is not an outdoor species within the definition used for archaeological material. The other more abundant species are Ptinus tectus Boieldieu, Atomaria spp., Ptenidium sp. and Lathridius minutus (Linnaeus) group. Of these, only P. tectus appears likely to have bred in the roof. Other than P. tectus and a single individual of Anobium punctatum (Degeer), the vast majority of the assemblage must be background fauna. A few specimens, for example some of the Carabidae, Scarabaeidae and Curculionidae, show breaks in the sclerites which are characteristic of beetles from bird droppings, but the bulk probably arrived in flight. Half the species are outdoor forms (OB) and they make up 28% of the assemblage. Aquatics and damp ground species are rare (together 2%), reflecting the great rarity of such habitats in the surroundings.

The second sample (Farrar Street, MOD FAS 3) also gave a fauna dominated by Anotylus tetracarinatus (30%). The assemblage is of modest size (170 individuals, 59 species). It is remininscent of the previous one, but the outdoor component is smaller (24% of species, 10% of individuals). There are many fewer species recorded as single individuals and the differences between the two assemblages may largely result from differences in the background fauna in the two areas; 1/1 came from a building quite close to parkland, allotments and a large cemetery, where there are diverse habitats, 4 from a house which for much of its life has been embedded in terraced streets. These roofs presented only very dry nest habitats and almost the whole of both assemblages must have originated in the flight component of the background fauna.

SAMPLE FROM AN OCCUPIED ROOF SPACE

This sample of dust and litter (MOD FAS 4) was collected from a roof space used as a workshop in a three-floored late Victorian terraced house in Portland Street, near the centre of York but adjacent to parkland. It gave a very large assemblage of moderate diversity (818 individuals, 111 species, Alpha = 35, SE = 2). The outdoor component is substantial (OB = 51%) and the condition of the specimens indicates that a majority originated in bird droppings. The most abundant outdoor forms are Sitona spp., Amara spp., Pterostichus spp., Bembidion properans Stephens, Agriotes spp., Clivina fossor (Linnaeus) and Simplocaria semistriata (Fabricius). Among the indoor forms, two species typical of roof spaces predominate; Ptinus tectus and Anobium punctatum. Apart from these two, none of the abundant species could have bred in the roof. There was no obvious flow of air through this roof space, but there was abundant evidence that birds had nested over a long period of time.

SAMPLES FROM A CELLAR

The samples were collected from the cellar floor and from the ledge of an unglazed window in the house in Portland Street. The material from the floor (MOD FAS 1) gave a large assemblage (703 individuals, 50 species) of low diversity (Alpha = 12, SE = 1) dominated by several abundant species likely to have bred in the cellar: Aridius nodifer (Westwood), Lathridius minutus group, Mycetaea hirta (Marsham), Euophryum confine (Broun), Pentarthrum huttoni Wollaston, Anobium punctatum, Cryptophagus sscutellatus Newman and C. acutangulatus (Gyllenhal). The outdoor component is small (2%), with a few specimens of a moderate number of species. The sample from the window ledge (MOD FAS 2) gave a quite small assemblage (157 individuals, 59 species, Alpha = 34, SE = 4) in which the only abundant species are Ptinus tectus and Lathridius minutus group. The outdoor component is much larger than in the previous sample both in proportion (20%) and in absolute numbers.

SAMPLE FROM UNDER FLOORBOARDS

This sample (MOD 120) was collected from under the hall floor of the house in Belle Vue Street. There was no nearby ground level access, but dust and insects may have entered through cracks in the floorboards near the door. The assemblage is surprisingly large (509 individuals, 130 taxa) and diverse (Alpha = 56, SE = 4). Spider beetles (Ptinus tectus, Niptus hololeucus (Faldermann) and Tipnus unicolor (Piller and Mitterpacher) are abundant, and some of the other species might have bred under the floor (for example Pentarthrum huttoni). Species associated with relatively dry decaying matter are predominant. There are, however, many specimens of `outdoor' species (%OB = 10.2) and the most abundant species is Anotylus tetracarinatus (14%), unlikely to have bred in the building.

DISCUSSION

Clearly large numbers of insect corpses from 'outdoor' habitats may become deposited within buildings. These insects may enter in three ways: human importation - believed to be negligible in the present cases, except perhaps for some wood-feeders in the cellar; as a result of the activities of birds and other insectivores - undoubtedly responsible for many corpses in MOD FAS 3; and through their own activity, principally in flight - as seems likely in the case of MOD FAS 1/1 and MOD FAS 4. Many of the outdoor species in MOD 120 also seem likely to have arrived on the wing, although in this case, and the window ledge sample, a proportion may have crawled.

Insects, especially small species, have little immediate control over the direction of their flight, being carried with the air flow. They are thus likely to be deposited in much the same way as inanimate particles, remaining suspended in fast air streams and being deposited in 'dead spaces' where air moves slowly. Roofs (and buildings in general) with a through-flow of air enclose such a dead space.

The assemblages from the roof spaces may be compared with those from external roof gutters (Kenward 1976 and unpublished). MOD FAS 1/1 and MOD FAS 4 gave assemblages which are subjectively very like the sample MOD 15, from a roof gulley in the centre of York, especially in the abundant *Anotylus tetracarinatus* (Kenward 1976); the differences are easily explained by differences in the surroundings (Kenward and Stone forthcoming). On the other hand, MOD FAS 3, from the occupied roof space, is very reminiscent of a sample from a roof in suburban York (Kenward, unpublished). This, too, gave an assemblage dominated by *Amara* and *Sitona* species, also apparently deposited by birds.

How applicable are these results to the interpretation of archaeological deposits? The present assemblages probably differ from most indoor archaeological ones in having accumulated over long periods of time (perhaps the whole life of the houses in the case of the roofs and underfloor cavity), and being preserved by a combination of dryness, soot and mortar rather than waterlogging. In addition, most contain recently introduced aliens in large numbers: *Aridius* spp., *Euophryum confine*, *Pentarthrum huttoni* and *Ptinus tectus*.

Human importation was probably a little more important in the past than in the deposits examined here, since there is abundant evidence that a variety of organic materials likely to contain live and dead insects were, in most periods, carried into buildings or incorporated into their structure. There are, however, few cases where specimens can confidently be said to have been imported in this way. Small numbers of individuals will be hard to distinguish from those which arrived by natural dispersal. Two cases may be cited: the flightless heathland weevil Strophosomus sus Stephens at the Lloyds Bank site (Hall et al. 1983); and some woodland species believed to have been imported in moss to a medieval site in Oslo (Kenward forthcoming a).

Insects originating in bird droppings are generally easily recognised; an archaeological example is cited by Girling (1977) and the characteristics of specimens of this kind outlined by Kenward (1976, 10). There are few specimens from urban deposits seemingly falling in this category, but the possible introduction of large numbers of specimens by birds must always be kept in mind. An example may be provided by the putlog holes in the masonry in the church of St Mary Bishophill Junior (Kenward forthcoming b). Obviously, the primary factor affecting this component will be the presence of insectivores; evidence of these will be useful in arriving at a description of the building in which deposition took place.

Transport by wind or flight leaves no distinguishing marks on fossils but death assemblage structure and composition may provide clues; on urban sites, very high diversity and a large concentration or percentage of OB is generally, but not rigidly, taken as evidence of a large background component.

Factors affecting the deposition of the flight component of background fauna are mentioned above. Its relative importance in archaeological assemblages is likely to have been determined by factors affecting, firstly, airflow and deposition and, secondly, the abundance of autochthones.

If there is no airflow through a building, flying insects can only enter through eddies or in directed flight and the numbers are not likely to be very large in most cases; this is supported by the samples from cellars (MOD FAS 1, MOD FAS 2), where the deposit by the opening contained many more species and individuals of outdoor insects than the deposit on the floor two metres away. Migrating insects seem unlikely to fly into dark, relatively cool places. It is, however, possible that some species found in archaeological deposits, for example Heterogaster urticae (Fabricius), entered buildings to hibernate. Without a through-flow of air, the numbers entering the building are likely to be proportional to the size of the holes available - whether doors, windows, smoke holes or cracks. Many insects, having entered, might of course leave again.

If there is a through-flow, the quantity of air passing and the magnitude of the velocity drop will be important determinants of settling rate. The times when the apertures are open will also have an effect; they must be open when the insects are on the wing. This may lead to selection in favour of particular species, for example those flying on very hot days or on muggy evenings. It is conceivable, but not thought likely, that the Carpelimusbilineatus Stephens found in quantity in many samples from at the Lloyds Bank site entered in this way.

The possibility that large components of background fauna may be deposited in buildings if the conditions are right casts doubt on our ability to determine whether deposits formed indoors or out. How relevant is this to the archaeological buildings so far examined? Few of them, if any, have large concentrations or percentages of OB or high diversity when compared with modern outdoor deposits, or with archaeological deposits believed on other evidence to have formed outdoors. This suggests that the conclusions drawn regarding the Lloyds Bank deposits were correct; the layers indeed formed indoors, in buildings which it now appears were kept well closed, or perhaps had only single small openings.

CONCLUSION

These observations emphasise, yet again, the need for caution in the interpretation of insect assemblages from archaeological deposits. In particular, more modern and ancient comparative material must be examined. Interpretations made without such comparisons are likely to be no more than guesses. Interpretation of death assemblages from within buildings presents a variety of special problems. However, there is no reason to despair where there is a good archaeological record: in such cases, we have a tool to determine, among other things, how well ventilated buildings were, surely an important factor in describing past human life.

ACKNOWLEDGEMENTS

I am grateful to Jim Spriggs and Patrick Ottaway for access to their homes to collect material; to Fionnagh Stone, who carried out the practical work and preliminary analyses of certain samples; to Frances Large, Alan Robertson and Philip Thomas, who processed and mounted the greater part of the material; and to my colleagues in the EAU for their comments on the manuscript.

REFERENCES

Addyman, P. V., and Hall, A. R. (forthcoming). Archaeology of York 8.

Fisher, R. A., Corbet, A. S. and Williams, C. B. (1943). The relationship between the number of species and the number of individuals in a random sample of an animal population. Journal of Animal Ecology 12, 42-58.

Girling, M. A. (1977). Bird pellets from a Somerset Levels Neolithic Trackway. The Naturalist 102, 49-52.

Hall, A. R., Kenward, H. K., Williams, D. and Greig, J.R.A. (1983). Environment and living conditions at two Anglo-Scandinavian sites. Archaeology of York 14 (4), 157-240 plus fiche 1. London, Council for British Archaeology.

Kenward, H. K. (1975). Pitfalls in the environmental interpretation of insect death assemblages. Journal of archaeological Science 2, 85-94.

Kenward, H. K. (1976). Reconstructing ancient ecological conditions from insect remains: some problems and an experimental approach. Ecological Entomology 1, 7-17.

Kenward, H. K. (1978). The analysis of archaeological insect assemblages: a new approach. Archaeology of York 19 (1), 1-68. London, Council for British Archaeology.

Kenward, H. K. (forthcoming a). Insect remains from Gamlebyen, Oslo.

Kenward, H. K. (forthcoming b). Insect remains from putlog holes. In Briden, C. and Buckland, P. C., 'The Church of St Mary Bishophill Junior'. Archaeology of York 8.

Kenward, H. K., Hall, A. R. and Jones, A. K. G. (forthcoming). Turf roofs and urban archaeological build-up.

Kenward, H. K. and Stone, F. A. (forthcoming). Experimental investigations of the 'background component' in urban insect death assemblages.

May, R. M. (1975). Patterns of species abundance and diversity. in M. L. Cody and J. M. Diamond (eds.), Ecology and Evolution of Communities (Cambridge, Mass.) 81-120.

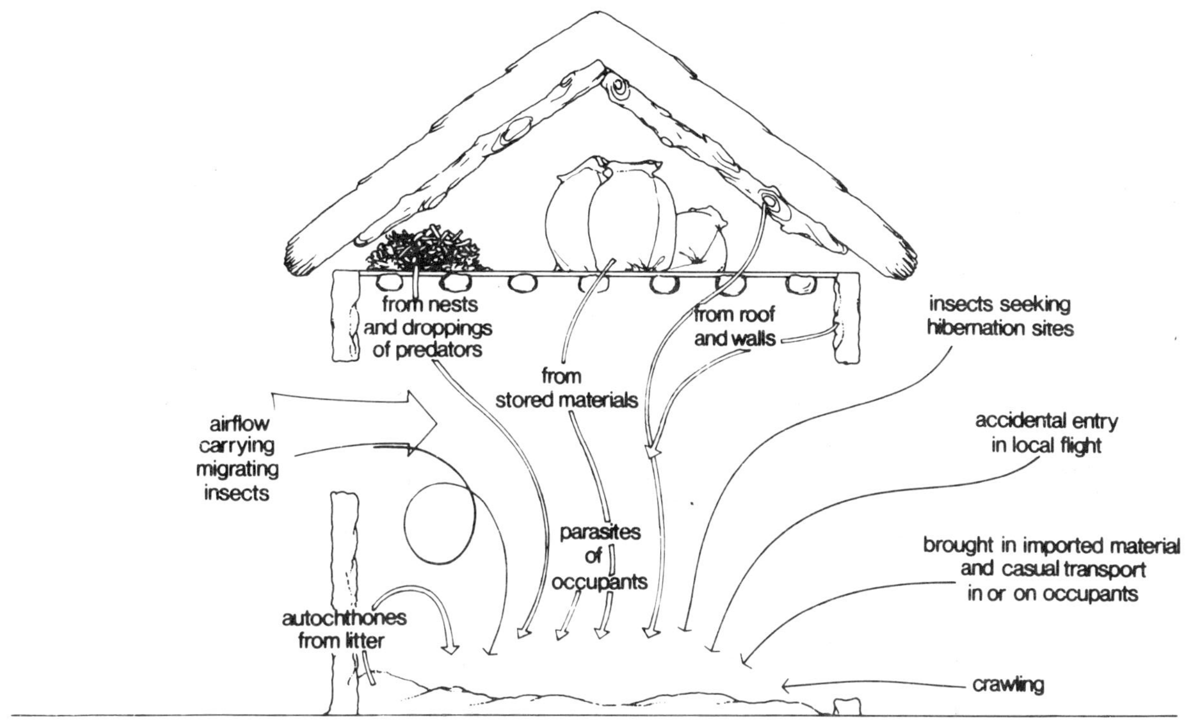

Figure 1. Likely sources of insect corpses in a deposit formed within a generalised building. Broad arrows - autochthones, originating in building; narrow arrows - allochthones, originating

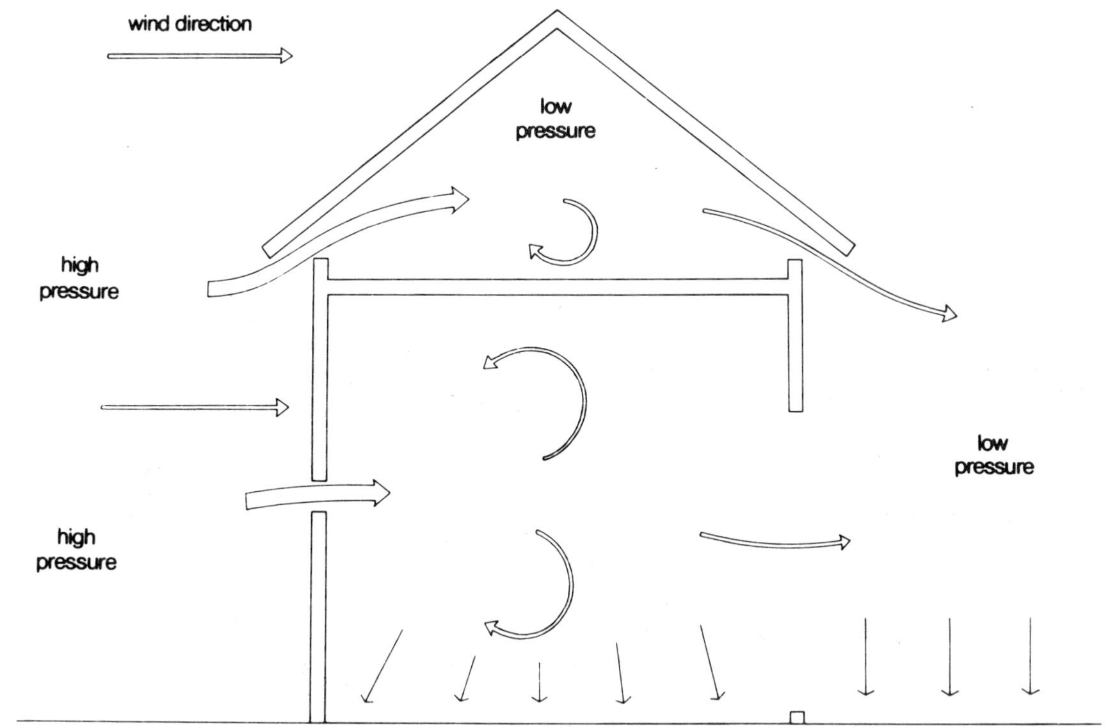

Figure 2. Depostion of flying insects within a generalised building. Broad arrows - airflow; narrow arrows - settling insects. (Drawing: A. R. Robertson.)

TRICHURID OVA IN ARCHAEOLOGICAL DEPOSITS: THEIR VALUE AS INDICATORS OF ANCIENT FAECES.

ANDREW K. G. JONES
Environmental Archaeology Unit,
Department of Biology,
University of York,
York YO1 5DD.

ABSTRACT

Ancient human intestinal parasite eggs (Trichuris and Ascaris spp.) have been found in many of the archaeological deposits so far examined in York, England. They occur over a wide range of concentrations. By considering modern egg-count data it is possible to determine whether a particular deposit is primarily faecal in origin, or if it is layer containing a small faecal component. Guidelines are set out for the interpretation of archaeological trichurid egg-count data.

INTRODUCTION

Eggs of the parasitic intestinal nematodes Trichuris (whipworm) and Ascaris (maw-worm) have been frequently reported in the archaeological and biological literature (Jones 1982). These ova are of great interest, for not only are they primary evidence of common human intestinal diseases in the past, but they also indicate the presence of ancient faeces.

It is usually possible to identify the trichurid ova to species by careful consideration of egg measurements. All the egg-bearing samples examined so far from deposits in York have proved to contain eggs of the human whipworm, Trichuris trichiura L. Today this is an extremely common parasite in communities having only primitive methods of faeces disposal; recent estimates suggest that 500 million people are infested with whipworms (Mahmoud 1979, 2157). People usually harbour only a small number of whipworms which cause comparatively slight clinical effects (Faust 1944). Children aged between five and 15 years have the highest prevalence and usually carry heavier worm burdens than adults.

Trichuris normally inhabits the lower part of the human digestive tract where it also breeds. It is a prolific egg producer, shedding into the gut lumen many thousands of ova which are passed with faeces. To complete the life cycle, an egg must embryonate outside the host before being ingested. Eggs then hatch in the stomach and continue their development in the intestine. Trichuris eggs need moisture to develop; they are readily desiccated and survive best in moist soil. *Ascaris* ova are more resistant to dry conditions and this genus is prevalent in hot dry climates (Spindler 1929). It is therefore not surprising that on British archaeological sites *Trichuris* ova are far more common than *Ascaris* eggs.

This paper examines trichurid egg-counts of samples from archaeological deposits in the light of modern evidence.

MODERN INVESTIGATIONS

To review the recent studies of human whipworm egg production, faecal egg-count data, and the distribution of parasite ova in and around buildings is beyond the scope of this paper. A comprehensive search for information relevant to archaeological investigations is urgently required. Much of the enormous medical and parasitological literature describes parasitic infestations in areas of the world (e.g. southern states of North America, Central America, Egypt) where the diversity of parasitic organisms is usually much greater than that found in archaeological deposits in Britain. Data gathered in temperate Europe are sparse. Consequently few data are directly relevant to the present investigations. A few papers, found by scanning the literature over the last four years have provided invaluable information and are summarized in Fig 01.

These are mainly studies giving the numbers of ova per gram faeces passed by people harbouring known numbers of worms. Papers showing the number of parasite ova in modern latrine deposits, soil and sweepings from buildings are particularly uncommon. Epidemiologists appear to have been more concerned with finding viable ova than with accurately determining the numbers of ova per gram present in samples collected around human habitation sites. Nevertheless, some data have come to light; doubtless many more will emerge.

When considering parasitological evidence it is important to bear in mind that workers have used a variety of different techniques to examine modern faecal and soil samples. There is a plethora of papers describing these techniques, many of which include procedures to allow for variation in recovery and stool consistency. Chandler and Read (1961) give details of the most frequently used methods e.g. those of Stoll and Telemann. In addition, the various teams of workers have used different units for presenting egg-count data. Most are presented as numbers of ova per gram faeces; others as numbers per cubic centimetre, while some are given as numbers per millilitre of faeces. These small differences in procedure and data presentation are unlikely to influence the overall trend that patients with small worm burdens produce less ova per unit faeces that people with heavy burdens. For present purposes it is assumed that numbers of eggs per gram, cubic centimetre, and millilitre faeces are equivalents.

It is important to stress the large range of numbers of trichurid ova per gram (opg) that can be passed with faeces. Numbers in excess of 100,000 opg have been recorded using the Stoll sedimentation method (Burrows 1950). The highest egg-count of Trichuris ova to come to light in the present survey is given by Swartzwelder, Frye, Muhleisen, Miller, Lampert, Pena Chavarria, Abadie, Anthony and Sappenfield (1957) who reported that a patient was passing the equivalent of 412,800 trichurid ova per millilitre faeces; this patient passed 3,162 adult T. trichiura after treatment with an anthelminthic. It seems that only a small number of cases are ever carefully investigated and these tend to be the exceptional cases where large worm burdens cause severe symptoms (e.g. Getz 1945 and Swartzwelder et al. 1957).

In a thorough study of patients in British mental hospitals, Lynch, Green, McFadzean and Pugh (1972) reported the egg-counts of 74 people infected with T. trichiura. This sample is of particular value as it is for British patients and therefore more relevant to British archaeological samples than data collected in other climatic zones. Although this study examined a number of patients with heavy worm burdens, it is unlikely to reflect accurately the pattern of medieval infestations. Medieval people were almost certainly more familiar with and tolerant of parasites. Faecal egg-counts, obtained by using a formol-ether concentration technique described by Allen and Ridley (1970), ranged from less than 10 opg to approximately 100,000 opg. Over 80% of the sampled population gave egg-counts of between 10 and 10,000 opg, with 34% giving 100-1,000 opg. Details of the number of worms producing these ova were not provided, but Burrows's (1950) paper does show that mean egg-counts of 1,100, 17,000 and 97,300 opg were produced by people harbouring 14, 466, and 2,544 worms respectively. Paine, Lower and Cooper (1960), examined faeces from 12 mental patients before and after treatment with the anthelminthic Dithiazanine and their results broadly concur with Burrows'. Their small sample of twelve cases gave mean egg-counts of between 100 and 9,200 opg for patients with between 6 and 131 worms.

Other useful recent data are available from parasitologists who have attempted to trace the life cycle of the worms in the field. Several papers describe the examination of samples of soil from gardens, housefloors, and compost heaps around human habitation sites where parasitic infections are endemic. Regrettably, few of the papers describing such investigations present data giving the numbers of ova per gram present in soil samples. Nevertheless, they show that parasite ova can resist decay given suitable environmental conditions, and that they frequently occur around and in human habitations. Such eggs form a background reservoir for infection. De Groot, Hoorweg, Kleevens and Rijpstra (1976), using the Telemann method, examined faecal and soil samples from a mental hospital in the Netherlands where human sewage was used for horticultural purposes. They discovered that 64% of the inmates were infected with T. trichiura and that water and 'compost' layers contained significant numbers of ova, often in excess of 50 opg. The only other data to emerge in a brief scan of the literature was gathered from Egypt and Alabama, U.S.A., where Ascaris infections were prevalent. In an Egyptian village, 5 g samples gathered from the floors of houses of infected people yielded up to 21 Ascaris eggs (Chandler 1954). A 100 g soil sample collected from the grounds of an Alabama mental hospital gave in excess of 5,000 Ascaris and Trichuris ova (Caldwell, Caldwell and Davis 1930).

To summarize, light infestation of T. trichiura may produce relatively small numbers of ova (often less than 1,000) per gram faeces while massive infestations give very large egg-counts. Parasite ova can be found in soil around habitation sites occupied by infested people using standard parasitological techniques.

TRICHURID EGG-COUNTS FROM ARCHAEOLOGICAL SOIL SAMPLES

METHODS AND MATERIALS

To date 114 samples of archaeological layers from medieval deposits from sites at Bedern and 16-22 Coppergate, York, England, have been examined for parasite ova. Much of the practical work was undertaken as student projects by Alison Cameron (Bedern) and Sandra Bicknell (Coppergate) while working in the Environmental Archaeology Unit, University of York.

The Bedern was the site of a medieval (12-15th century) college of Vicars Choral - a group of clerics who sang masses in York Minster. All the samples from this site were taken from pits; some from within small buildings, others from areas which appeared at the time of excavation to have been open yards.

By contrast, only a small number of pit samples have been examined from excavations at Coppergate, the majority being from layers in and around buildings. The Coppergate site comprised a series of Viking Age (9th-11th century) tenements. With the help of the excavators, it has been possible to divide these samples into context groups. Nineteen samples were taken from dump layers above or below floor deposits. Thirteen were taken from in situ occupation layers, while sixteen were from paths and yards immediately around the buildings.

While waterlogged organic materials were preserved on both sites, organic preservation was better at Coppergate. Soil samples were examined moist, containing approximately the same amount of water as when excavated. Nevertheless, considerable variation in water content was observed. Some were rather crumbly and dry, while free water could be squeezed from some of the Coppergate pit samples. Generally, pit fills were moist and contained little inorganic material while the samples from around the buildings on Coppergate tended to be somewhat drier and contained more sand particles.

A modified version of the `Stoll' method for examining faecal samples was used to examine all soil samples. A three gram sub-sample of each layer was disaggregated in 42 ml water and filtered through a 250 micron mesh to remove coarse particles. 0.15 ml aliquots of the filtrate were examined in their entirety for parasite ova and the number of ova per gram calculated by multiplying the number of observed ova by 100. In cases where trichurid ova were particularly common a 0.05 ml aliquot was examined and the number of ova per gram calculated by multiplying the number of ova observed by 300. Two counts were made for all samples from Coppergate while at Bedern the number of counts per sample varied from one to seven.

RESULTS

These investigations have shown that trichurid ova occur in most soil samples, and that they may occur in small numbers or they may be extremely abundant. Results for the Bedern and Coppergate pit samples are presented in Fig 02. Egg-counts for samples collected around and in buildings on Coppergate are given in Fig 04.

The Pit Samples from Bedern and 16-22 Coppergate

A small number of samples failed to produce any parasite ova, but 90% contained *Trichuris* ova. Numbers in excess of 1,000 opg were frequently obtained from pit fills which also yielded abundant fragments of cereal spermoderm (part of the bran of wheat, *Triticum*, or rye, *Secale*), fruitstones and other food refuse (A. R. Hall personal communication).

Pit 1359 from the Bedern was extensively sampled (six subsamples were examined), and provides a opportunity to investigate the distribution of faeces-associated animal and plant remains. The occurrence of `bran' and seeds of exotic food plants (fig, *Ficus carica* and grape, *Vitis vinifera*) were scored using an abundance rating. Results are summarized in Fig 03 and show that there is an increase in the amount of faeces-associated material towards the centre of the pit until sample 40, below which faecal matter becomes less abundant.

Pits on the Coppergate site yielded particularly high concentrations of trichurid ova, often 10 times more ova per gram than pits on the Bedern site.

Samples from Inside and Around Buildings at 16-22 Coppergate

All the deposits examined yielded trichurid ova. Figure 04 shows their distribution. The dump layers mainly contained small numbers of ova per gram: most gave counts of 100-300, though one gave 4,700. The floor levels gave counts in the range 250-800 opg, slightly more than the dumps. The yard and path deposits produced a wide range of egg-counts, ranging from 50-2,700 opg. Although the numbers of samples examined is small, there does appear to be a consistency in the results, with dumps generally producing very low egg-counts, floors generally producing higher egg-counts, while paths and yard deposits (like pits) are rather variable. Some egg-counts from samples in and around buildings were of the same order of magnitude as those from some of the Bedern pits.

DISCUSSION

The Pits from Bedern and 16-22 Coppergate

There can be little doubt that all but a minority of the pits examined in this study were used as latrine pits at some stage. Overall, the evidence from pits confirms results of Greig (1981), Hall, Jones and Kenward (1983) and others who have examined faecal deposits for plant macrofossils and nematode ova, and suggest that faecal layers are extremely common in medieval urban pit fills. It is also clear that egg-count data should not be considered in isolation; evidence from other archaeological investigations and modern parasitological studies are necessary before interpretation begins. Examination of other finds (animal bones, pottery etc.) from the pits shows that a range of domestic refuse was discarded into them. The mixing of human excreta with other kinds of domestic waste may account for the gradual increase in faeces-associated finds in pit 1359 at the Bedern.

Unfortunately, inferences about the size of the worm burdens which people carried cannot be drawn from egg-count data obtained from pit fills. The addition of non-faecal material to cess pits may have reduced egg-counts, while decay may have occurred, resulting in some concentration of resistant parasite ova. However, it is apparent that whipworm infestations were common both in the religious community of the Bedern during the 12-15th century and in the families of artisans who lived in Coppergate during the Anglo-Scandinavian occupation of the town.

It is salutary to note that, once biological investigations were completed on the extensively sampled Bedern pit 1359 (see Fig 03), the remaining untreated sediment was carefully examined using field techniques. All samples contained a large amount of fine organic matter but none appeared to be different from adjacent samples on grounds of colour, moisture content or texture. This example demonstrates how difficult it is to tell the origin of a layer without the use of a microscope, and illustrates the need for rigorous sampling.

Other sets of samples from individual pits on the Bedern show considerable variation in the numbers of parasite ova, bran and seeds of food plants. Rarely were more than two or three samples taken from most of these features and, while excavators attempted to identify and sample those layers which were likely to have been laid down while the feature was used, it is possible that some of the richer faecal layers were not collected.

Samples from inside and around Buildings from 16-22 Coppergate

The most remarkable fact to emerge from these investigations is the concentrations of ova distributed in layers in and around buildings at Coppergate. The data suggest that both dumped layers and yard/path layers contained a very variable number of trichurid ova, several giving in excess of 1,000 opg. Floor layers tended to give slightly lower egg counts. It is tempting to suggest that faeces were scattered unevenly around buildings while a small amount was deposited indoors. However, to generalize about living conditions in the buildings from such a small number of samples would be both unwise and premature for a number of reasons. The samples might not be fully representative of the whole site.

A number of depositional processes may have lead to such a build up of faecal matter in and around buildings. The most probable, given the large numbers of ova, is that people, most likely children, were defecating indiscriminately around and within buildings.

The implications of this suggestion are unlikely to shock anyone with recent experience of people living crowded together with primitive means of sanitation. In several papers describing parasitic infestations in people living in mental hospitals in Europe, the mountains of Virginia, U.S.A., Panama, India, Australia and Nigeria, the most important factor in the spread of ascariasis and trichuriasis is indiscriminate and casual defecation within and around habitation sites.

While the results unequivocally show that substantial amounts of human faeces were present in most deposits so far examined, it is not yet apparent if all were freshly passed or if they were secondarily deposited. Excavations at Coppergate revealed a complex sequence of intercutting pits, many of which appeared to be cess pits. Inevitably pit digging would have brought to the surface long buried human faecal remains, and this `soil' is likely to have been trampled over the site, so introducing parasite ova into occupation layers.

A further explanation is that faeces were laid down during brief periods of abandonment, i.e. between building phases. Today, derelict buildings are used as latrines, particularly by vagrants. In addition, it is possible that ova were concentrated into the buildings by rainwater flooding over the site.

To conclude this brief and essentially preliminary survey of the Viking Age and medieval town, there can be little doubt that peoples' living conditions, standards of hygiene and cleanliness fell very far short of those that generally obtain today in Britain. Some confirmation of this can be seen in the recently published account of investigations at 6-8 Pavement (Hall, Kenward, Williams and Greig, 1983) where an unusually well preserved coprolite was recovered intact. The mineralized coprolite (almost certainly human) gave egg-counts of 60,000 opg which, when corrected to allow for the mineralization, would have been in the region of 120,000 opg fresh faeces (Jones 1983).

CONCLUDING OBSERVATIONS

As more workers examine soil samples for parasite ova, there emerges a clear picture that medieval urban populations frequently harboured nematode worms. What scant evidence there is for the Roman period suggests a similar pattern of parasitic infestation. Thus, reports of parasite ova in archaeological deposits will soon cease to be mere novelties, but, like other biological records, will become abundant.

Further research should be directed into at least two areas. First, to complete a survey of periods and site types to see the prevalence of parasite populations through time and space; and second to examine rigorously a small number of sites in great detail to study spatial and temporal differences on a small scale.

While it is important to remember that this paper discusses only a small number of samples from the highly organic deposits of central York, it is hoped that it will be of value to other workers investigating similar waterlogged urban deposits in Britian and north west Europe. In an attempt to make this paper of practical use I wish to suggest some guidlines, based on experience of archaeological material and modern published data, for interpreting the numbers of trichurid ova which can be found in archaeological layers.

Necessarily, the figures presented below must be used cautiously and will almost certainly be revised in the light of future work. At all times a conservative estimate has been given. It is suggested that any sample giving egg-counts of 20,000 opg or over be regarded as primarily faecal in origin. Samples of 2,000-19,999 are almost certainly faecal but may be contaminated with other materials. Egg-counts of 500-1,999 opg are probably from layers containing a substantial amount of faeces. Layers giving egg-counts of between 200 and 500 opg are the most difficult to interpret. The ova are probably best regarded as part of the urban background fauna for

Britain, although some ancient faeces is present. Little significance can be attached to egg-counts of less than 200 opg.

ACKNOWLEGDEMENTS

I am particularly grateful to Alison Cameron and Sandra Bicknell (Bradford University) who carried out most of the practical work for this study. In addition, I wish to thank Alan Robertson who carefully drafted the figures and Allan Hall (EAU), Richard Hall (York Archaeological Trust), Harry Kenward (EAU) and Terry O'Connor (EAU) who read and made helpful comments on the text of this paper.

REFERENCES

Allen, A. V. H. and Ridley, D. S. (1970). Further observations on the formol-ether concentration technique for faecal parasites. Journal of Clinical Pathology **23**, 545-547.

Burrows, R. D. (1950). On the estimation of Trichuris worm burdens in patients. Journal of Parasitology **36**, 227-231.

Caldwell, F. C., Caldwell, E. L. and Davis G. E. (1930). Some aspects of the epidemiology of infestation with Trichuris and Ascaris revealed in a study at the hospitals for the insane and the home for mentally defective children in the state of Alabama. American Journal of Hygiene **11**, 619-651.

Chandler, A. C. (1954). A comparison of helminthic and protozoan infections in two Egyptian villages two years after the installation of sanitary improvements in one of them. American Journal of Tropical Medicine and Hygiene **3**, 59-73.

Chandler, A. C. and Read, C. P. (1961). Introduction to parasitology. New York: John Wiley.

Faust, E. C. (1944). Diseases due to helminths and arthropods. In (J. Brennemann, Ed.) Practice of Pediatrics. Hagerstown: W. F. Prior, vol. 2, p 19.

Hall, A. R., Jones A. K. G. and Kenward, H. K. (1983) Cereal bran and human faecal remains - some preliminary observations. In Proudfoot, B. (ed.) Site, Environment and Economy. Symposia of the Association for Environmental Archaeology No. 3. British Archaeological Reports International Series 173, p 85-104.

Getz L. (1945). Massive infection with Trichuris trichiura in children. American Journal of Diseases in Children **70**, 19-24.

Greig J. R. A. (1981). The investigation of a medieval barrel-latrine from Worcester. Journal of Archaeological Science **8**, 265-282.

De Groot, J. F. M., Hoorweg, L., Kleevens, J. W. L. and Rijpstra, A. C. (1976). Trichiuriasis, amoebiasis en andere parasitaire darminfecties ineen zwakzinnigeninrichting. Nederlands Tijdschrift Voor Geneeskunde **120** (51), 2263-2267.

Jones, A. K. G. (1982). Human parasite remains: prospects for a quantative approach. In (A. R. Hall and H. K. Kenward Eds.) Environmental Archaeology in the Urban Context. Council for British Archaeology Research Report No. 43, 66-70.

Jones, A. K. G. (1983). A coprolite from 6-8 Pavement. In (A. R. Hall, H. K. Kenward, D. Williams and J. R. A. Greig) Environment and living conditions at two Anglo-Scandinavian sites. The Archaeology of York 14/4. London: Council for British Archaeology, pp. 225-229.

Lynch, D. M., Green, E. A., McFadzean, J. A. and Pugh, I. M. (1972). Trichuris trichiura infestations in the United Kingdom and treatment with Difetarsone. British Medical Journal **4**, 73-76.

Mahmoud, A. A. F. (1979). Intestinal nematodes (roundworms). p 2157-2159. In (Mandell, G. L., Douglas, R. G. Jn., Bennett, J. E. Eds.). Principles and practice of infectious disease. New York: Wiley.

Paine, D. H. D., Lower. E. S. and Cooper, T. V. (1960). Treatment of Trichuriasis with Dithiazanine in a hospital for mental defectives. British Medical Journal **1**, 770-774.

Spindler, L. A. (1929). The relation of moisture to the distribution of human Trichuris and Ascaris. American Journal of Hygiene **10**, 476-496.

Swartzwelder, J. C., Frye, W. W., Muhleisen J. P., Miller, J. H., Lampert, R., Pena Chavarria, A., Abadie, S. H., Anthony, S. O. and Sappenfield, R. W. (1957). Dithiazanine, an effective broad spectrum anthelminthic. Journal of the American Medical Association **165**, 2063-2067.

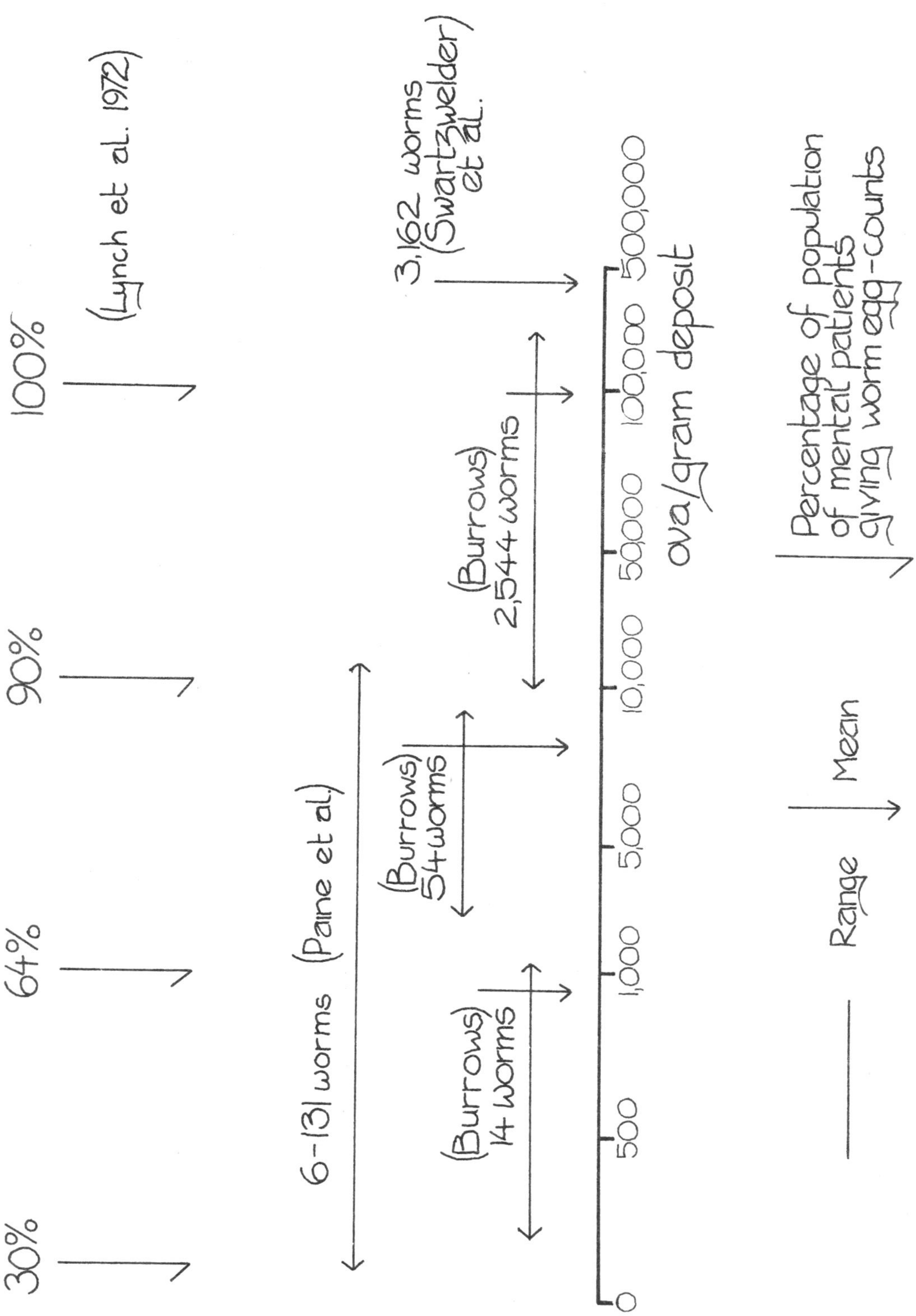

FIGURE 1: The numbers of trichurid ova per gram from recent studies. Horizontal axis shows the number of ova per gram. In order to represent the huge range of egg-counts, the data are presented using a condensed linear scale. Initially counts jump in units of 100, i.e. 100, 200, etc. to 1,000. Here the scale leaps in units of thousands, i.e. 1,000, 2,000 etc until 10,000 where it leaps in units of 10,000, and so on as necessary. The same scale have been used on all figures showing modern or ancient egg-counts. Data from Lynch <u>et al.</u> (1972) shows the cumulative percentages of British mental patients with <u>Trichuris</u> egg counts i.e. 64% of the population gave egg-counts of 1,000 opg or less. The other data, Paine <u>et al.</u> (1960), Burrows (1950) and Swartzwelder <u>et al.</u> (1957) show the number of eggs per unit faeces passed by people harbouring known numbers of worms.

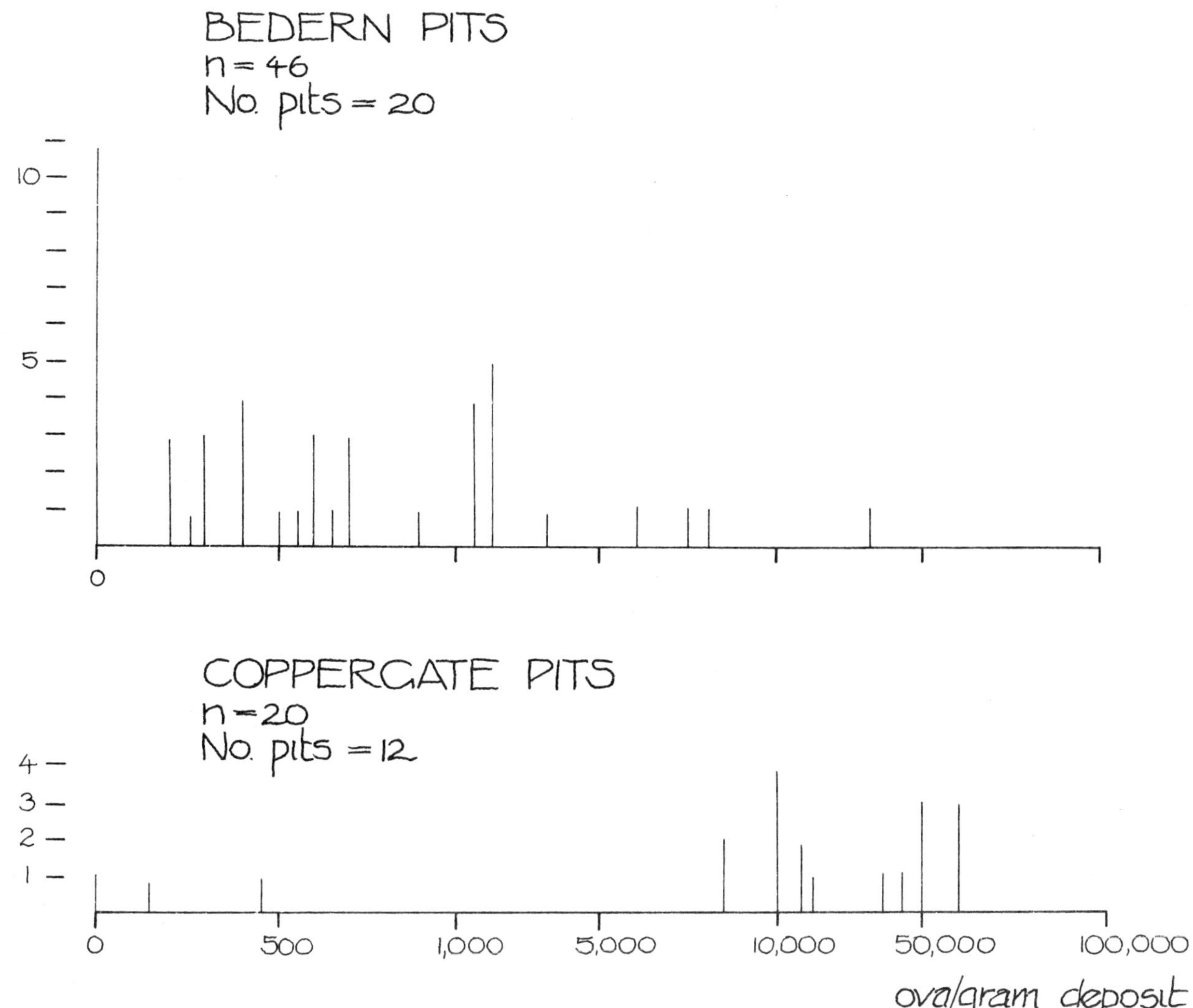

FIGURE 2: The numbers of trichurid ova per gram from soil samples collected from pits a) The Bedern, York. b) Anglo-Scandinavian levels at 16-22 Coppergate, York. n = number of samples examined.

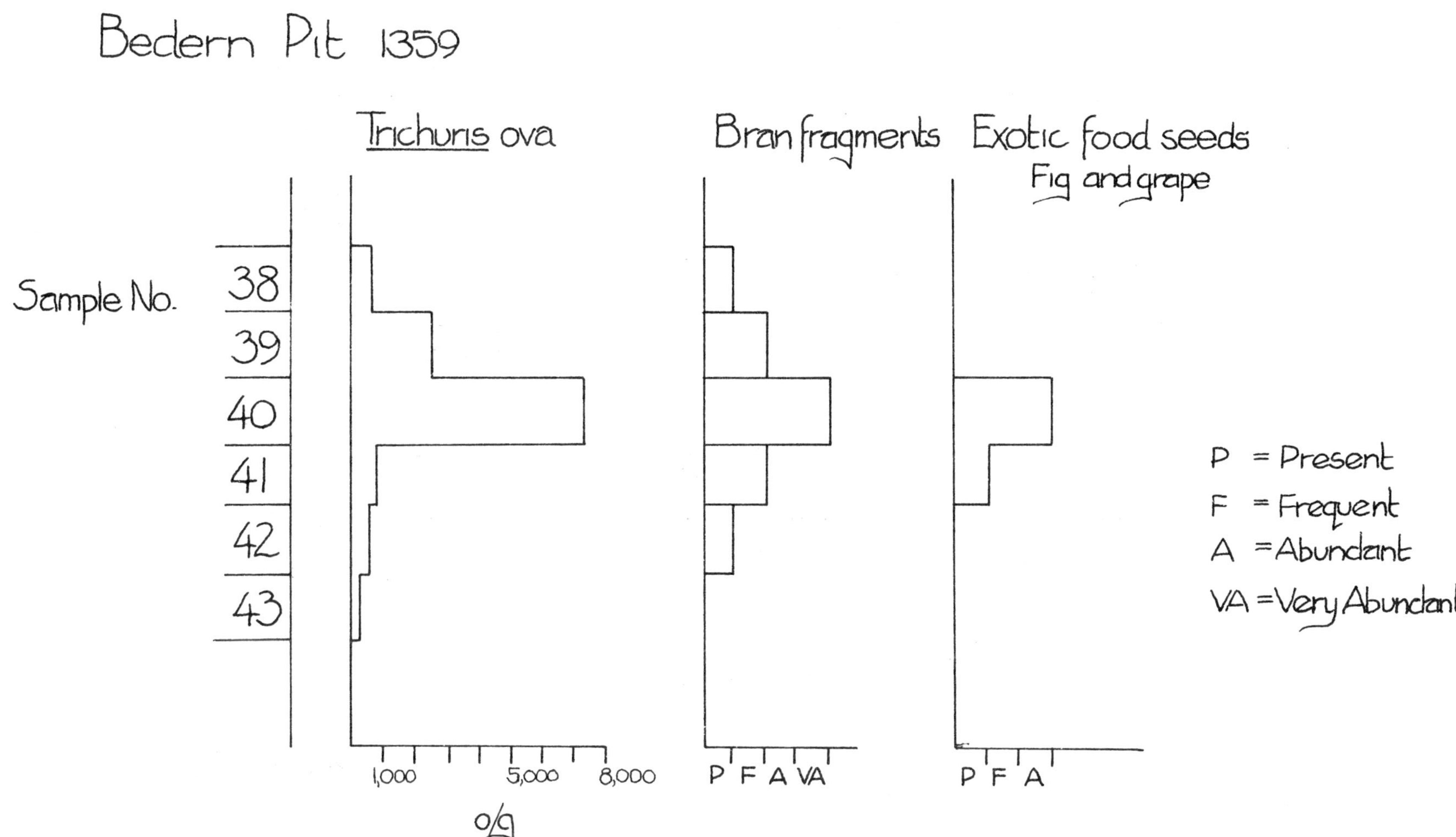

FIGURE 3: The distribution of trichurid ova, cereal spermoderm (bran) fragments and fig and grape seeds in pit 1359 from the Bedern, York.

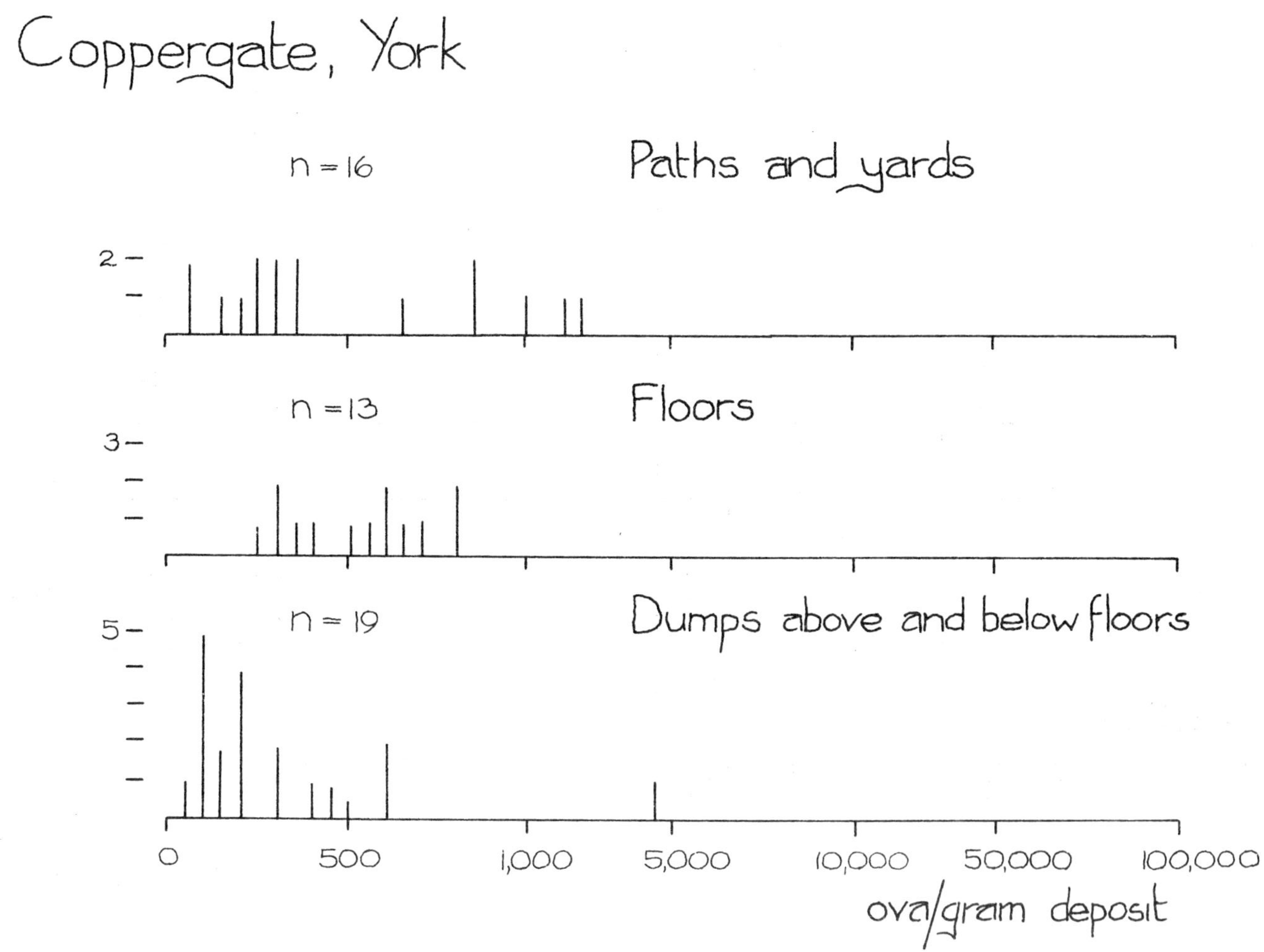

FIGURE 4: The number of trichurid ova from layers associated with Anglo-Scandinavian tenements at 16-22 Coppergate, York.

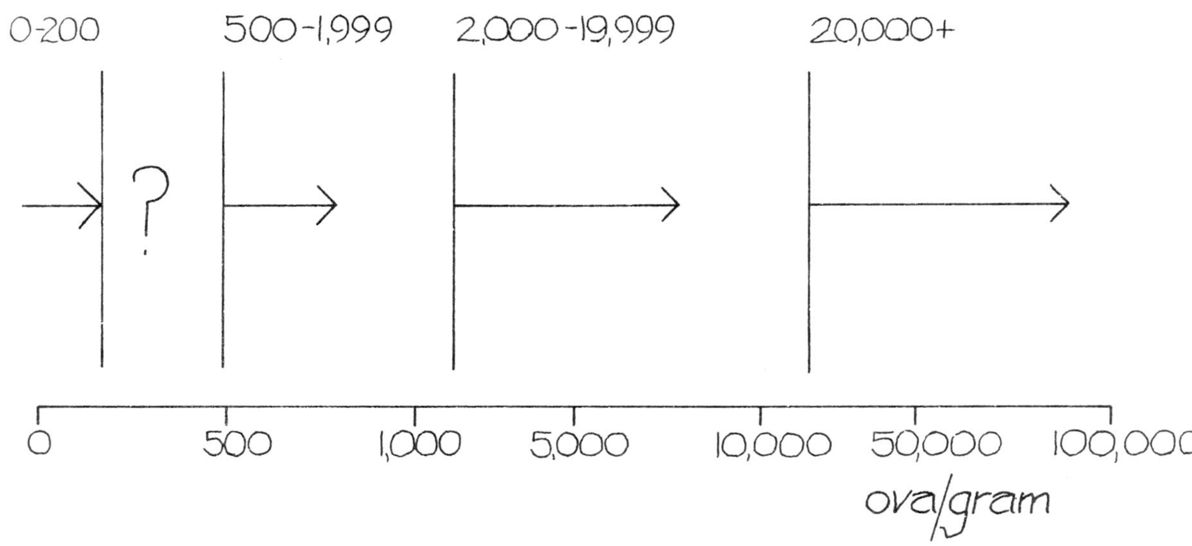

FIGURE 5: Suggested guidelines for the interpretation of egg counts of Trichuris trichiura from medieval waterlogged urban archaeological deposits.

Molluscs

SEASONALITY FROM SHELLS: AN EVALUATION OF TWO TECHNIQUES FOR SEASONAL DATING OF MARINE MOLLUSCS.

Margaret R. Deith
University of Cambridge
The Godwin Laboratory
Free School Lane
Cambridge

The potentialities of two techniques for seasonal dating on marine shells, growth-line analysis and oxygen isotope analysis, are investigated. Both the possibilities and the limitations of each technique are assessed in relation to different species and different habitats. The two techniques are seen to be largely complementary.

INTRODUCTION

The reconstruction of settlement pattern, of annual cycles of movement in response to seasonal changes in resource availability, is one of the acknowledged aims of hunter-gatherer archaeology (Welinder 1978). Such an objective requires an understanding of the seasonal use of the sites within the settlement system. Direct evidence for such seasonality may come from bone and tooth studies, molluscs, plant remains, or the presence of migratory species; inferences concerning the seasons when sites would have been able to support human groups can be drawn from reconstructions of the contemporary environment (Jochim 1976).

Marine molluscs have certain advantages as seasonal indicators, being less at the mercy of taphonomic processes than are bones. They are available all the year round, they are often well-preserved, and there is little question that they have been brought to the site by human beings. They are of no interest to other mammalian predators and their relative immobility ensures that natural deaths occur within the intertidal zone. The kind of complex problems that beset the interpretation of bone remains in cave sites (McGuire 1980), such as apportioning responsibility to different predators or to the movements of the prey itself, consequently do not apply to them. Nor are they likely to have been stored for use at other times of the year. The samples are therefore as reliable as it is possible for human food remains to be.

Nevertheless, there is no simple correspondence between the seasonal consumption of shellfish and the season of site occupation, any more than there is between the seasonal consumption of any other resource that is available throughout the year and the span of occupation. The role of shellfish in the diet is liable to fluctuate according to the availability of other resources (Meehan 1982). Indeed, in certain circumstances, such as those of 'complex hunters' who occupy their sites throughout the year, shellfish may be kept in reserve for periods of hardship (Rowley-Conwy 1980, Deith 1983 a). Bailey (1983) has pointed out that patterns of shell discard behaviour vary and that this variation is sometimes a direct response to seasonal changes. What remains in the midden is not necessarily a complete record of shellfish consumption. With these caveats on how the shellfish data should be interpreted in terms of settlement, the seasonal information which can be retrieved from shells is relatively 'fine-grained' and may add much to the understanding of economy and settlement.

This paper examines two techniques for obtaining seasonal information from shells, growth-line analysis and oxygen isotope analysis. The aim is to assess the circumstances in which each functions most effectively, rather than to discuss the methodology, which has already been described in detail elsewhere (Farrow 1971, Shackleton 1973). The major areas to be considered are the species and habitats for which each is especially suitable, some of the problems involved in retrieving seasonal information, and an appreciation of the relative strengths and weaknesses of each technique. Shell growth is a complex phenomenon, subject to influences from a wide range of environmental factors. It is therefore not appropriate to attempt to draw up lists or tables from which the archaeologist can read off the column of information relevant to his data. It is preferable to offer some insights into the complexities of shell growth, especially those aspects which have a bearing on the deciphering of seasonal information.

GROWTH-LINE ANALYSIS

The principle underlying the growth-line technique is that of the 'biological clock' (Cloudsley-Thompson 1980). In brief, many behavioural and physiological processes in both plants and animals are rhythmic, with a natural, innate ('endogenous') periodicity. The periodicities of the endogenous rhythms approximately match those of the dominant rhythms of the environment which, for marine molluscs, are the daily, tidal and seasonal cycles. The endogenous rhythms are 'entrained' or regulated by the environmental rhythms, with which they are synchronised (Saunders 1977).

Shell growth, like that of many tree species, is a rhythmic physiological process, whereby new material is deposited incrementally. If the rhythm is an entrained one, keyed in to seasonal, daily or tidal rhythms, the incremental sequence has a chronological significance and can be read in temporal terms. If daily or tidal sequences can be placed within the context of an annual cycle which is apparent in the shell structure, a date of death can be determined.

Figure 1B is a stylised drawing of growth lines and increments in the section of a valve of Cerastoderma edule Linne. Each increment constitutes material deposited during one period of tidal immersion and each line represents material deposited during a low-tide emersion period (Richardson, Crisp and Runham 1979). Growth ceases in the winter in this species, when a groove is formed in the surface of the shell (Figure 1A). The groove is also evident in the section, where it is emphasised by a gradual reduction in the size of the increments and a similarly gradual increase at the beginning of the new spring growth.

There are, therefore, three fundamental requirements for doing seasonal dating by growth-line analysis: the lines between the increments must be visible, they must have a chronological significance, and there must be a seasonal marker from which to measure the final growth section.

A. Line Visibility

The visibility or otherwise of the growth lines is partly a function of crystal structure, which is species-specific, and partly of habitat. The outer shell layer in some families, such as the Pectinidae (scallops), is of foliated calcite (Taylor, Kennedy and Hall 1969), in which the arrangement of the crystals is such as to obscure any divisions between increments (Pannella and MacClintock 1968).

Some habitats are not ideal for producing clear lines. In the subtidal zone, for example, shells of the same species as those in the intertidal zone have much fainter lines (Rhoads and Pannella 1970, Bourget 1980), while shells from abyssal depths have none at all (Berry and Barker 1975). Striking differences in the clarity of the lines have been observed in shells which have been transferred from one zone to the other (Richardson *et al*. 1979). Shells from the Mediterranean, with its very limited intertidal zone, such as the cockle, *Cerastoderma glaucum* Poiret, present problems of line visibility. In some parts of the shells, the lines are faint, while in others they disappear altogether (Deith, unpublished observations).

B. Seasonal Markers

The second essential condition for seasonal dating is that there should be a recognisable point in the annual sequence. The winter groove in the cockle, with its associated characteristic zone of winter growth, illustrated in Figure 1, is ideal. It provides a point of reference from which to start a line count. A date for the beginning of the new spring growth can be established by control work with modern specimens (Deith, 1983 b).

In some circumstances, such as where a species is in the middle of its geographical range and conditions are therefore optimal, the shell may continue to grow throughout the year. Where there is no sharp seasonal contrast in temperature or food supply, as in tropical zones, there may be no perceptible slowing down of growth at any point in the annual cycle. Other events which may affect shell growth and sometimes provide a 'marker', such as spawning, often take place sporadically, several times a year, in warmer waters. For example, spawning in the Mediterranean C. glaucum may take place four or five times a year (Ivell 1979).

In the common limpet, Patella vulgata Linne, occurring in the waters around Britain, there is no sign of winter in the shell microstructure, despite a winter stoppage in growth (Lewis and Bowman 1975). Nor is increment width sufficiently variable to be any guide. However, the outer surface of the shell changes its orientation as the faster, new spring growth begins (Figure 2), producing a 'step'-like section. The effect is most dramatic in limpets from the upper shore, where growth is normally slow and the typical shell shape is tall with a narrow base. The low-shore limpets are generally flatter because of their higher growth rate, and the spring increase in growth rate consequently has less impact on the shape of the shell and may, in fact, be difficult to distinguish (Figure 2). The almost horizontal, fast growth that occurs in spring has also been produced artificially during marking experiments (Ekaratne and Crisp 1982). It appears that the severity of the stress is responsible for this phenomenon in the shell of P. Vulgata. In default of any other possible seasonal marker, the change of slope has been used as a base-line from which to attempt to establish a method for determining seasonality in limpets (Deith 1982).

C. Chronological Correlation

The third prerequisite for using growth lines in seasonal dating is that they should have a chronological significance. A tidal periodicity of line and increment formation has been amply demonstrated for the cockle, C. edule (Richardson et al. 1979).

Evans' (1975) analyses of two bivalve species on the Pacific coast of the United States revealed a remarkably close correlation between tidal cover and increment size in one of them (Clinocardium nuttalli Keen) and a complete lack of congruence between the growth patterns of the other (the rock borer, Penitella penita Conrad) and any rhythmic events in the environment. It has been suggested that some species are sensitive in recording environmental events and others complacent, using terminology developed in dendrochronology (Pannella 1972).

Line counts on the shells of Patella Vulgata collected from high- and low-shore levels of Robin Hood's Bay, Yorkshire, display a high degree of variability from the marker to the growing edge of shells collected on the same day. The number of lines between two 'seasonal' markers is also very variable (if the orientation of the outer surface of the shell is used to indicate the beginning of new growth) (Table 1). There are three possible explanations here. Either the 'annual' markers include other stress events in addition to the spring 'shell shoot' or the timing of the beginning of new growth varies greatly from one individual to another or, thirdly, the increments have no chronological significance. Experimental work by Ekaratne and Crisp (1982) demonstrates that, as in the cockle, growth lines in the limpet are laid down during each tidal emersion. From the viewpoint of interpreting archaeological shells, the question of whether there is a great deal of variability in the beginning of the growing season or whether the choice of stress effects as seasonal markers is too erratic to form an adequate basis for seasonality work is largely an academic one.

The periodicity of line formation is also dependent on habitat. Animals growing in the intertidal zone are subjected to sudden and extreme changes, of oxygen levels, partial pressure of CO_2, pH and temperature, on immersion and emersion. It is these changes in conditions which give rise to the slightly different forms of shell material in the increments and in the lines (Deith 1982). For animals growing in subtidal habitats, although such violent changes are not experienced externally, conditions within the microenvironment of the shell interior are subject to variations through alternating periods of activity and quiescence, controlled by the endogenous rhythms of the animal. Because the changes in conditions within the shell during periods of quiescence are not as severe as those imposed on the intertidal animals, the lines are fainter, less distinct from the increments. They are also under endogenous control, which implies that the periodicity may be less precise. Shells with zones of growth lacking growth lines are clearly not susceptible to precise chronological analysis.

If the conditions needed for the successful implementation of the growth-line technique are met, it can be a very powerful tool, capable of a high degree of seasonal resolution. Shells of Cerastoderma edule from the mesolithic site of Morton, south-east Scotland, provide an illustration. During the main growing season, collection dates with a precision of $\pm$ ten days can be calculated from growth-line counts. The site has been shown to have been visited at different times of year (Deith, 1983 b).

OXYGEN ISOTOPE ANALYSIS

The second technique for seasonal dating is oxygen isotope analysis, which approaches the question from a different angle, that of measuring temperature variation by analysing the ratio of ^{16}O to ^{18}O in the shell carbonate. If samples of calcium carbonate are taken at close intervals along the direction of growth, the sequence of isotopic readings can be plotted to give a temperature-related curve. Two factors determine the ratio of ^{16}O to ^{18}O in calcium carbonate, the isotopic ratio of the water in which the carbonate was precipitated and a fractionation coefficient which is dependent on the temperature of the water to a known degree (Shackleton 1973). If the first variable, the isotopic composition of the water, is constant, as it is for example in sea water, any variation in the signal should be attributable to temperature variations. A plot of isotopic ratios taken through a shell is thus the equivalent of a record of the fluctuations in water temperature through the period represented by the shell growth. That is, annual growth cycles are detectable in the isotope profile. (The procedure is to compare the isotope ratio in each sample with that in a standard carbonate. The deviation of the ratio in the sample from that in the standard is expressed as a delta value, in parts per thousand. An increase of $1^{o}C$ in seawater temperature produces an approximate decrease in delta value of 0.2‰.)

The upper graph in Figure 3 charts mean monthly sea temperatures for the coastal region of the Tavoliere in southeastern Italy. The lower one is a record of ^{18}O values through a razor shell, Solen vagina Linne, collected near Manfredonia in the Tavoliere, at the end of September 1982. The samples were taken at regular 2mm intervals. They have been plotted, however, in such a way as to demonstrate the relationship between the shape of the isotope profile and that of the sea temperature curve by telescoping the fast summer growth into a relatively short time period and drawing out that part of the curve that represents the slow winter and early spring growth. The fluctuations in ^{18}O values in the summer part of the graph are at least partly due to the variations in actual seawater temperatures, which are masked by the smoothed curve of mean monthly values.

Analyses of archaeological razor shells from the nearby neolithic site of Masseria Valente show that shellfish were being collected throughout the year at this site (Deith 1983 c). The role of molluscs in the economy of this site is in sharp contrast to that in the hunter-gatherer sites of the Cantabrian plain, where shellfish were collected in the winter and autumn only (Deith in press a), although these sites are also considered to have been occupied on a year-round basis (Clark 1983). As with the growth-line technique, there are certain requirements for the successful use of the isotope technique in seasonality studies.

A. Water Temperature Variation

A fundamental requirement is that there should be sufficient annual variation in water temperature to produce a clear differential between shell deposited in summer and shell deposited in winter. It should also be sufficient for shorter-term variations not to be confused with the major seasonal ones. This effectively excludes the majority of shells from tropical waters, where there is little variation in temperature through the year. It also excludes shells which have a relatively short growing season from the restricted determinations which might be possible by using other techniques, such as growth-line analysis.

B. Salinity

If temperature is the significant variable, then the delta value of the water needs to be constant. For this reason, sea water is ideal. Estuarine water is liable to pose a problem, however. Fresh water is isotopically lighter than sea water, that is, it has a lower ratio of ^{18}O to ^{16}O. The mixture of the two in estuaries and the variation in their relative proportions appears to preclude estuarine shells from seasonal analysis. In practice, in shells from some estuaries, the isotopic profile is still recognisably temperature-related, although the range of delta values has shifted towards the lighter values associated with lowered salinity. Figure 4 shows the isotope profiles of three specimens of Cerastoderma glaucum from Masseria Valente. These shells have come from the estuary of the Candelaro river, as the range of delta values indicates when it is compared with that in contemporary razor shells (which are fully marine) from the same deposits (Figure 5). In fact, each of these shells has come from a slightly different part of the estuary. The variations produced by temperature fluctuations are still strongly marked in specimens from this estuary, although this is not true of all situations. For example, the temperature effect is entirely swamped by huge variations in salinity in neolithic winkles (Littorina littorea Linne) from the Danish site of Norsminde at the Baltic end of the Limfjord (unpublished data).

C. Isotopic Exchange

A problem which is of particular importance to archaeological material is whether any secondary changes have taken place in the shells since their deposition at the site. Shell material may be dissolved and reprecipitated, with isotopic exchange between the shell carbonate and the ground water. This water is isotopically much lighter than that in which the carbonate was originally formed, a fact which often helps in its detection. How far isotopic exchange is likely to occur depends very much on the conditions at a site. It is much more likely where the site is badly drained and damp and where temperatures are warm.

Certain species are also more prone to recrystallization than others. The outer surface of the limpet is peculiarly susceptible to attack from endolithic blue-green algae, which bore into the shell extensively, producing a network of cavities. In places where sea temperatures are high, the holes are subsequently filled by secondary carbonate, but in colder waters the holes remain empty (Alexandersson 1972). It is not clear how long it takes for this process to begin, but it has been suggested that the decomposing plant material may help to create a favourable environment for carbonate formation and so speed up the process (Margolis and Rex 1971).

The network of holes, whether partially filled with secondary carbonate or not, opens up the interior of the shell, making it vulnerable to dissolution and recrystallization processes. This has made seasonality work on the limpets Patella vulgata and P. intermedia Jeffreys, from Asturian (Mesolithic) and Upper Palaeolithic levels at the cave site of La Riera, on the Cantabrian coast of Spain, impossible, because the outer surfaces of the shells, and the growing edges in particular, are heavily recrystallized at this site (Deith and Shackleton in preparation).

The practicability of using limpets for seasonal dating by the isotopic technique clearly depends on their origins and on the conditions at the site. Shackleton's work on Patella tabularis Kraus from Nelson Bay Cave, South Africa, shows that not all limpets are necessarily afflicted in this way (Shackleton 1973).

A COMPARISON OF THE TWO TECHNIQUES

1. Both growth-line analysis and isotope analysis can operate only within the limits set by the growing season. If a shell stops growing for six months in the year, there is clearly no technique which can determine the date of collection within that period.

2. Intertidal and subtidal specimens are equally suitable for isotope analysis. Mediterranean shells are particularly effective because growth rates are fast enough to give good seasonal resolution and because many species grow throughout the year. Intertidal shells are preferable to subtidal ones for growth-line work, although there are some species which inhabit the subtidal zone and have been found to display regular, rhythmic growth patterns.

3. Fully marine species are preferable for isotope work, although not all estuarine situations are unsuitable. Growth-line analysis can be applied to shells from either marine or estuarine situations.

4. The potential seasonal resolution is greater from growth-line analysis than it is from isotopes. Fluctuations in seawater temperatures from year to year and within the seasons, especially in the summer, make the resolution of the isotope technique somewhat coarser than that of growth-line analysis. There is also a practical difficulty in that, unless growth is exceptionally rapid, it is impossible to produce large enough samples to represent precise points in time. This is particularly a problem in the winter when growth may be very slow. A sample represents the mean value over a period of time.

5. The range of species on which isotope analysis can be used successfully is, however, much larger than that open to growth-line analysis.

6. Finally, the potentiality for retrieving other information about the environment and hence about other aspects of human activity, is greater from the isotope technique, largely because the signal is made up of two variables, the delta value of the water and its temperature. If the temperature component is allowed for by including the entire range of values within each shell, comparisons between ranges can be used to establish variations in the isotopic composition of the water in which the shells grew. This can be applied either to variation through time, which might indicate climatic change, or to variation in space, for example collecting areas along an

estuary.

To conclude, the two techniques are in many respects complementary. The situations in which each is most effective differ and the problem areas rarely overlap.

ACKNOWLEDGEMENTS

Funding for the growth-line work referred to in this paper came from the Department of Education and Science and for the isotope work from the Science and Engineering Research Council. I should also like to thank Su Wanek for drawing figures 1 and 2, and Geoff Bailey and Dave Gilbertson for their helpful comments on the manuscript, and Rosemary Bowman for useful talks on limpet growth.

REFERENCES

Alexandersson, T. (1972) Micritization of carbonate particles: processes of precipitation and dissolution in modern shallow-marine sediments. Bulletin of theGeological Institute of the University of Uppsala New Series 3, 201-236.

Bailey, G.N. (1983) Problems of site formation and the interpretation of spatial and temporal discontinuities in the distribution of coastal middens. In Quaternary Coastlines and Marine Archaeology, P.M. Masters and N.C. Flemming (eds), pp. 559-582. London: Academic Press.

Berry, W.B.N. and Barker, R.M. (1975) Growth increments in fossil and modern bivalves. In Growth Rhythms and the History of the Earth's Rotation. G.D. Rosenberg and S.K. Runcorn (eds), pp. 9-25. London: John Wiley and Sons Ltd.

Bourget, E. (1980) Barnacle shell growth and its relationship to environmental factors. In Skeletal Growth of Aquatic Organisms. D.C. Rhoads and R.A. Lutz (eds), pp. 469-491. New York: Plenum Press.

Clark, G.A. (1983) Boreal phase settlement/subsistence models for Cantabrian Spain. In Hunter-Gatherer Economy in Prehistory. G. Bailey (ed), pp. 96-110. Cambridge: Cambridge University Press.

Cloudsley-Thompson, J.S. (1980) Biological Clocks: their Functions in Nature. London: Weidenfeld and Nicholson.

Deith, M.R. (1982) Seasonality from shells: growth-line studies of the cockle, Cerastoderma edule L., and the limpet, Patella vulgata L., as a guide to prehistoric shell-collecting activities. Unpublished Ph.D. dissertation, University of Sheffield, Department of Prehistory and Archaeology.

Deith, M.R. (1983a) Seasonality of shell collecting determined by oxygen isotope analysis of marine shells from Asturian sites in Cantabria. In Animals and Archaeology 2. Shell Middens, Fishes and Birds. C. Grigson and J. Clutton-Brock (eds), pp. 67-76. Oxford: British Archaeological Reports.

Deith, M.R. (1983b) Molluscan calendars: the use of growth-line analysis to establish seasonality of shellfish collection at the Mesolithic site of Morton, Fife, Journal of Archaeological Science 10, 423-440.

Deith, M.R. (in press) The role of shellfish in the economy of neolithic sites in Apulia. Origini.

Deith, M.R. and Shackleton, N.J. (in prep.) Seasonal exploitation of marine molluscs: a report on the oxygen isotope analysis of shells from La Riera. To be published as part of a monograph on La Riera, edited by G.A. Clark.

Ekaratne, S.U.K. and Crisp, D.J. (1982) Tidal Micro-growth bands in intertidal gastropod shells, with an evaluation of band-dating techniques, Proceedings of the Royal Society of London B 214, 305-323.

Evans, J.W. (1975) Growth and micromorphology of two bivalves exhibiting non-daily growth lines. In Growth Rhythms and the History of the Earth's Rotation. G.D. Rosenberg and S.K. Runcorn (eds), pp. 119-134. London: John Wiley and Sons Ltd.

Farrow, G.E. (1971) Periodicity structures in the bivalve shell: experiments to establish growth controls in Cerastoderma edule from the Thames estuary, Palaeontology 14, 571-588.

Ivell, R. (1979) The biology and ecology of a brackish lagoon bivalve, Cerastoderma glaucum Bruguiere, in Lago Lungo, Italy, Journal of Molluscan Studies 45, 364-382.

Jochim, M.A. (1976) Hunter-Gatherer Subsistence and Settlement: a Predictive Model. London: Academic Press.

Lewis, J.R. and Bowman, R.S. (1975) Local habitat-induced variations in the population dynamics of Patella vulgata L. Journal of Experimental Marine Biology and Ecology 17, 165-203.

Margolis, S. and Rex, R.W. (1971) Endolithic algae and micrite envelope formation in Bahamian oolites as revealed by SEM, Bulletin of the Geological Society of America 82, 843-852.

McGuire, K.R. (1980) Cave sites, faunal analysis, and big-game hunters of the Great Basin: a caution, Quaternary Research 14, 263-268.

Meehan, B. (1982) Shell Bed to Shell Midden. Canberra: Australian Institute of Aboriginal Studies.

Pannella, G. (1972) Palaeontological evidence on the earth's rotational history since Early Precambrian, Astrophysics and Space Sciences 16, 212-237.

Pannella, G. and MacClintock, C. (1968) Biological and experimental rhythms reflected in molluscan shell growth, Journal of Palaeontology 42, 64-79.

Rhoads, D.C. and Pannella, G. (1970) The use of molluscan shell growth patterns in ecology and palaeoecology, Leithaia 3, 143-161.

Richardson, C.A., Crisp, D.J. and Runham, N.W. (1979) Tidally deposited growth bands in the shell of the common cockle, Cerastoderma edule (L.), Malacologia 18, 277-290.

Rowley-Conwy, P.A. (1980) Continuity and change in the Prehistoric Economies of Denmark 3700 bc - 2300 bc. Unpublished PhD dissertation, University of Cambridge, Department of Archaeology and Anthropology.

Saunders, D.S. (1977) An Introduction to Biological Rhythms. Glasgow: Blackie.

Shackleton, N.J. (1973) Oxygen isotope analysis as a means of determining season of occupation of prehistoric midden sites, Archaeometry 15, 133-141.

Taylor, J.D., Kennedy, W.J. and Hall, A. (1969) The shell structure and mineralogy of the Bivalvia. Introduction. Nuculacea-Trigonacea. Bulletin of the British Museum of Natural History (Zoology) Supplement 3, 1-125.

Welinder, S. (1978) The concept of 'ecology' in Mesolithic research. In The Early Postglacial Settlement of Northern Europe. P.A. Mellars (ed). pp. 11-25. London: Duckworth.

Table 1A. Numbers of growth lines between 'annual increments in shells from high- and low-shore localities from Robin Hood's Bay, Yorkshire, showing a high degree of variance (mean 201.3 ± 106.3).

Shell number (shore level and date of collection)	Number of growth lines 1977	1976	1975	1974	1973
HL 22.2.77		164	140	116	130
HL 17.5.77		129	118	202	
LL 17.5.77	43	92			
HL 14.6.77	67	330	338		
LL 14.6.77	149	150	144	155	
HL 10.7.77	192	333			
LL 10.7.77	113	202			
LL 6.10.77	146	428			

B.

	Size of annual increment in mm				
HL 22.2.77		1.53	1.40	1.16	1.34
HL 17.5.77		1.11	1.47	2.09	
LL 17.5.77	0.44	1.47			
HL 14.6.77	0.70	2.72	2.62		
LL 14.6.77	1.82	1.34	1.25	1.35	
HL 10.7.77	1.47	3.46			
LL 10.7.77	0.86	1.97			
LL 6.10.77	3.00	5.75			

Table 1B. Size of 'annual' increments in the same shells, measured normal to the growth lines. (Spearman Rank correlation coefficient between size of annual increment and number of growth lines $r_s = 0.992$.)

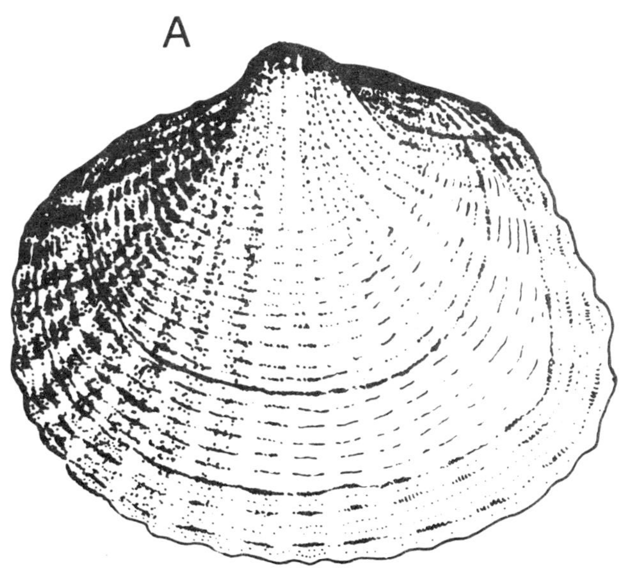

Figure 1A. Valve of *Cerastoderma edule* L., with two winter rings.

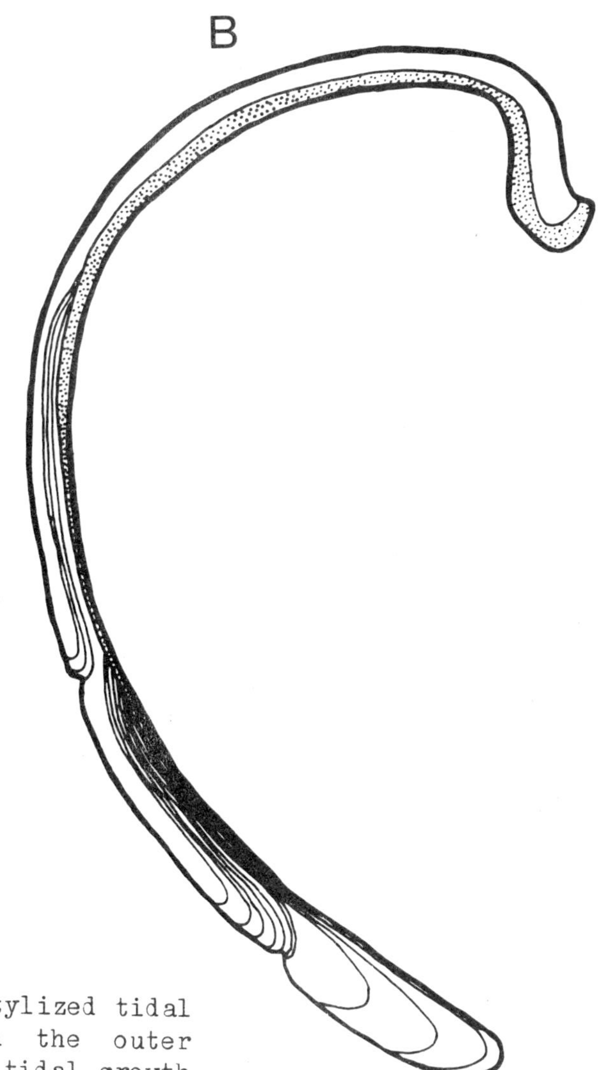

Figure 1B. Section of valve of *C. edule*, showing stylized tidal growth lines. Winter grooves can be seen in the outer surface of the shell; at these points, the tidal growth lines are closer together.

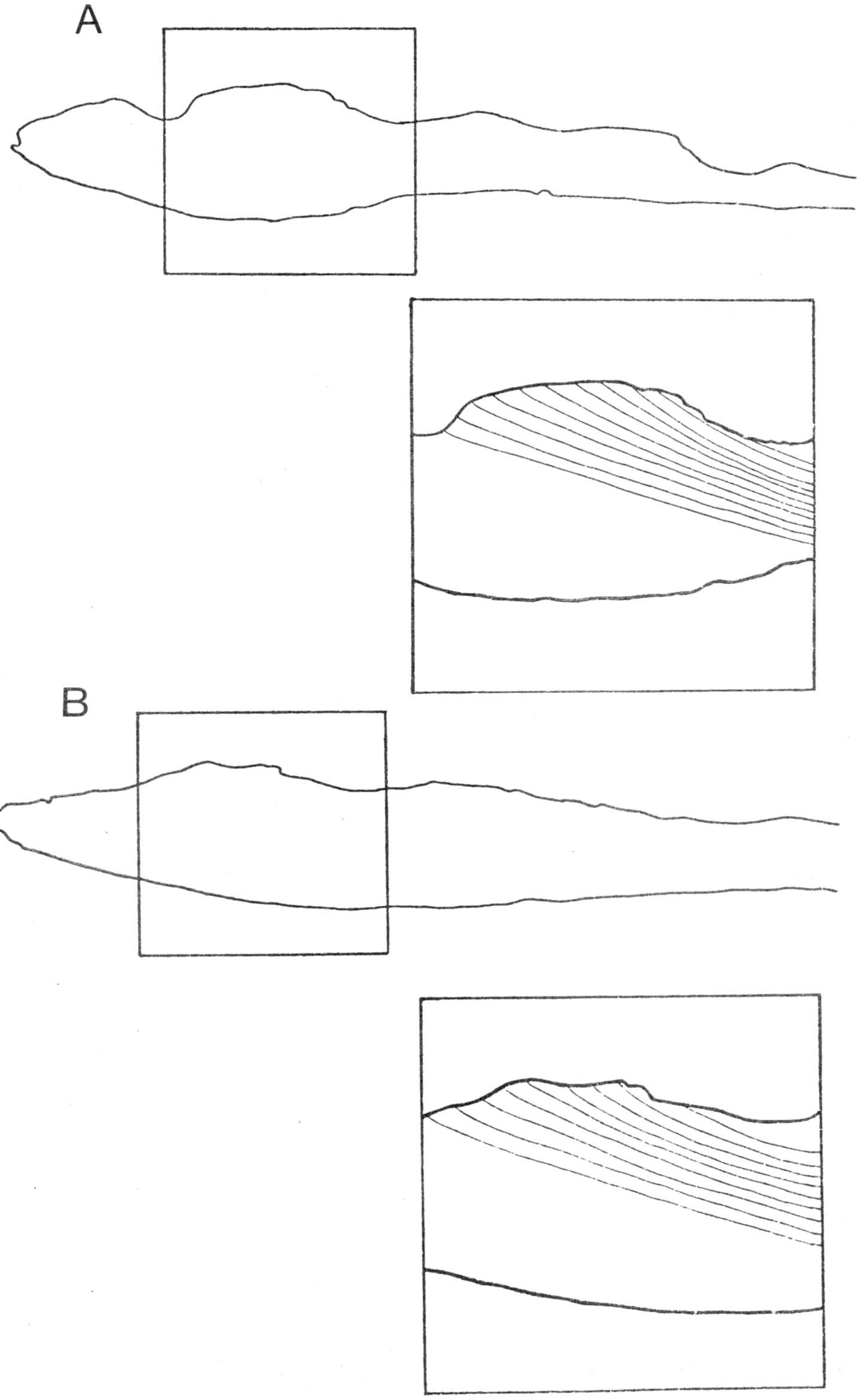

Figure 2. Profiles of a high-shore (A) and a low-shore (B) specimen of Patella vulgata L., demonstrating the smoother outline of the low-shore specimen and the orientation of the growth lines in relation to the outer surface of the shells.

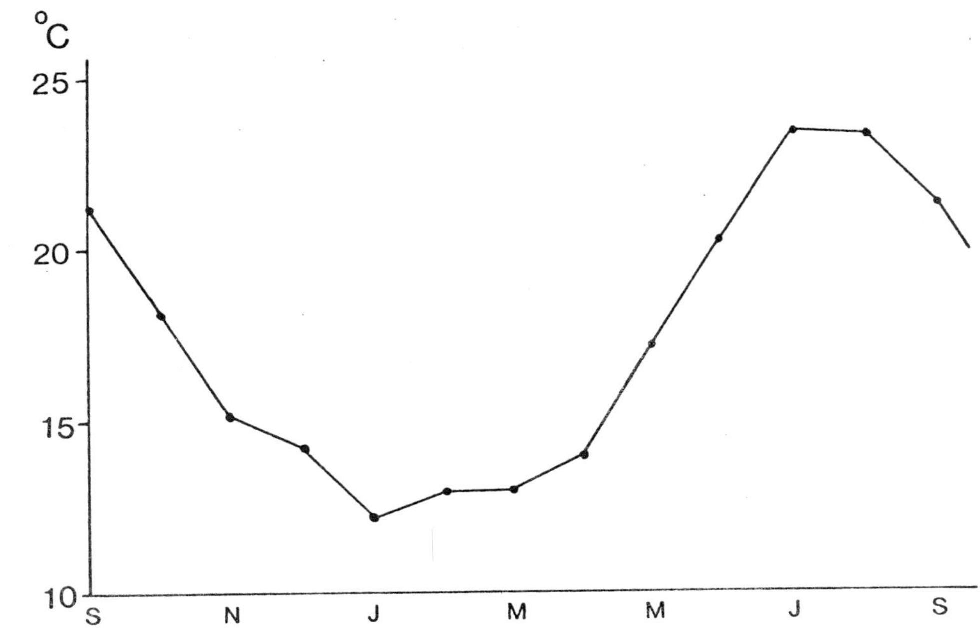

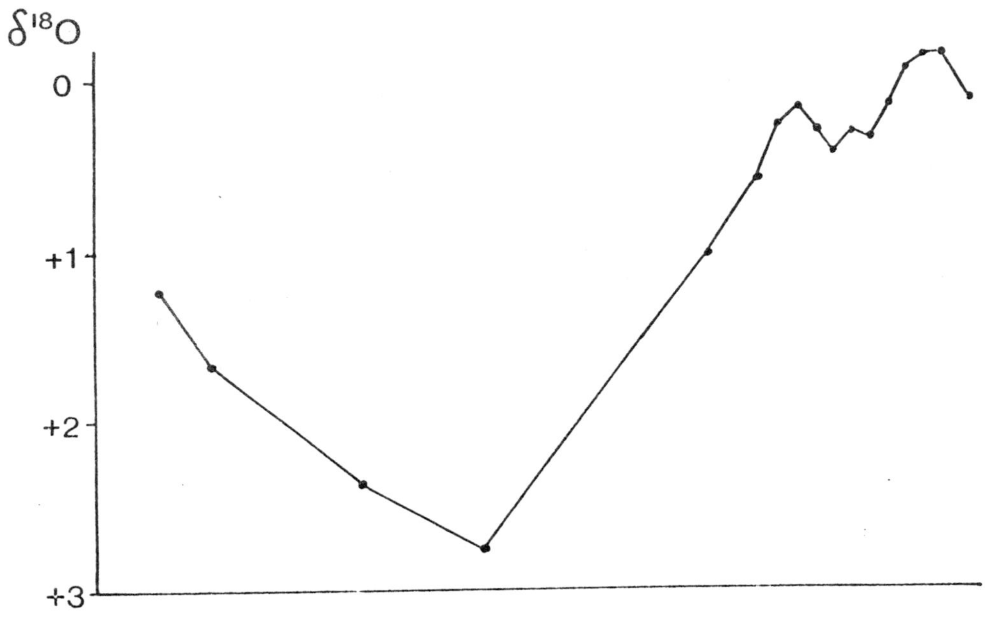

Figure 3A. Mean monthly sea temperatures for inshore waters in the Tavoliere, southeastern Italy.

Figure 3B. Oxygen isotope profile of a modern specimen of Solen vagina L., collected at the end of September 1982. Samples were taken at regular 2mm intervals.

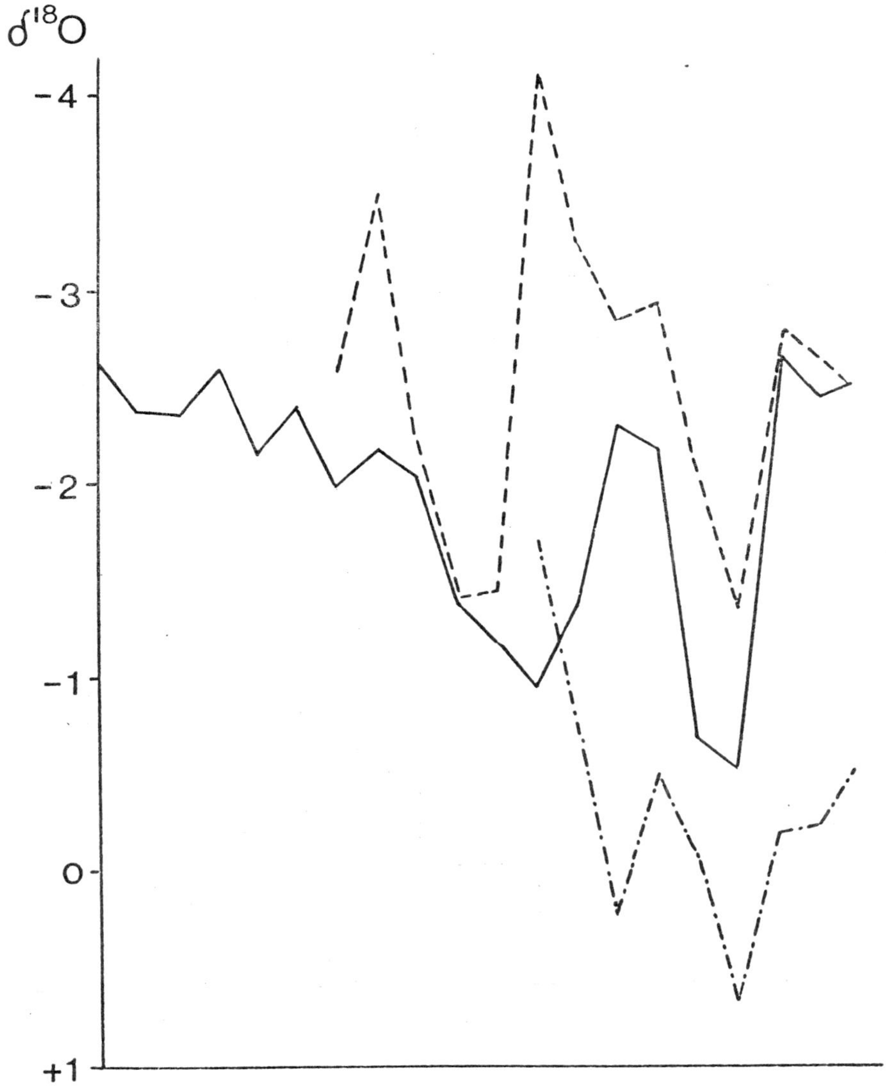

Figure 4. Isotope profiles of three neolithic specimens of Cerastoderma glaucum Poiret from Masseria Valente, demonstrating the variation in isotopic range. Edge values are on the right.

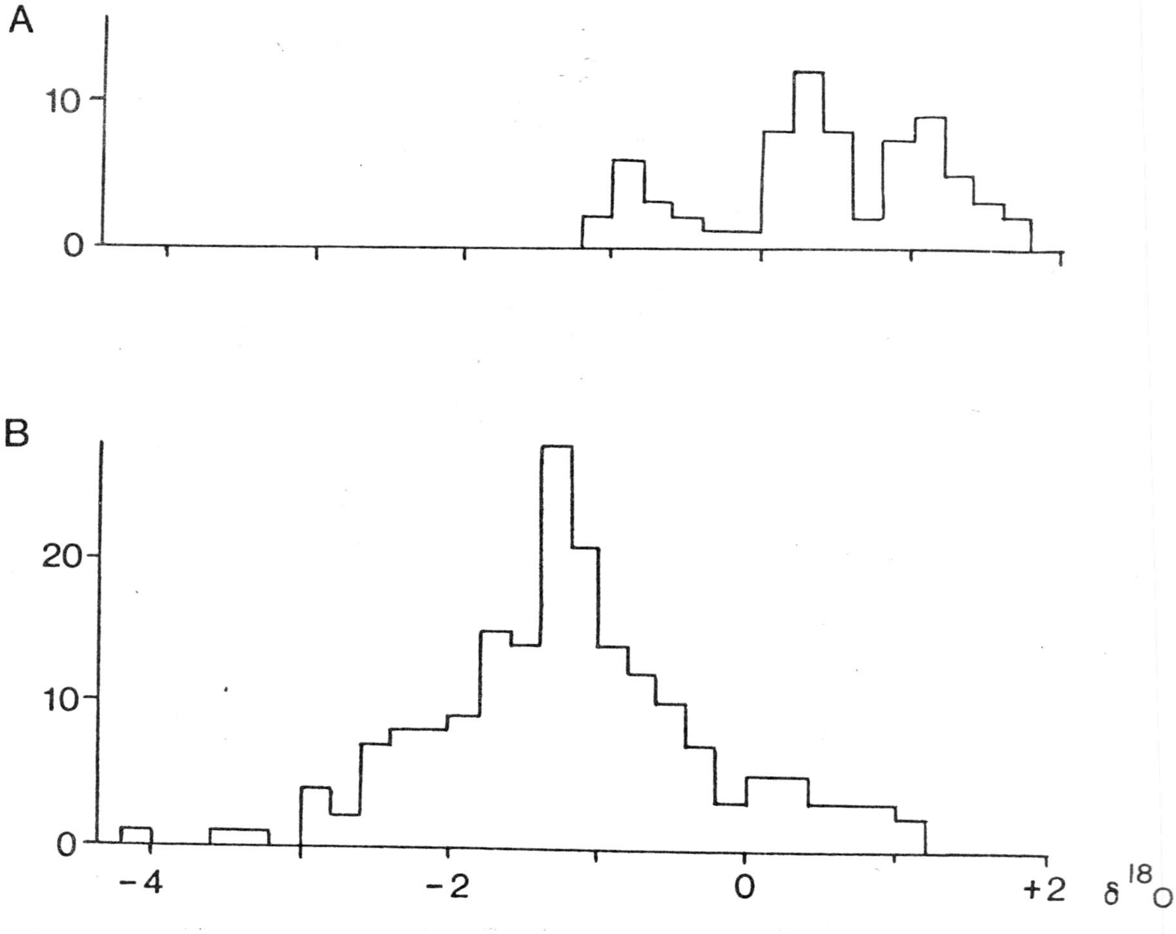

Figure 5. Histograms of all 18O values through neolithic specimens of S. vagina and C. glaucum sampled from Masseria Valente.

LAND SNAIL ANALYSIS IN ARCHAEOLOGY: THEORY AND PRACTICE.

K.D. Thomas
University of London
Institute of Archaeology
Department of Human Environment
31-34 Gordon Square
LONDON WC1H OPY

ABSTRACT

Land-snail analysis in archaeology has 'worked' because the data generally make ecological sense; however some of the explicit and implicit assumptions of this empirical approach are rather dubious on theoretical grounds. Various problems are highlighted in this paper and are discussed with reference to palaeoecological studies undertaken in Britain.

i. Quantification of assemblages; the use of percentage frequencies is criticised and suggestions are made for improving absolute estimates of species abundance.

ii. The problems and potential of using autecological data for key indicator species is discussed; it is suggested that the use of absolute frequencies gives a more reliable picture of the past dynamics of indicator species. This is illustrated with reference to *Pomatias elegans* (Müller) from Neolithic assemblages.

iii. It is suggested that the analysis of sub-fossil assemblages should not only include the use of ecological groups but also be more quantitative. Much can be learned about the past ecology of species by examining species-associations in assemblages.

iv. Simple models are used to illustrate the problems of interpreting rates and scales of responses. The value of using biogeographic models in interpreting assemblages of molluscs is examined. Particular attention is paid to the problems of studying assemblages from buried soils; studies on modern soils are needed to give estimates of time-depth functions and of the degree of ecological resolution which could potentially be achieved.

v. Studies on intra-specific variation (polymorphisms and metrical attributes) will probably add to the power of existing approaches and give information on ecophysiological stress in various species.

vi. Closer collaboration with pedologists and sedimentologists is urged.

INTRODUCTION

The analysis of land snails from archaeological contexts has produced valuable data on past environments and of man's activities in them. The technique has not, however, been subjected to the same level of critical analysis as, say, pollen analysis. Recent palynological reviews (e.g. Delcourt, Delcourt and Webb 1983, Edwards 1979) have stressed the value of using models to aid interpretation on various spatial and temporal scales. Experimental studies of pollen-catchment areas (Edwards 1979:257) are providing valuable, and often salutary, information on the origins of pollen assemblages. Stimulated by such work, I attempt here to discuss some of the practical and theoretical difficulties of land snail analysis in the context of archaeology. The aim is not to present a thorough review of the state of the art (if that is an appropriate term) or its many achievements. Instead, the paper is intended as a foil against the naive environmentalism (or 'habitatualism') which many students and archaeologists adopt when discussing land snail data. The ecological relationships of animals are very complex and land snails, being animals (this might surprise some 'zooarchaeologists'), are no exception to this generalisation. By using simple models to illustrate highly complex situations, I hope to show that the interpretation of assemblages of snail shells is a procedure fraught with dangers for the unwary.

Many of the points made here are certainly not unique to the land Mollusca; they could apply to most of the classes of biological data recovered from archaeological sites. Most workers involved with the analysis of biological materials from archaeological contexts will already be aware of these problems but there is a need to discuss them for the benefit of potential users of the data (Edwards 1979:255). As Evans (1978) pointed-out, archaeologists can seriously mis-interpret the evidence (e.g. Barker and Webley 1978).

BACKGROUND

The study of land snails has had varied applications in archaeology and will continue to make valuable contributions in the areas of palaeoeconomy (e.g. Evans 1969a, Lubell, Hassan, Gautier and Ballois 1976, Reed 1962), relative dating (Evans 1969b), absolute dating by radiocarbon (Burleigh and Kerney 1982), biostratigraphy (Burchell 1961, Ellis 1983, Kerney 1977, Kerney, Preece and Turner 1980), archaeological zonation (Bottema and Ottaway 1982), palaeotemperature analysis by oxygen-isotope determinations (Burleigh and Kerney 1982), reconstructions of past climates (e.g. Burchell and Davis 1957; see Evans 1970 and 1972 for a critical review) and the reconstruction of past environments and ecosystems (e.g. Evans 1970, 1972, Sparks 1964, 1969). Palaeoecological reconstruction is the subject of this discussion.

Environmental reconstructions may be 'instantaneous', to answer questions like "What was the environment around this site when it was occupied?", or 'sequential', yielding a history of landscape change - or stability. Often, land-snail data will yield both types of reconstruction; this depends on the contexts available for study. Snail assemblages may be obtained from a wide range of contexts, each with their own problems of sampling and interpretation. One major limitation is preservation; land snail analysis is restricted to those contexts which are sufficiently calcareous to permit the preservation of shells. In practice this means restriction to soils developed from calcareous parent materials, to calcareous deposits and to low-lying situations where ground-waters are charged with calcium carbonate. Assemblages may be extracted from soils buried under archaeological monuments (Evans 1972), under colluvium (Evans 1972), under and within tufa or travertine (Evans 1972) and beneath blown sand (Evans 1971b, Spencer 1978). They may also be incorporated into sediments such as the fills of pits and ditches on archaeological sites (Evans 1972), in lynchets (Fowler and Evans 1967), in colluvium and the fills of dry-valleys (Bell 1981, 1982,Evans and Valentine 1974, Kerney 1963, Kerney, Brown and Chandler 1964) and re-deposited in alluvium (Shotton 1978). Thus, although Sparks (1964:91) has suggested that land-snail assemblages are more truly representative of once-living communities than assemblages of fresh water molluscs, it is clear that mixed assemblages can occur in many terrestrial contexts.

Buried soils might be expected to contain the most reliable assemblages with relatively low frequencies of allochthonous elements. Evans' (1972) studies on the snail assemblages from buried soils suggest that they are probably age-stratified and therefore have great potential for fine environmental resolution. This is discussed further below. Unfortunately, well-preserved buried soils are usually only found under archaeological sites which are also well-preserved and not under any serious threat from present-day land use. In recent years, rescue (salvage) archaeology has dominated fieldwork in Britain with work focussed on sites which are under threat or which have already been almost destroyed. As a result, few buried soils have been exposed for analysis by malacologists, who have instead worked with assemblages from ditches, pits and other sedimentary environments.

METHODOLOGY

The main stages of a land-snail analysis are shown in Figure 1; detailed discussions of the methods are available in the works of Evans (1972: chapter 3), Sparks (1961, 1964) and Kerney (1963). A fundamental step is the recognition of the context, or contexts, to be sampled; the interpretation of the context will influence the sampling strategy (location of samples, size of samples, sampling interval, etc.). Equally, the type of sample required,such as an autochthonous assemblage or one 'most contemporary' with the archaeological features, will determine which of the available contexts will be sampled. Sampling represents a considerable practical and theoretical problem which cannot be fully discussed here. However, many of the questions raised in this paper have important implications both for sampling designs and the on-site implementation of sampling strategies.

The laboratory extraction and identification of snail shells involve essentially practical problems; these are continuously being minimised by refinements in technique (e.g. Preece 1981) and improvements in the skills of individual investigators.

The major problems discussed below are:

i. Quantification of data, particularly the use of percentages to indicate the abundance of species.

ii. The use of assemblages and of individual species of snails in environmental reconstruction.

iii. The ways (rates and trajectories) in which communities and individual species of snails respond to environmental changes.

iv. Scales of discrimination in environmental reconstruction.

QUANTITATIVE ANALYSIS AND DATA PRESENTATION

To date, surprisingly little use has been made of statistical and other mathematical devices (such as diversity indices) in the manipulation of data on assemblages of snails from archaeological sites, although a few biometrical studies have been made (Burleigh and Kerney 1982, Preece 1980). This situation is in marked contrast to the work on sub-fossil assemblages of insects in which diversity indices, association analyses and other approaches have been used to great effect (e.g. Coope and Sands 1966, Kenward 1978, 1982).

Assemblages of land snails are usually analysed in terms of the percentage frequencies of the component species and of various ecological groups (Evans 1972, Kerney et al. 1964, Sparks 1961, 1964). There are, however, various difficulties associated with the use of percentage frequencies. Severe distortions will arise where some species are superabundant (see Figure 2). More fundamentally, the use of percentages implies certain assumptions about the ecology of land snails and the functional relationships between species. It is unlikely that mollusc communities saturate their environments (Kerney 1963:231). As a consequence, individual species of molluscs may fluctuate in abundance independently of other species. This is not to deny that inter-species interactions do not occur; competition and prey-predator relations may be important factors in the ecology of many species. However, such ecological relationships might occur not only between various species of snails, but also between snails and other taxonomic groups such as mites (Acari), woodlice (Isopoda), millipedes (Diplopoda), centipedes (Chilopoda), springtails (Collembolla), beetles (Coleoptera), as well as mammals and birds. In short, we should not assume that land-snail communities (and even less so, sub-fossil assemblages) are unified ecologically; the '100 percent assemblage' concept is only valid at the level of taxonomy (all the elements of a snail assemblage are gastropods).

Kerney (1963:231) advocated the use of absolute values of abundance for quantifying assemblages of land snails; he used the numbers of shells per unit weight of soil. Evans and Jones (1979) calculated absolute abundances for species of snails extracted from ditch deposits at the site of Mount Pleasant, a neolithic henge monument. They made allowance for the dilution effect of the ditch rubble by excluding the weight of those stones coarser than 0.5mm from their calculations. Such simple numerical devices have two major advantages: they allow the independent quantification of species without *a priori* assumptions about their ecological relationships and provide scope for the application of mathematical analyses, ranging from simple tests such as chi-squared to multivariate analyses of species associations. If only things were so simple in reality. A major problem is that the abundance of shells in a deposit is in part a function of the rate of accumulation of the deposit (or, in the case of a buried soil, the rate at which snails were incorporated into the soil). Where such processes are rapid, snail shells will have low densities and low numbers per unit of sample. Obviously, absolute snail frequencies, comparable to absolute pollen frequencies (e.g. Pennington 1975), can only be meaningful when assemblages can be assigned to discrete units of time. It is not likely that radiocarbon dating of the shells of land-snails will provide sufficiently accurate dates for fine stratigraphic discrimination (Burleigh and Kerney 1982:36) and, unless adequate samples of charcoal can be found stratified with assemblages of land snails, the problem of absolute quantification may prove to be intractable.

The only practical way forward is to use both percentage and quasi-absolute frequencies in interpretation and data-presentation, much as Dimbleby (1957, 1961) has done in his analyses of soil pollen. The percentage values give a general idea of fluctuations or trends over time while the absolute values enable the analyst to see if a species is genuinely increasing or declining in abundance. This is illustrated in Figure 3 for various key species from a lynchet at Bishopstone in Sussex (National Grid Reference TQ 46750072). The basic data (from Thomas 1977) are given in Table 1 and the relative and absolute frequencies of the six selected species are plotted in Figure 3. It is clear that the low percentage frequencies of such species as *Pomatias elegans* (Müller) and *Helicella itala* (L.) in the buried soil (samples 28 to 30), compared with the body of the lynchet, are due to the super-abundance of *Pupilla muscorum* (L.). This example will be returned to later.

INDICATOR SPECIES, AUTECOLOGY AND ENVIRONMENTAL RECONSTRUCTION

Most of the species of snails which occur in archaeological contexts in Britain are still living today. We can therefore hope to use modern ecological data on these species to reconstruct past conditions. If we knew enough about the key limiting factors determining the distribution and abundance of a particular species, these could be extrapolated into the past in order to 'explain' the fossil occurrence of that species. Such 'substantive uniformitarianism' (Rymer 1978) is, however, very dubious on ecological grounds.

We could measure in the laboratory the survival responses of any species of animal to a range of physical and chemical factors and thereby define the upper and lower limiting values, beyond which the species would not be able to survive, for each of the factors. Taking all of these values together, we could define a multi-dimensional space within which the species would be able to live (the so-called 'fundamental niche'). In the real world these factors will interact and over-lap in various ways and reduce the size of the fundamental niche to the 'realised niche'. The size of this niche will vary according to local conditions: the ways in which the various physical factors combine and the presence of other living organisms. The size of the niche will also vary over time as the values of the environmental factors change. If the values change beyond certain limits, the habitat will become untenable for that species. In simple terms, the distribution and abundance of a species is determined by the complex interaction of a large number of physical and biotic variables which are not readily predictable. Thus, simplistic substantive uniformitarianism is only possible if we can be certain that all factors in the past were equal to those of the present day. We must also assume that the present-day is 'typical' in terms of the combinations of factors for any given species. Obviously we cannot accept either of these assumptions. Even today, a species may respond quite differently to environmental factors within quite small geographic areas (e.g. Cameron 1978, Evans 1972:89). In practical terms, the use of land snails in palaeoecology is made difficult by the paucity of knowledge of the ecology of many species and by the problem of adaptive responses to environmental changes (Cheatum and Allen 1964).

Evans (1972) has discussed the various factors which can affect the distribution and abundance of land snails. Availability and retention of water are obviously important (e.g. Cameron 1970a, Machin 1967) and temperature may have an important impact on both geographic and local distributions (e.g. Cameron 1970b). Soil pH and the availability of calcium may be critical (e.g. Atkins and Lebour 1923) and although land snails may survive in rather acid environments (Bishop 1976), the chances of preservation are low. Some species may be able to acclimatize to shifts in the environment during the life-span of an individual (Segal 1961), but not all can do so (Mason 1971); this ability is related to the rates and degrees of change. Many species may respond to shifts in the environment by genetic (trans-generation) changes, but there is a limit to this for individual species - or so we believe.

To what extent food is a limiting factor to land snails is unknown. We know little of the food preferences of most species of land snails; some are herbivores and a few are carnivorous, but the majority appear to be facultative omnivores and detritivores. Mason (1971) has shown that snails in a deciduous wood in Oxfordshire ingest less than one percent of the annual litter production. This would suggest that total quantities of available food are rarely, if ever, limiting. Other workers (Grime, Blyth and Thornton 1970, Wolda, Zweep and Schuitema 1971) have suggested that the snail Cepaea nemoralis(L.) may be regulated in abundance in some habitats by the quality of its food.

The interaction of physical and biotic factors in regulating distribution and abundance is complex. Some species may co-exist without serious interaction (Figure 4.I), others may mutually exclude one another from particular parts of their niches (Figure 4.II) while one species might totally exclude another from that part of the niche they share in common (Figure 4.III). The outcome of species-interactions may be determined by factors in the physical environment (with varying results according to variation in the various factors), by the presence of alternative resources, or by various threshold factors associated with relative numbers or the presence of other interacting species. Mordan (1977) has highlighted the complex of physical and biotic factors which influence the distribution of two species of land snails in grassland and woodland habitats in Huntingdonshire. Aegopinella nitidula (Draparnaud) appears to be limited by climatic or micro-climatic factors, being intolerant of dry conditions, while Nesovitrea hammonis (Ström) appears to be partly limited by predation by A. nitidula.

Implicit in the use of land snails to reconstruct past environments is the assumption that their ecological tolerances have not changed (see Evans 1972:30). Modern populations of all out-breeding species possess enormous genetic variability which may give them the ability to respond adaptively to environmental change. Such responses may be physiological and leave no trace in the morphology of the species. In most palaeoecological studies it is assumed that no gross adaptations have biased the assemblages; such assumptions are usually justified since the associations of sub-fossil species resemble present-day associations.

Occasionally, genetic responses to the environment may be detected in the phenotypes of modern and sub-fossil snails. For example, the land snail Cepaea nemoralis is polymorphic for shell colour and banding patterns and these polymorphisms appear to be related to selective pressures in various environments. In some parts of the range of the species, colour and banding patterns appear to confer camouflage against various vegetational backgrounds so that some forms are predated less heavily by song-thrushes (Cain and Sheppard 1954). It would be tempting to infer past types of vegetation from the frequencies of banding and colour morphs in sub-fossil assemblages of C. nemoralis, were it not for the fact that other selective factors are also important. In some areas at the present day the selective pressure of predation is over-ridden by other pressures associated with climate and other factors; in these areas there is no obvious correlation between morph-frequencies and vegetation type (Arnold 1971; Cain and Currey 1963).

Lastly, there is a need for more data on the factors which affect the distribution and abundance of species at the present-day. Much of our knowledge of the autecology of British land snails derives from the compilation of Boycott (1934) of the range of habitats in which various species have been recorded. There have been a number of important studies since the publication of Boycott's paper (see Evans 1972 and Cameron 1978a), but the gaps in our knowledge of the ecology of all but a few species remain large. This will continue to be a fundamental constraint for the foreseeable future; as Kenward (1982:71) has pointed out, it is unlikely that time and funds will be available to collect adequate ecological data for most species. We should therefore focus our attention on those species which are potentially most interesting.

Pomatias elegans as an Indicator Species

Evans (1971a, 1972:272) has suggested that peaks in the percentage frequency of Pomatias elegans in mollusc sequences from various neolithic sites in Wessex probably indicate loose, broken soils which could have been created during woodland clearance phases. In this section the value of P. elegans as an indicator species will be examined with reference to three of the sites studied by Dr Evans:

i. South Street long barrow, near Avebury in Wiltshire (National Grid Reference SU 091693). Data from Evans (1971b, 1972).

ii. Avebury henge monument in Wiltshire (National Grid Reference SU 101698). Data from Evans (1972).

iii. Durrington Walls henge monument, near Stonehenge in Wiltshire (National Grid Reference SU 152435). Data from Evans (1971a).

Well preserved buried soils were revealed by the excavations at these sites and they yielded numerous assemblages of land snails which enabled Dr Evans to make detailed reconstructions of environmental changes and Neolithic land-use patterns. Marked environmental changes were attested at each site (we are only considering the evidence from the buried soils here), from habitats dominated by shade-loving species of snails to more open conditions with evidence of disturbed ground; see Evans (1971a, 1971b, 1972) for detailed discussion. At each site, the transition from shaded to open conditions was marked by an increase in the frequency of P. elegans, a species which favours loose and friable soil (Kerney 1972). The percentage frequency of P. elegans rose sharply as the frequency of shade-loving elements in the assemblages declined and then fell again, rather slowly, as open-country elements asserted themselves. This effect is shown in Figure 5. However, if we calculate the absolute abundance of P. elegans in the assemblages (see above), the picture changes (Figure 5). The changes in abundance of the species become less asymmetrical and the peak abundance becomes spread over a longer time-span. In general, the modal peak occurs later in the sequence than when relative values are used. The use of absolute data to study the population dynamics of P. elegans suggests that not only does the species increase in abundance as a result of clearance, it maintains these high abundances through phases interpreted as periods of ploughing (see the papers by Evans, op. cit.). [Since I wrote this paragraph, Dr Evans has informed me that "shells of P. elegans in the upper (grassland) levels are usually small, worn and corroded apices, while those in the "clearance" horizons are large and fresh-looking" (Evans in litt.). This might indicate the presence of old or relict shells in the upper soil horizons (in which case the stratigraphic integrity of the assemblages must come into question). Alternatively, broken and worn shells might be expected to occur in soil horizons in which biological activity is intense and where tillage has been attested. Such problems illuminate the need for studies on land-snail taphonomy (see below)].

Figure 6 shows these data plotted in a more comparable way. Three estimates of abundance are used: percentage frequency, numbers of shells per kg of soil and numbers per kg per cm of soil. The highest value of each estimate is taken to be 100 percent; all other values in each series are expressed as a proportion of this. Clearly, there are marked differences in the patterns of fluctuation of *P. elegans* revealed by the different methods of estimation. In the ensuing discussion I will compare the percentage estimates with those derived from numbers per kg per cm.

i. South Street long barrow

The absolute frequencies of *P. elegans* (Figure 6.I) show peaks in those horizons associated with Evans' ploughing, or soil-disturbance, phases. One peak occurs near the base of the soil profile (from 17 to23cm) and the other from 6 to 13cm. Between these peaks is a decline in abundance associated with the relatively stone-free horizon, which Evans attributes to worm-sorting under stable soil conditions - perhaps grassland. The abundance of *P. elegans* also declines near the surface of the soil, again in association with a stone-free horizon. Even here, the species remains fairly abundant; the dramatic decline shown by the percentage estimates is an artifact produced by the large numbers of open-country species in the samples.

ii. Avebury henge monument

Here the clearance episode is interpreted (Evans 1972) from changes in the assemblages which occur just below the old turf line, at about 6 to 9cm. The absolute abundance of *P. elegans* (Figure 6.II) reveals a more dramatic picture of change than the percentage estimates, with a relatively huge increase in abundance from 26cm to the peak at 4cm. As at South Street, the absolute estimates suggest that *P. elegans* was not only favoured by the relatively transient conditions created by the initial disturbance of the soil (presumably during woodland clearance) but that it continued to thrive during subsequent phases of tillage in the Neolithic period. Evans (1972:273) suggested that the sample from 0 to 1cm is not in fact from the buried soil but represents a trampled disturbance layer, probably associated with the earliest stages of the construction of the henge. The abundant fragments of crushed chalk would support this (we might also speculate if the chalk was used to mark out the monument on the ground). Again, this interpretation is supported by the absolute abundance of P. elegans which increases in frequency in this zone. No doubt the species was taking advantage of the loose chalky soils which existed during this short-lived phase.

iii. Durrington Walls henge monument

The absolute estimates for P. elegans again reinforce some of Evans' (1971a) interpretations. The percentage frequencies (Figure 6.III) are severely distorted by dramatic changes in the abundance of other species; the rapid decline in the proportion of shade-loving species following clearance was in reality followed by a slow progressive rise in the abundance of P. elegans which reaches a peak in the 3 to 6cm horizon (coincident with Evans' postulated phase of middle neolithic cultivation) and falls away slightly in the upper soil horizons - possibly indicating a change of land-use to pasture.

iv. Conclusions

These analyses in no way invalidate any of the interpretations made by Evans (op cit.) but do add support to them and give a clearer picture of the various events. The use of percentage data alone tends to obscure both the pattern and the timing of the observed responses of Pomatias elegans. The data also suggest that, at least as far as the techniques of cultivation in the Neolithic period are concerned, P. elegans was able to thrive in arable environments.

SUB-FOSSIL ASSEMBLAGES, COMMUNITIES AND ENVIRONMENTS

We have seen that the ecology of individual species (autecology; although some ecologists would limit this term to the ecology of individual populations of a species) is a complex mix of interactive variables, the relative importance of any of which may vary both spatially and temporally in response to changes in the other physical, chemical or biological variables. Fortunately, we do not have to rely on individual species to make judgements about past environments (i.e. we are not forced to undertake some sort of 'ecology-by-summation' exercise) because observations of molluscs in modern habitats show that species tend to occur in recognisable associations. Unfortunately, the data on modern snail communities in Britain are far from adequate. Much of our knowledge is still based on the work of Boycott (1934), but recent work by, inter alia, Bishop (1976), Cameron and Morgan-Huws (1975), Evans and Jones (1973), Mason (1970), Paul (1975, 1978a, 1978b) and Wöreborn (1969), in Britain and elsewhere, has added to our knowledge of species associations and the quantitative relationships between species in a range of habitats.

However, whether we will be able to "match the species compositions of ancient and modern faunas", as suggested by Cameron (1978a:21), is open to question because we do not always know if we are dealing with an ancient _fauna_ (in the strict sense of this term) when studying assemblages of land snails from archaeological contexts. The study of archaeological assemblages presents a number of problems:

i. The assemblages are death assemblages and are therefore not exact replicas of once-living communities; they have suffered losses by differential destruction and gains by the incorporation of allochthonous elements.

ii. We do not know if such archaeological assemblages are derived from equilibrium communities. Many certainly are not: they represent communities in a state of flux, in which the representation of various species might result from chance effects or from the peculiar conditions created by man's past activities.

iii. We do not know if equivalent biomes at the present day invariably have quantitatively similar associations of species of snails. Historical accidents, associated with relative rates of migration, colonisation, ecological succession and the importation of alien species by man, could have a marked impact on the species structures of supposedly equilibrium communities (if such exist).

iv. We cannot be sure that any present-day environments and their associated snail faunas provide adequate models for prehistoric environments - particularly those in which man was having a significant impact. The environment of a neolithic or bronze age field may have no adequate counterpart in the modern British landscape, therefore the land-snail faunas of such fields may have been quite unlike any living in Britain today. We have already suggested, for example, that _Pomatias elegans_ may have been a successful species on tilled land in neolithic times.

Assemblages of land snails are often analysed and interpreted by the percentage frequencies of their contained 'ecological groups' (Sparks 1961, Evans 1972). Species are classified, according to their habitat preferences, into 'shade-loving', 'catholic' and 'open-country' groups (other categories, such as 'marshland', can be added as needed); the proportions of these groups, along with information provided by certain key indicator species, is then used to make inferences about the past environment or environmental mosaic. Recent work by Cameron (1978a), Cameron and Morgan-Huws (1975) and Evans and Jones (1973) has led to a re-evaluation of some of these groups, particularly the composition of the shade-loving category (which is sometimes used to infer wooded conditions in the past). Cameron and Morgan-Huws (1975) showed that some long-grassland habitats located in open-country contained no species of the 'open-country' ecological group but had up to 60 per cent snail species which would be classified as 'shade-loving'. _Carychium tridentatum_ (Risso) occurred as a dominant species in some of these assemblages. It should be obvious from all this that interpretations of past woodland from assemblages dominated by shade-loving species should be done with great care. A statement such as: "The soil contained a woodland mollusc fauna dominated by _Carychium_ *tridentatum* (Risso)" (Murphy 1983:56) should be treated with due scepticism until further data emerge on the other elements of the 'woodland fauna'.

The use of ecological groups in the palaeoenvironmental interpretation of assemblages of land snails has given us important insights into past environments and man's activities but, as we have seen, there are problems. The majority of land snails in Britain fall into the 'catholic' ecological group and have little interpretative value as such (Cameron 1978a:21). By sorting species into apparently exclusive groups we are losing potentially valuable information; many 'catholic' species are ecologically compatible with 'shade-loving' or with 'open-country' species. Species should not be assigned to ecological groups by some rule-of-thumb procedure; they should be assigned according to their association with other species in the sub-fossil assemblage. A possible procedure is as follows:

i. Assign the various species to their 'exclusive' ecological groups in the usual way.

ii. Assess the ecological balance which is suggested by the resulting frequencies.

iii. Re-assign 'catholic' elements according to their known ecological tolerances (i.e. if the overall balance of an assemblage is toward 'open-country', those 'catholic' elements which are quite compatible with this interpretation should be assigned to the open-country group).

This is not just juggling with the data, or cooking the books; it is an attempt to increase the ecological reality of using habitat groupings in land snail palaeoecology. A similar approach was used by Coope and Sands (1966) in their analysis of sub-fossil assemblages of Carabidae (Coleoptera) from Devensian deposits in the Tame Valley in Warwickshire, central England (various sites were sampled with National Grid References SP 168918, SP 212924 and SP 201907). Using the known distribution of species of carabids in the various vegetation zones of the Scandinavian mountains as a model, Coope and Sands gave each species in their assemblages a score according to the ecological zones in which it could be found today: '0' if it is never found in a particular zone, '1' if often found and '2' if commonly found. These scores were then summed for each possible type of habitat (data from ecotones were also used) and the resulting distributions compared with similar ones for the beetles from these present-day zones. In this way a reasonably objective assessment could be made of the likely climatic variables in the past. This approach could be used in the analysis of assemblages of land snails and could be further refined by weighting the habitat scores by the abundance of each species in an assemblage.

Kenward has also suggested that some of the problems of interpreting species-associations in sub-fossil assemblages could be resolved in part by a more rigorous analysis of the assemblages themselves (Kenward 1982:71). I wish to return now to an example discussed earlier in this paper. It was shown that the use of absolute frequencies for various species of snails from an iron age lynchet at Bishopstone in Sussex (data from Thomas 1977; see Table 1 and Figure 3) produced a rather different picture of species-fluctuations than when percentage frequencies were used. The major problem with using absolute frequencies is that they depend on the rates at which the deposits accumulated. However, it might be possible to use such data to obtain information about the relative behaviour of different species in the sequence. In Table 2 the data from Bishopstone are presented in a form which permits comparisons of the patterns of fluctuations of the selected species. Using the numbers of shells per kg per cm (Table 2, rows (c)) as our unit of abundance, correlation coefficients were calculated for each pair of species. The results are shown in the matrix in Figure 7.A. Of the 15 coefficients only 5 were statistically significant at the formal level of significance and, interestingly, all of these had positive values. This could be taken to indicate that the various species were fluctuating in abundance relatively independently of each other (in the sense of biotic interactions between them), although the results could be an artifact of small sample sizes. Figure 7.B is a simple constellation diagram showing the associations between the various species. Interestingly, Pomatias elegans, Pupilla muscorum and Helicella itala form an integrated group, with the other species out on a limb. These three species appear to have responded in parallel to whatever factors were responsible for their fluctuations. The processes leading to the formation of lynchet deposits are complex and it is difficult to unravel the past spatial relationships between species. Some could have been living at the site of accumulation of soil; others could have been transported to the site of deposition. However, it would be tempting to interpret the fluctuations in the three species as indicating alternate phases of tillage and fallow. H. itala is rarely found today associated with arable agriculture - it is a species of dry grassland (Evans 1972:180); Pupilla muscorum similarly tends to avoid arable environments today (Evans 1972:146) but can be very abundant in dry grasslands where there are patches of bare earth. P. elegans, as we have seen, prefers disturbed calcareous soils and might have been tolerant of prehistoric tillage practices. Taking this information together, we can postulate that this mini-assemblage would have been favoured by the conditions of a tilled field allowed to go fallow, perhaps under a regime of grazing (H. itala) would not tolerate rank vegetation). I would not wish to push this example too far; I use it here to illustrate the potential of other forms of analysis of assemblages of land snails.

Palaeontologists often use diversity indices in their analyses of fossil assemblages (e.g. Gould 1976, Lasker 1976) and such mathematical devices have also been used in environmental archaeology - notably by Kenward (e.g. 1978). Such indices are easy to calculate and reduce sets of complex data to simple numbers. They may have some potential value in land-snail palaeoecology although the ecological interpretation of diversity indices is difficult (May 1975, Thomas and Foin 1982). In autochthonous assemblages, low diversity indices could be interpreted in two quite different ways:

habitats of relatively low stability which do not favour molluscan life <u>or</u> habitats of high stability which favour a few competitively-superior species. On the other hand, high indices could lead us to postulate habitats of intermediate stability, or of high micro-habitat diversity - or even prompt us to consider if the assemblages are as autochthonous as originally thought. Used in an empirical way, without any theoretical ecological assumptions, diversity indices might be a valuable device, particularly in the analysis of allochthonous assemblages (as Kenward, op cit., has shown with great success for sub-fossil assemblages of insects). The potential value of diversity indices is considered by Ellis and Gordon (1984, this volume) and J.G. Evans is analysing their application to the interpretation of assemblages of land snails from buried soils and ditch-fills (personal communication).

RATES OF RESPONSE

Both the nature of responses shown by land snails to environmental change, and the rates at which such responses occur, can fundamentally affect the interpretation of sequences of sub-fossil assemblages. Some possible effects are shown in some simple models in Figure 8, where only two key indicator species are considered - although the model could equally apply to two ecological groups. The various models are described in the legend to Figure 8. These models deal with an environmental change which happens at a rapid rate; equally, we could devise models of responses to slow environmental changes, involving: i. little numerical response, due to adaptation, ii. slow numerical responses which reflect the rate of environmental change or iii. rapid responses due to ecological threshold effects.

The rates at which communities of land snails respond to changes in their environment depend upon a number of factors:

i. The degree of change will affect not only the degree of response, but the rate at which it can occur (due to threshold effects).

ii. Rates of local extinction and replacement. If an environment becomes unsuitable for a species it will die out fairly rapidly. This is apparent in some of the neolithic buried soils discussed above, where shade-loving elements decline rapidly in frequency following woodland clearance, giving rise to the apparently high percentage frequencies of *Pomatias elegans*. Rates of replacement will depend upon successful colonisation; in the Neolithic soils this appears to have occurred slowly enough to distort the relative abundance of *P. elegans* in the sequences.

iii. The presence of species at low frequencies in the community which can take advantage of newly-created opportunities; for example, some open-country species are able to survive in suitable refuges in woodland.

iv. Successful colonisation. This is a complex process which has a number of determining variables:

(a) Dispersal: land snails are not known to have special dispersive phases in their life-cycles. Some species are known to have phases of aggregation and dispersal during the year (Baker 1968) but few studies have been made on the distances moved by snails over long periods of time. Recent studies by Cowie (1980) and Paul (1978a) deal only with short-term dispersal which could be reversible (there is some evidence to suggest that some species of land snails are territorial).

In general, dispersal can occur either rapidly as a result of chance events - accidental transport by man or in birds' plumage, for example (Kerney 1966, Rees 1965, South 1974), or slowly as a result of a wave of advance of a population front which is gradually extending into suitable habitats.

(b) Chances of reaching the new habitat: these will depend on a range of factors (see MacArthur and Wilson 1967):

(i) the nearness of source areas with potential colonists

(ii) the population sizes or densities of potential colonists in the source areas

(iii) the presence of refuges or suitable micro-habitats near the new area to provide 'stepping stones' for colonists during migration.

(iv) the presence of adequate migration routes for colonists

(v) the nature of potential barriers to effective dispersal; how much 'hostile territory' exists between the source area and the new area? To what extent will this prevent colonisation?

(vi) chance effects: suitable habitats are not always colonised by species even if they are accessible (Ayal and Safriel 1983)

(vii) successful colonisation entails not only arriving in a suitable area but establishing a viable population. Sufficient numbers of colonists must arrive for this to happen; most species of land snails in Britain are hermaphroditic and are potentially well-adapted for establishing populations.

Some of these concepts have been used by Thomas (1982) to test various hypotheses regarding the palaeoenvironmental interpretation of assemblages of land snails from neolithic enclosures in south-east England; he concluded that these enclosures were probably constructed in localised temporary woodland clearings which were colonised successfully by only a few pioneering open-country species (Pupilla muscorum, Vallonia costata (Müller) and Vertigo pygmaea (Draparnaud)), which could anyway be regarded as 'fugitive' species exploiting temporary clearings within the natural woodland ecosystem.

Figures 9 and 10 show possible trajectories of change in land-snail communities resulting, on the one hand, from the expansion of woodland clearance (with the potential for rapid colonisation of newly-cleared areas from contiguous open habitats) and, on the other, from the creation of a temporary clearing in a wooded landscape. See the legends of these Figures for details. Clearly, a wide range of alternative trajectories of change are possible (but not necessarily equally probable) and these can give rise to different 'histories' of change as manifested in the age-stratified assemblages of molluscs incorporated into the soil profiles. Where newly created habitats are colonised slowly by suitable species various elements of the 'old community' might be able to survive in the absence of competition. Many of these relict species would be classed in the 'catholic' ecological group and the nature of the environmental change would be difficult to deduce from the resulting sub-fossil assemblages.

Time Resolution and Assemblages of Snails from Buried Soils

Time resolution is a major problem in the analysis and interpretation of assemblages of snails - and of other bioarchaeological materials (e.g. Behrensmeyer 1982, Edwards 1979). Some of the problems have been discussed above (see Figure 8).

Evans (1972:228) has discussed the possibility of estimating the total time-span represented by land-snail assemblages in the weathering zones of buried soils. He suggests that shell sequences within such soils may only represent the last few hundred years before burial and not the total history of the soil (although assemblages of snails may be recovered from sub-soil features which relate to earlier phases). We have noted that buried soils are particularly valuable contexts for the palaeoecologist because they preserve autochthonous assemblages of snails which appear to be age-stratified, presumably as a result of regular rates of incorporation of shells into the soil by, among other processes, the activities of earthworms. The precise time-depth relations of such stratified assemblages are not clear (Figure 11). A linear model of incorporation (Figure 11.I) is unlikely; the rates of pedological processes generally decline with increased depth, so a decreasing power function is more appropriate (Figure 11.II). Snail shells have some finite life span in active soils and will eventually be destroyed by weathering processes. The life-span will vary between different types of shells, according to size, shell thickness, surface texture, etc. Therefore the apparent diversity of species will decline with depth (i.e. age).

Observations on the rendzina soils of the Chalklands suggest that earthworm sorting under stable soil conditions (grassland) leads to the formation of a stone-free zone some 20-25cm thick at most (Evans 1972:228). This implies that snail assemblages may be lowered through the soil profile to about this level and probably not much further. Therefore, assemblages found in the lower parts of stone-free zones will be time-compressed and possibly mixed (Figure 11.III). In addition, we must doubt that assemblages of snails are incorporated into soils in a neat and orderly way; earthworms and moles may mix assemblages in the soil by bringing older shells up and by creating channels down which more recent shells may fall (Figure 11.IV and V). In shallow soils with intense biological and physical mixing, assemblages of snails may become thoroughly mixed (Figure 11.VI); such a situation appears also to arise with pollen in many soils on the Chalk (Dimbleby and Evans 1974).

The situation is complicated even further by post-burial compression of soils by the weight of the over-burden. This will be most marked in the surface horizons of the soil, which have to bear the weight directly and which have a loose crumb structure with numerous voids. Such compression will lead to the effective compression of age-stratified assemblages near the surface. If we add to this the effects of man, who may cultivate the soil and mix the assemblages and perhaps remove the turf horizon before constructing the mound or bank which will preserve the soil, our picture of despair is almost complete! Given all of these problems, it is astonishing that any ecological sense can be made from assemblages of snails from buried soils. Yet, as Evans (1972, and elsewhere) has shown, dramatic environmental changes can be detected from assemblages of snails from thicknesses of soil of 10cm or less, even when past cultivation is attested. This must imply that the processes tending to produce age-stratification by the sequential incorporation of assemblages into soils are so effective that other disruptive processes have a relatively small impact.

In order to develop further the potential of land-snail analysis of archaeological palaeosols we need to have a better idea of the time-depth relations of assemblages. This will permit better absolute quantification of the various species (see above) and, by permitting improved sampling techniques (sample intervals), will lead to improved time-discrimination of events. This in turn will enhance the degree of ecological resolution which can be achieved by the technique. To further these aims, work on modern soils with known land-use histories and on recently buried soils, using pedological and biological techniques, is in progress.

SPATIAL SCALES

Snails are both proverbially and actually stenotopic and are therefore influenced strongly by local environmental factors. There arises the question: to what extent are land-snail assemblages indicative of the general ecosystem? (Cheatum and Allen 1964:32).

Figure 12 illustrates some of the problems associated with making spatial inferences about environments based upon snail assemblages derived from particular loci of preservation. The sampling frame for the land-snail assemblage is readily defined by the contexts available for sampling, but the catchments from which the snails in the assemblage are derived are less readily deduced. The area of possible or probable catchments (shown simply as I to IV in our model in Figure 12) will depend upon the over-all area being considered, the topographic relationships between the context and the catchments, and the local processes which might have been operating in the past. Clearly, assemblages of snails cannot be adequately interpreted without due reference to the archaeological, pedological and geomorphological data.

Evans (1972:112) has argued persuasively that because samples of land snails do not represent discrete populations of species (in the spatial or standing-crop sense used by animal ecologists) and because snails are continually moving from one micro-habitat to another (thereby blurring, over a number of years, differences in highly localised distributions), a snail assemblage will reflect the range of micro-habitats which were available for exploitation and therefore give us a good idea of the past ecosystem.

Some of the problems of interpretation at the spatial level are intimately related to other problems discussed above, especially those of the rates of movement of species and in particular those species which are colonising new areas of suitable habitat. The use of simple biogeographic models led Thomas (1982) to infer that some neolithic enclosures in southern England were built in localised woodland clearings; it was not possible to deduce the size of these postulated clearings, but it was suggested that they were certainly not contiguous with any nearby cleared areas.

MacArthur and Wilson (1967) point out that there is a distinct relationship between the area of habitat islands and the numbers of species which inhabit them (the so-called 'species-area curve'). Perhaps diversity indices could be used to give some idea of past areas of habitats? This is unlikely to be possible. With snails we are dealing with stenotopic and small animals whose diversity is probably related to local micro-habitat diversity. The diversity of local micro-habitats is not necessarily simply related to the total area of a habitat.

Another more practical, and proven, approach is to analyse assemblages from as many contemporaneous contexts as are available within a given - preferably restricted - area. Evans (1972) has done this for many archaeological sites in Wessex and has convincingly demonstrated the widespread creation of woodland clearings by man in the neolithic period.

Lastly, there is, as ever, a need for more experimental work on modern assemblages and, in particular, for comparative samples to be taken from modern contexts in which land-snail assemblages are accumulating.

PROSPECTS

Land-snail analysis in archaeology has 'worked' because the data generally make ecological sense, even if some of the assumptions of the empirical approach are rather shaky. There are some fundamental limitations to the application of the technique: shells are only preserved in calcareous environments and snail species are not sufficiently stenoecious to indicate, say, the presence of particular species of plants in an environment. Other limitations of the technique which have been discussed in this paper could be partly overcome - or at least be better understood. In this way we can hope to realise the potential of the technique: to understand its true limits and the degree of ecological resolution that the technique can potentially achieve.

Specific areas for future development include:

i. Improvements in the basic quantification of species abundance to avoid ecological and logical pitfalls and to allow greater use of mathematical approaches to data analysis.

ii. The collection of data on rates of responses of land snail communities in recent environments with known ecological histories.

iii. Improvements in sampling designs, with experimentation to test the limits of resolution for assemblages in buried soils.

iv. The analysis of intra-specific attributes of sub-fossil land snails, particularly genetic variation, through the study of polymorphisms (e.g. Cain 1971, Currey and Cain 1968) and metrical variation of size and shape (e.g. Burleigh and Kerney 1982, Cameron 1978b, Thomas 1978). Snail shells preserve a record of growth patterns through life and can yield longitudinal data on growth which might cast light upon subtle environmental changes which are manifest through, say, ecophysiological stress.

v. Greater integration of studies on land snails with other palaeoenvironmental techniques (particularly palaeopedology and sedimentology), in the field, in the laboratory and at the level of interpretation.

ACKNOWLEDGEMENTS

I thank numerous undergraduate and graduate students for their keen scepticism and for posing many of the questions which I have discussed in this paper. My debt to the stimulating writings of Dr J.G. Evans will be apparent to all who have read this paper. C. Ellis, J.G. Evans and D. Gordon made valuable comments on a draft of this paper.

REFERENCES

Arnold, R.W. (1971) Cepaea nemoralis on the East Sussex South Downs, and the nature of area effects, Heredity 26, 277-298.

Atkins, W.R.G. and Lebour, M.V. (1923) The hydrogen ion concentration of the soil and of natural waters in relation to the distribution of snails, Scientific Proceedings of the Royal Dublin Society 17, 233-240.

Ayal, Y. and Safriel, U.N. (1983) Does a suitable habitat guarantee successful colonisation? Journal of Biogeography 10, 37-46.

Baker, R.E. (1968) The ecology of the wrinkled snail Helicella caperata (Mont.) on the Braunton Burrows sand dune system, Proceedings of the Malacological Society of London 38, 41-54.

Barker, G. and Webley, D. (1978) Causewayed camps and early neolithic economies in central southern England, Proceedings of the Prehistoric Society 44, 161-168.

Behrensmeyer, A.K. (1982) Time resolution in fluvial vertebrate assemblages, Paleobiology 8, 211-227.

Bell, M. (1981) Valley sediments and environmental change. In The Environment of Man: the Iron Age to the Anglo-Saxon Period M. Jones and G. Dimbleby (eds) pp. 75-91. Oxford: British Archaeological Reports British Series No. 87.

Bell, M. (1982) The effects of land-use and climate on valley sedimentation. In Climatic Change in Later Prehistory A.F. Harding (ed.) pp 127-142. Edinburgh: University Press.

Bishop, M.J. (1976) The Mollusca of some poor fens in Co. Cork, Ireland, Journal of Conchology 29, 57-60.

Bottema, S. and Ottaway, B.S. (1982) Botanical, malacological and archaeological zonation of settlement deposits at Gomolava. Journal of Archaeological Science 9, 221-246.

Boycott, A.E. (1934) The habitats of land Mollusca in Britain, Journal of Ecology 22, 1-38.

Burchell, J.P.T. (1961) Land shells and their role in dating deposits of post-glacial times in south-east England, Archaeological Newsletter 7, 34-38.

Burchell, J.P.T. and Davis, A.G. (1957) The molluscan fauna of some early post-glacial deposits in north Lincolnshire and Kent, Journal of Conchology 24, 164-170.

Burleigh, R. and Kerney, M.P. (1982) Some chronological implications of a fossil molluscan assemblage from a Neolithic site at Brook, Kent, England, Journal of Archaeological Science 9, 29-38.

Cain, A.J. (1971) Colour and banding morphs in subfossil samples of the snail Cepaea. In Ecological Genetics and Evolution R. Creed (ed), pp 65-92. Oxford: Blackwell Scientific Publications.

Cain, A.J. and Currey, J.D. (1963) Area effects in Cepaea, Philosophical Transactions of the Royal Society of London B 246, 1-81.

Cain, A.J. and Sheppard, P.M. (1954) Natural selection in Cepaea, Genetics 39, 89-116.

Cameron, R.A.D. (1970a) The survival, weight-loss and behaviour of three species of land snail in conditions of low humidity, Journal of Zoology London 160, 143-157.

Cameron, R.A.D. (1970b) The effect of temperature on the activity of three species of Helicid snail (Mollusca: Gastropoda). Journal of Zoology London 162, 303-315.

Cameron, R.A.D. (1978a) Interpreting buried land-snail assemblages from archaeological sites - problems and progress. In Research Problems in Zooarchaeology D.R. Brothwell, K.D. Thomas and J. Clutton-Brock (eds), pp. 19-23. London: University of London Institute of Archaeology Occasional Publication No. 3.

Cameron, R.A.D. (1978b) Differences in the sites of activity of coexisting species, Journal of Conchology 29, 273-278.

Cameron, R.A.D. and Morgan-Huws, D.I. (1975) Snail faunas in the early stages of a chalk grassland succession, Biological Journal of the Linnean Society 7, 215-229.

Cheatum, E.P. And Allen, D. (1964) Limitations in palaeoecological reconstruction utilising data from fossil non-marine molluscs. In The Reconstruction of Past Environments J.J. Hester and J. Schoenwetter (eds) pp. 31-33. Fort Burgwin Research Center Publication No. 3.

Coope, G.R. and Sands, C.H.S. (1966) Insect faunas of the last glaciation from the Tame Valley, Warwickshire, Proceedings of the Royal Society of London B 165, 389-412.

Cowie, R.H. (1980) Observations on the dispersal of two species of British land snail, Journal of Conchology 30, 201-208.

Currey, J.D. and Cain, A.J. (1968) Studies on Cepaea. IV. Climate and selection of banding morphs in Cepaea from the Climatic Optimum to the present day, Philosophical Transactions of the Royal Society of London B 253, 483-498.

Delcourt, H.R., Delcourt, P.A. and Webb, T. III (1983) Dynamic plant ecology: the spectrum of vegetational change in space and time. Quaternary Science Reviews 1, 153-175.

Dimbleby, G.W. (1957) Pollen analysis of terrestrial soils. The New Phytologist 56, 12-28.

Dimbleby, G.W. (1961) Soil pollen analysis, Journal of Soil Science 12, 1-11.

Dimbleby, G.W. and Evans, J.G. (1974) Pollen and land snail analysis of calcareous soils, Journal of Archaeological Science 1, 117-133.

Edwards, K.J. (1979) Palynological and temporal inference in the context of prehistory, with special reference to the evidence from lake and peat deposits, Journal of Archaeological Science 6, 255-270.

Ellis, C. (1983) Dry-valley fills at Asham and the Devils Dyke. Excursion Guide, pp. 9-23. Brighton: 4th International Flint Symposium.

Evans, J.G. (1969a) The exploitation of molluscs. In The Domestication and Exploitation of Plants and Animals. P.J. Ucko and G.W. Dimbleby (eds), pp. 479-484. London: Duckworth.

Evans, J.G. (1969b) Land and freshwater Mollusca in Archaeology: chronological aspects, World Archaeology 1, 170-183.

Evans, J.G. (1970) Interpretation of land snail faunas, Bulletin of the University of London Institute of Archaeology Nos. 8 and 9, 109-116.

Evans, J.G. (1971a) Durrington Walls: the pre-henge environment. In Durrington Walls: Excavations 1966-1968 G.J. Wainwright and I.H. Longworth, pp. 229-337. Reports of the Research Committee of the Society of Antiquaries of London No. 29. London: Society of Antiquaries.

Evans, J.G. (1971b) Habitat change on the calcareous soils of Britain: the impact of Neolithic man. In Economy and Settlement in Neolithic and Bronze Age Britain and Europe D.D.A. Simpson (ed), pp. 27-73. Leicester: University Press.

Evans, J.G. (1972) Land Snails in Archaeology. London: Seminar

Press.

Evans, J.G. (1978) Comment (upon paper by Barker and Webley 1978, op. cit.) Proceedings of the Prehistoric Society 44, 185-186.

Evans, J.G. and Jones, H. (1973) Sub-fossil and modern land-snail faunas from rock-rubble habitats. Journal of Conchology 28, 103-129.

Evans, J.G. and Jones, H. (1979) Mount Pleasant and Woodhenge: the Land Mollusca. In Mount Pleasant, Dorset: Excavations 1970-1971 G.J. Wainwright, pp. 190-213. Reports of the Research Committee of the Society of Antiquaries of London No. 37. London: Society of Antiquaries.

Evans, J.G. and Valentine, K.W.G. (1974) Ecological changes induced by prehistoric man at Pitstone, Bucks. Journal of Archaeological Science 1, 343-351.

Fowler, P.J. and Evans, J.G. (1967) Plough-marks, lynchets and early fields. Antiquity 41, 289-301.

Gould, S.J. (1976) Palaeontology plus ecology as palaeobiology. In Theoretical Ecology: Principles and Applications, R.M. May (ed), pp. 218-236. Oxford: Blackwell Scientific Publications.

Grime, J.P., Blythe, G.M. and Thornton, J.D. (1970) Food selection by the snail Cepaea nemoralis L. In Animal Populations in Relation to their Food Resources, A. Watson (ed), pp. 73-99. British Ecological Society Symposium No. 10. Oxford: Blackwell Scientific Publications.

Kenward, H.K. (1978) The Analysis of Archaeological Insect Assemblages: a new approach. The Archaeology of York 19/1.

Kenward, H.K. (1982) Insect communities and death assemblages, past and present. In Environmental Archaeology in the Urban Context, A.R. Hall and H.K. Kenward (eds), pp. 71-78. London: Council for British Archaeology Research Report No. 43.

Kerney, M.P. (1963) Late-glacial deposits on the Chalk of South-east England. Philosophical Transactions of the Royal Society of London B 246, 203-254.

Kerney, M.P. (1966) Snails and man in Britain. Journal of Conchology 26, 3-14.

Kerney, M.P. (1972) The British distribution of Pomatias elegans (Müller). Journal of Conchology 27, 359-361.

Kerney, M.P. (1977) A proposed zonation scheme for late-glacial and Post-glacial deposits using land Mollusca, Journal of Archaeological Science 4, 387-390.

Kerney, M.P., Brown, E.H. and Chandler, T.J. (1964) The late-glacial and Post-glacial history of the Chalk escarpment near Brook, Kent. Philosophical Transactions of the Royal Society of London B 248, 135-204.

Kerney, M.P., Preece, R.C. and Turner, C. (1980) Molluscan and plant biostratigraphy of some late Devensian and Flandrian deposits in Kent. Philosophical Transactions of the Royal Society of London B 291, 1-43.

Lasker, H. (1976) Effects of differential preservation of the measurement of taxonomic diversity, Paleobiology 2, 84-93.

Lubell, D., Hassan, F.A., Gautier, A. and Ballois, J.-L. (1976) The Capsian escargotières, Science 191, 910-920.

MacArthur, R.H. and Wilson, E.O. (1967) The Theory of Island Biogeography. Princeton: University Press.

Machin, J. (1967) Structural adaptations for reducing water-loss in three species of terrestrial snail, Journal of Zoology London 152, 55-65.

Mason, C.F. (1970) Snail populations, beech litter production, and the role of snails in litter decomposition, Oecologio 5, 215-239.

Mason, C.F. (1971) Respiration rates and population metabolism of woodland snails, Oecologia 7, 80-94.

May, R.M. (1975) Stability and Complexity in Model Ecosystems. 2nd edition. Princeton: University Press.

Mordan, P.B. (1977) Factors affecting the distribution and abundance of Aegopinella and Nesovitrea (Pulmonata: Zonitidae) at Monks Wood National Nature Reserve, Huntingdonshire, Biological Journal of the Linnean Society 9, 59-72.

Murphy, P. (1983) Studies of the environment and economy of a Bronze Age Fen-edge site at West Row, Mildenhall, Suffolk: a preliminary report, Circaea 1, 49-60.

Paul, C.R.C. (1975) The ecology of Mollusca in ancient woodland. 1. The fauna of Hayley Wood, Cambridgeshire, Journal of Conchology 28, 301-327.

Paul, C.R.C. (1978a) The ecology of Mollusca in ancient woodland. 2. Analysis of distributions and experiments in Hayley Wood,

Cambridgeshire. Journal of Conchology 29, 281-294.

Paul, C.R.C. (1978b) The ecology of Mollusca in ancient woodland. 3. Frequencies of occurrence in west Cambridgeshire woods, Journal of Conchology 29, 295-300.

Pennington, W. (1975) The effect of Neolithic man on the environment in north-west England: the use of absolute pollen diagrams. In The Effect of Man on the Landscape: the Highland Zone J.G. Evans, S. Limbrey and H. Cleere (eds), pp. 74-86. London: Council for British Archaeology Research Report No. 11.

Preece, R.G. (1980) The biostratigraphy and dating of the tufa deposit at the Mesolithic site at Blashenwell, Dorset, England. Journal of Archaeological Science 7, 345-362.

Preece, R.C. (1981) The value of shell microsculpture as a guide to the identification of land Mollusca from Quaternary deposits. Journal of Conchology 30, 331-337.

Reed, C.A. (1962) Snails on a Persian hillside: ecology, prehistory and gastronomy. Postilla, Yale Peabody Museum 66, 1-20.

Rees, W.J. (1965) The aerial dispersal of Mollusca, Proceedings of the Malacological Society of London 36, 269-282.

Rymer, L. (1978) The use of uniformitarianism and analogy in palaeoecology, particularly pollen analysis. In Biology and Quaternary Environments, D. Walker and J.C. Guppy (eds), pp. 245-257. Canberra: Australian Academy of Science.

Segal, E. (1961) Acclimation in molluscs, American Zoologist 1, 235-244.

Shotton, F.W. (1978) Archaeological inferences from the study of alluvium in the Lower Severn/Avon valleys. In The Effect of Man on the Landscape: the Lowland Zone, S. Limbrey and J.G. Evans (eds), pp. 27-32. London: Council for British Archaeology Research Report No. 21.

South, A. (1974) Changes in composition of the terrestrial mollus fauna. In The Changing Flora and Fauna of Britain, D.L. Hawksworth (ed), pp. 255-274. Systematics Association Special Volume No. 6. London and New York: Academic Press.

Sparks, B.W. (1961) The ecological interpretation of Quaternary non-marine Mollusca, Proceedings of the Linnean Society of London 172, 71-80.

Sparks, B.W. (1964) Non-marine Mollusca and Quaternary ecology. Journal of Animal Ecology 33, 87-98. (Jubilee Symposium Supplement).

Sparks, B.W. (1969) Non-marine Mollusca and Archaeology. In Science in Archaeology (2nd edition), D. Brothwell and E.S. Higgs (eds), pp. 395-406. London: Thames and Hudson.

Spencer, P.J. (1975) Habitat change in coastal sand-dune areas: the molluscan evidence. In The Effect of Man on the Landscape: the Highland Zone, J.G. Evans, S. Limbrey and H. Cleere (eds), pp. 96-103. London: Council for British Archaeology Research Report No. 11.

Thomas, K.D. (1977) A preliminary report on the Mollusca from the lynchet section. In Excavations at Bishopstone, M.G. Bell. Sussex Archaeological Collections 115, 258-264.

Thomas, K.D. (1978) Population studies on molluscs in relation to environmental archaeology. In Research Problems in Zooarchaeology D.R. Brothwell, K.D. Thomas and J. Clutton-Brock (eds), pp. 9-18. London: University of London Institute of Archaeology Occasional Publication No. 3.

Thomas, K.D. (1982) Neolithic enclosures and woodland habitats on the South Downs in Sussex, England. In Archaeological Aspects of Woodland Ecology, M. Bell and S. Limbrey (eds), pp. 147-170. Oxford: British Archaeological Reports International Series No. 146.

Thomas, W.R. and Foin, T.C. (1982) Neutral hypotheses and patterns of species diversity: fact or artifact? Paleobiology 8, 44-55.

Würeborn, I. (1969) Land molluscs and their environments in an oligotrophic area in southern Sweden. Oikos 20, 461-479.

Wolda, H., Zweep, A. and Schuitema, K.A. (1971) The role of food in the dynamics of populations of the landsnail Cepaea nemoralis. Oecologia 7, 361-381.

Sample number:	1	3	7	10	14	18	22	26	28	29	30
Sample interval (cm):	10-14	26-30	50-55	65-70	85-90	105-110	125-130	145-150	155-158	158-161	161-167
Sample weight (kg):	0.96	0.71	0.96	0.90	0.90	0.93	1.00	0.95	0.86	0.78	0.83
Pomatias elegans (Müller)	22	7	8	10	29	41	41	65	35	45	11
Carychium tridentatum (Risso)	-	-	-	-	-	-	-	-	-	-	-
Cochlicopa lubricella (Porro)	-	-	-	-	-	-	-	-	1	-	7
Cochlicopa sp.	5	3	1	1	3	-	2	16	3	7	14
Truncatellina cylindrica (Ferussac)	-	-	-	-	-	-	-	-	-	4	1
Vertigo pygmaea (Draparnaud)	3	1	-	-	-	-	-	2	2	4	6
Pupilla muscorum (Linne)	240	121	41	44	45	12	31	191	920	694	421
Vallonia costata (Müller)	32	65	39	16	65	27	13	23	58	57	82
Vallonia excentrica Sterki	108	68	20	26	53	39	16	18	16	24	10
Vallonia spp.	8	-	-	-	13	-	-	-	-	3	-
Ena obscura (Müller)	-	1	-	-	-	-	-	-	-	-	-
Punctum pygmaeum (Draparnaud)	-	-	-	-	-	-	-	1	-	1	-
Discus rotundatus (Müller)	-	-	-	1	-	2	-	-	-	-	-
Vitrea contracta (Westerlund)	-	1	-	-	-	-	-	-	-	-	-
Nesovitrea (Perpolita) hammonis (Ström)	-	-	-	-	-	-	-	-	1	-	-
Limacidae	3	3	1	-	6	10	4	-	-	1	1
Cecilioides acicula (Müller)	2	5	6	-	-	-	-	-	-	-	-
Clausilia bidentata (Ström)	3	1	1	1	5	6	12	9	12	7	2
Candidula intersecta (Poiret)	7	-	-	-	-	-	-	-	-	-	-
Cernuella virgata (Da Costa)	66	4	-	-	-	-	-	-	-	-	-
Helicella itala (Linne)	-	-	4	13	15	15	15	13	33	21	13
"Helicella" spp.	39	-	-	-	-	-	-	-	-	-	-
Monacha cartusiana (Müller)	-	4	12	10	10	24	11	4	37	5	4
Trichia hispida (Linne)	21	158	96	32	72	50	23	15	53	45	61
Cepaea nemoralis (Linne)	-	-	-	-	-	-	-	-	+	-	-
Cepaea sp.	-	-	4	2	2	16	8	8	6	8	2
Helix aspersa Müller	-	2	-	-	-	-	-	-	-	-	-

Table 1

The numbers of shells of various species of land snails extracted from samples from the Iron Age lynchet at Bishopstone, Sussex. (From Thomas 1977).

Sample:		1	3	7	10	14	18	22	26	28	29	30
Pomatias elegans												
	(a)	16.7	6.9	14.6	27.5	39.1	72.5	100.0	76.4	12.9	21.0	7.3
	(b)	33.8	14.7	11.8	16.2	47.0	64.7	60.3	100.0	60.3	85.3	19.1
	(c)	31.1	12.9	8.3	11.4	14.5	45.6	42.5	70.5	71.0	100.0	11.4
Pupilla muscorum												
	(a)	54.8	34.9	22.5	36.1	18.0	6.3	22.5	66.9	100.0	95.8	84.8
	(b)	23.4	15.9	4.0	4.6	4.7	1.2	2.9	18.8	100.0	83.2	47.4
	(c)	17.5	11.9	2.4	2.7	2.8	0.8	1.7	11.3	100.0	83.2	23.7
Vallonia costata												
	(a)	27.9	71.6	81.9	50.0	100.0	54.4	36.3	30.9	24.0	29.9	63.2
	(b)	33.3	92.9	41.4	18.2	72.7	29.3	13.1	24.2	67.7	73.7	100.0
	(c)	33.9	94.6	33.7	14.8	59.3	23.9	10.7	19.8	95.9	100.0	67.9
Vallonia excentrica												
	(a)	100.0	79.3	44.5	86.5	86.5	83.4	47.1	25.4	7.2	13.5	8.3
	(b)	100.0	85.7	18.7	25.9	52.7	37.5	14.3	17.0	17.0	27.7	10.7
	(c)	100.0	85.7	15.0	20.7	42.1	30.0	11.4	13.6	22.5	36.8	7.2
Helicella itala												
	(a)	-	-	20.0	97.6	55.3	72.9	100.0	42.3	32.9	27.0	24.7
	(b)	-	-	10.5	36.8	44.7	42.1	39.5	36.8	100.0	71.1	42.1
	(c)	-	-	6.3	22.0	26.8	25.2	23.6	45.7	100.0	70.9	21.3
Trichia hispida												
	(a)	8.9	86.4	100.0	49.7	54.8	50.2	31.6	9.9	10.9	11.9	23.3
	(b)	9.9	100.0	45.0	16.2	36.0	24.3	10.4	7.2	27.9	26.1	32.9
	(c)	9.9	100.0	36.0	13.0	28.8	19.5	8.3	5.8	39.3	34.8	22.0

Table 2

The relative frequencies of selected species of land snails from the Iron Age lynchet at Bishopstone, Sussex (see Table 1). Three methods have been used to calculate abundance: (a) percentage frequency in the total assemblage extracted from each soil sample, (b) numbers of shells per kg of soil, (c) numbers of shells per kg of soil per cm of soil. In each case, the highest estimate has been taken to be 100 per cent; all other values in each series are expressed as a proportion of this.

Figure 1

Simplified flow-chart showing the main stages in the analysis and interpretation of sub-fossil land-snail assemblages.

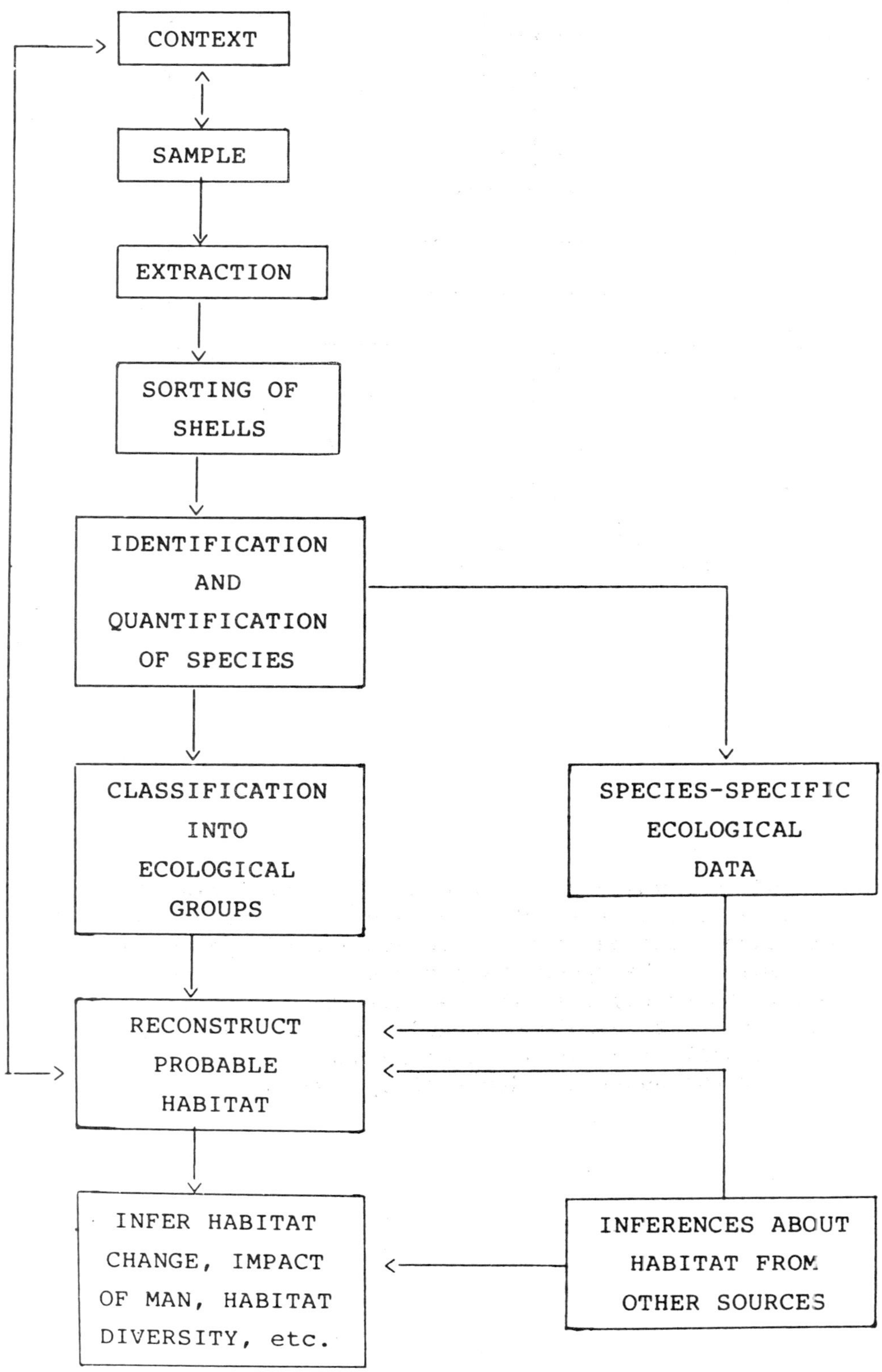

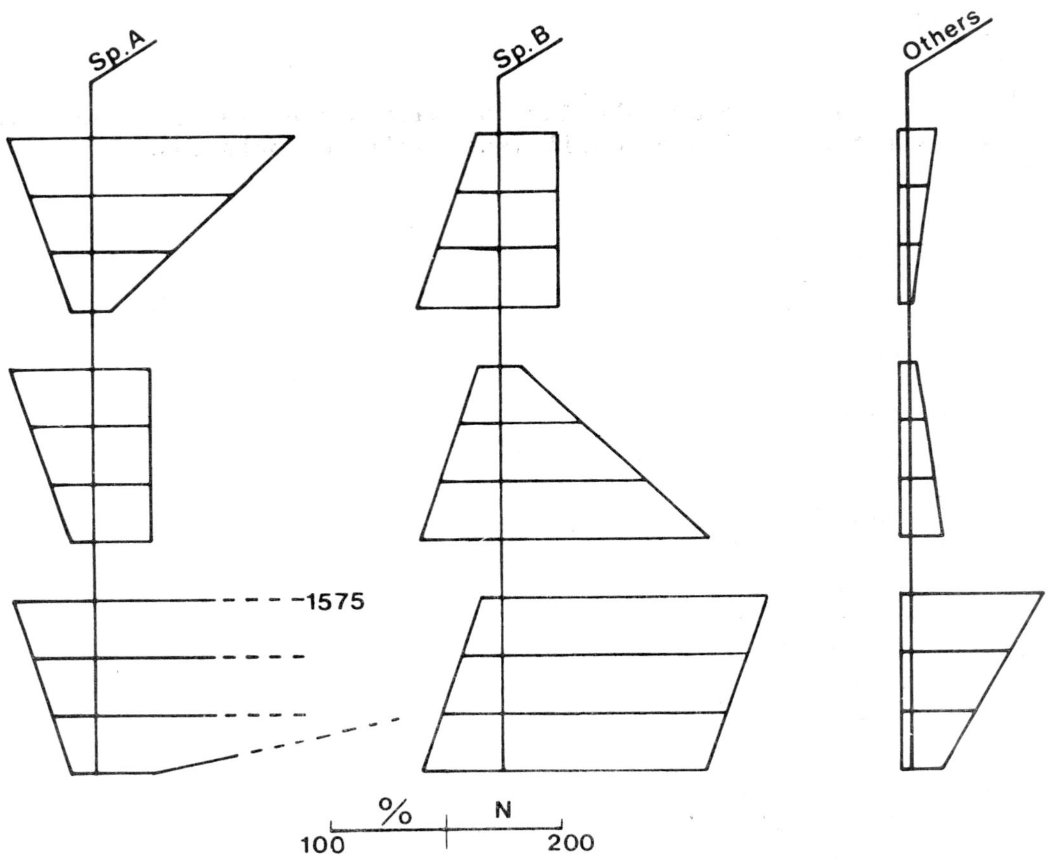

Figure 2

The quantification of assemblages of land snails: three examples showing the distorting effect of absolute abundance on percentage frequencies. In all cases, the percentage frequencies of species A, B and Others is the same. In the upper example species B appears to decline in frequency (percent) through a series of samples yet it remains constant in terms of absolute abundance. In the lower example species B is actually increasing in abundance but appears to be in decline because of the super-abundance of species A.

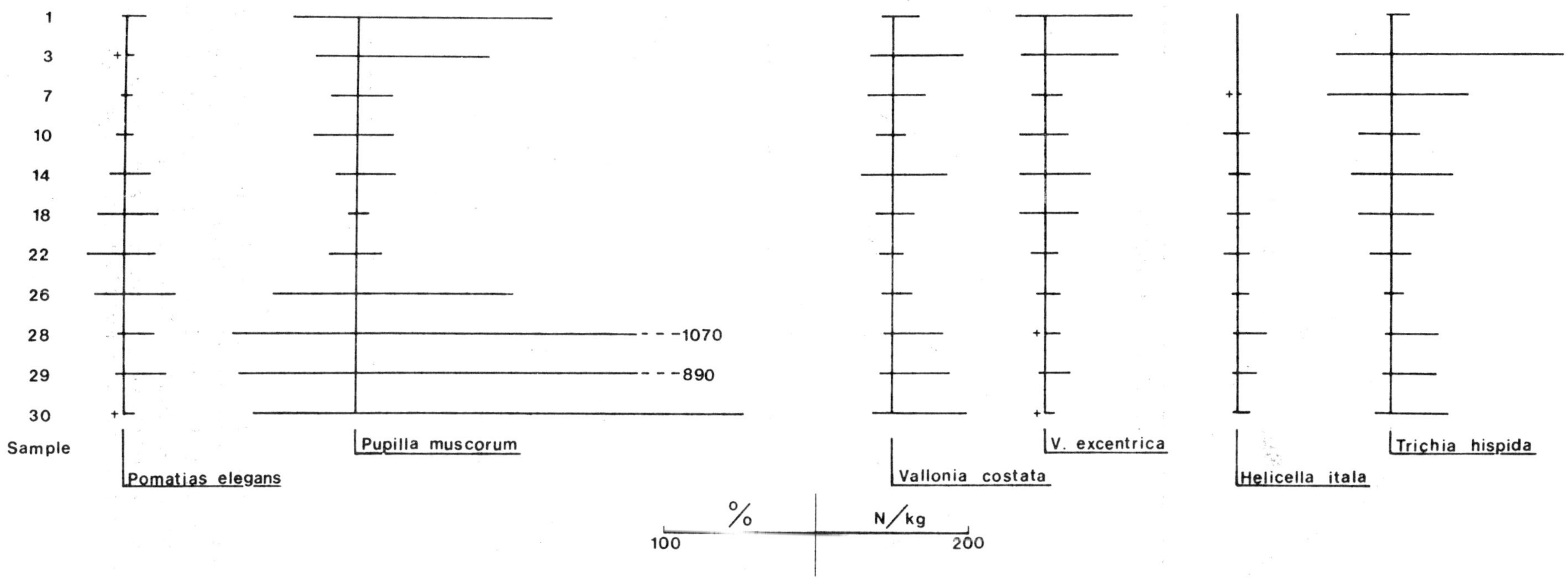

Figure 3

Bar histograms of species-abundance for selected species of land-snails from the Bishopstone lynchet (see Table 1). The data are plotted as relative (percentage) and absolute frequencies.

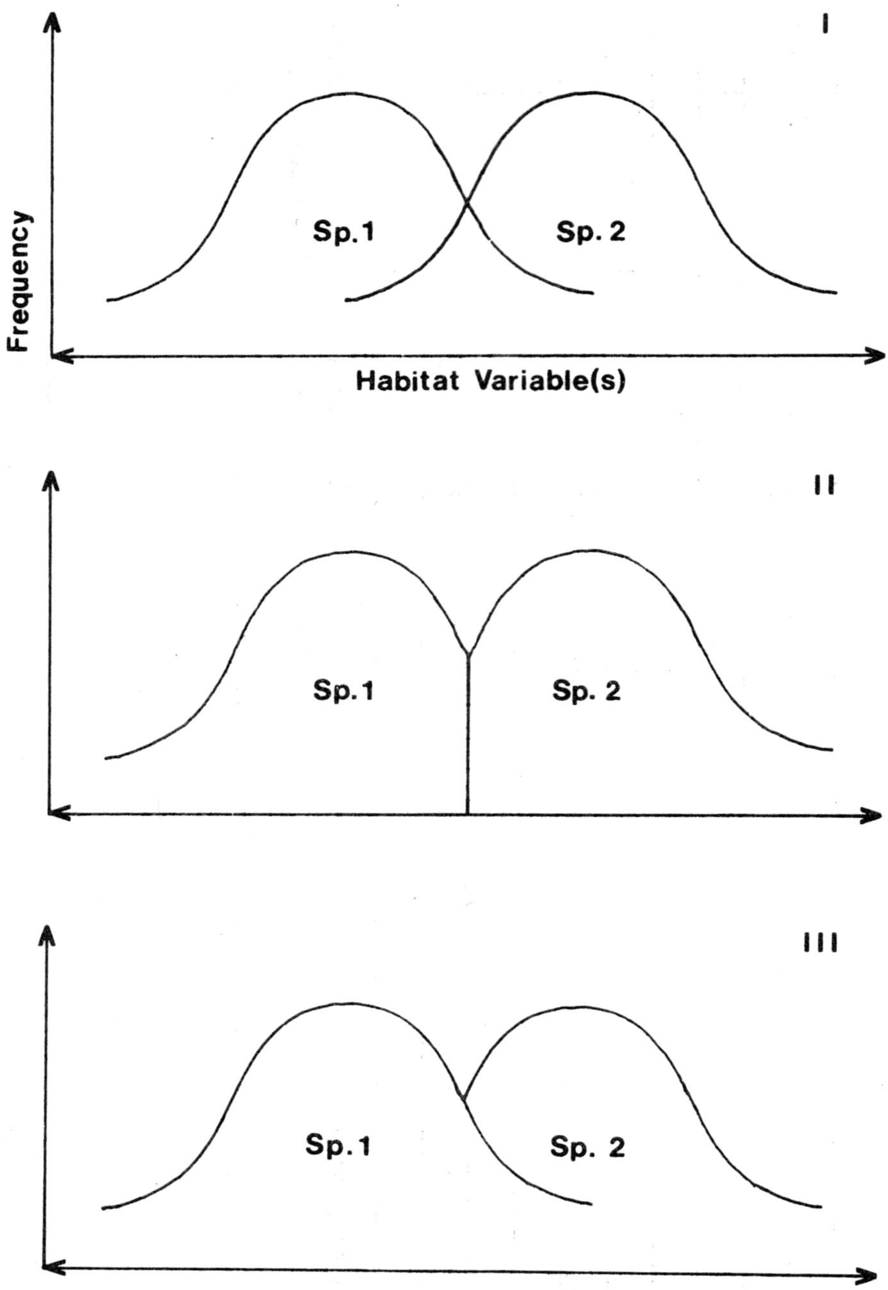

Figure 4

Limiting factors and species abundance. I: the range of habitat variables tolerated by two species which are either allopatric or non-interactive sympatric species. II: interactions between the two species causes mutual exclusion from parts of the species ranges. III: species 1 is competitively superior and excludes species 2 from part of its range when the two species are sympatric.

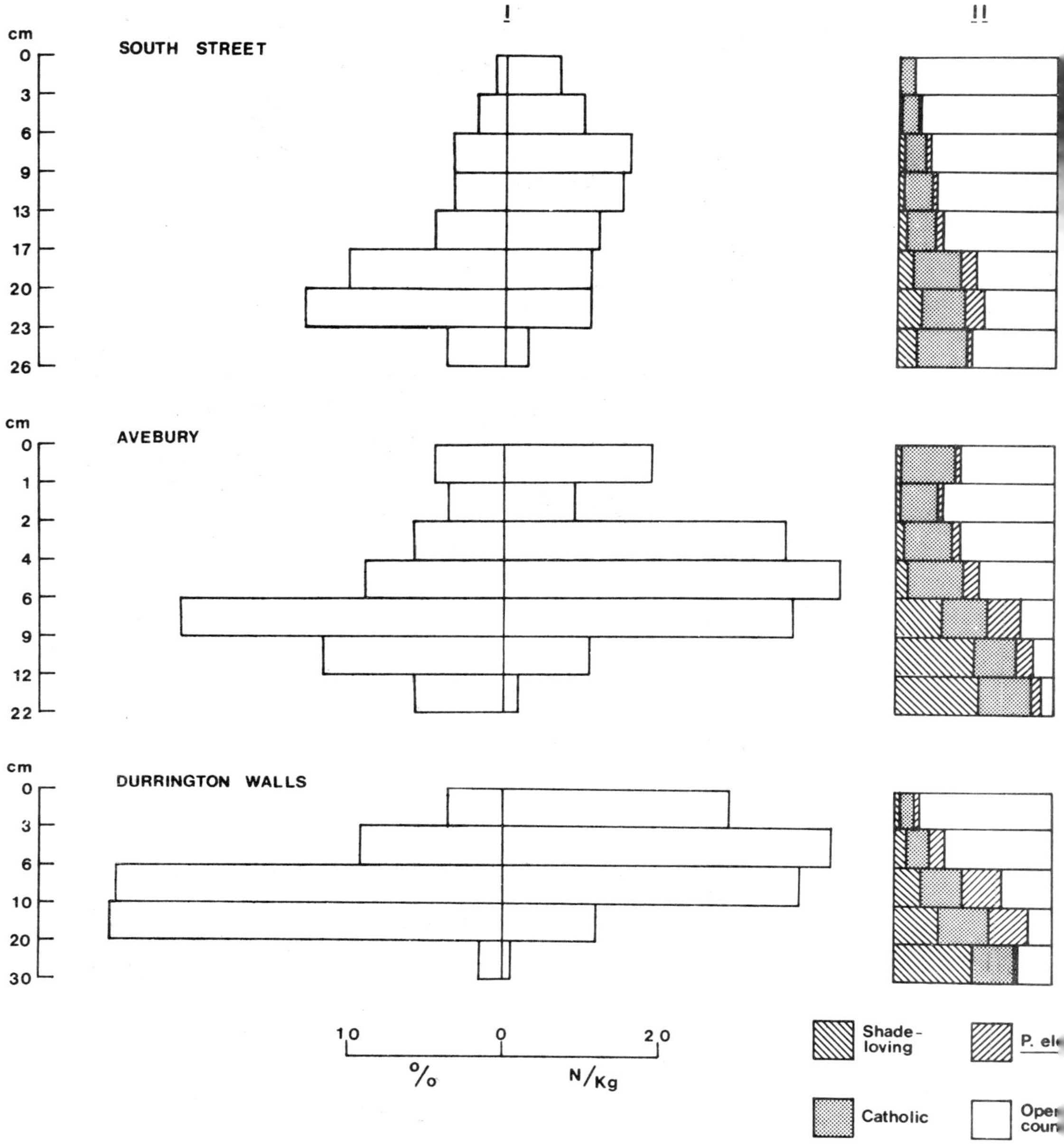

Figure 5

Mollusc histograms for the buried soils from the South Street long barrow, Avebury henge and Durrington Walls henge. (Data from Evans 1971a, 1971b, 1972).
I. percentage and absolute frequencies of Pomatias elegans. II. percentage frequencies of various ecological groups.

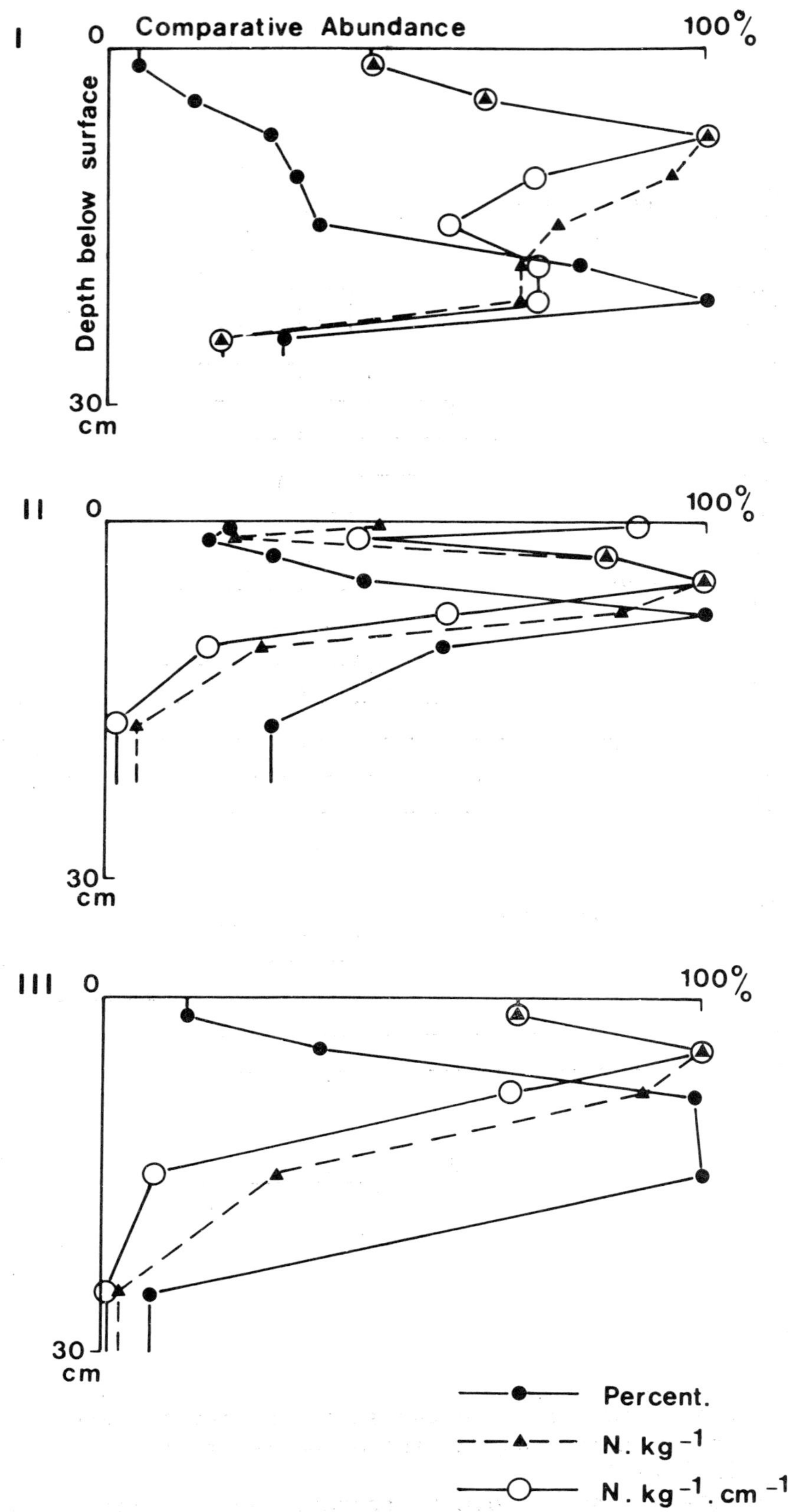

Figure 6

The relative frequencies of Pomatias elegans from the buried soils from: I: South Street long barrow; II: Avebury henge; III: Durrington Walls henge (see Figure 5). Three methods of quantification have been used: percentage frequencies; numbers of shells per kg of soil; numbers of shells per kg of soil per cm of soil. The highest value in each series has been taken as 100 percent; all other values have been adjusted to this. For clarity, the values have been plotted to the mid-point of each sampling interval.

A

	2	3	4	5	6
1	+0.703[a]	+0.278	-0.140	+0.788[b]	-0.179
2		+0.734[a]	-0.034	+0.858[c]	+0.171
3			+0.267	+0.453	+0.695[a]
4				-0.386	+0.439
5					-0.094

1 *Pomatias elegans*

2 *Pupilla muscorum*

3 *Vallonia costata*

4 *V. excentrica*

5 *Helicella itala*

6 *Trichia hispida*

[a] $P < 0.05$
[b] $P < 0.02$
[c] $P < 0.01$

B

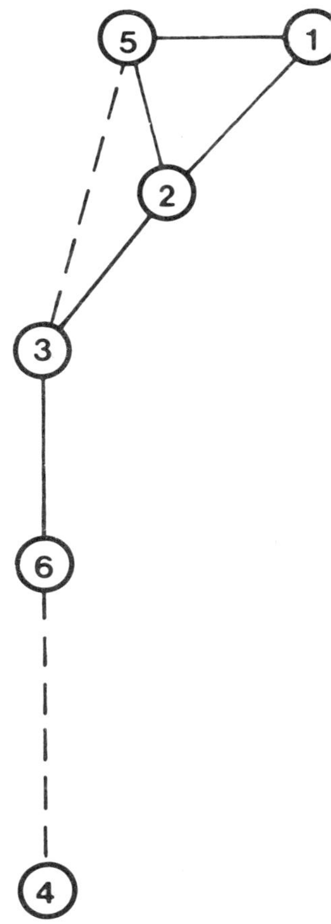

Figure 7

Analysis of the data from the Bishopstone lynchet (see Tables 1 and 2). A: Correlation matrix showing the correlation coefficients calculated for each pair of species, using the data in Table 2. Probabilities have been calculated from Fisher's z-transformation. B: Constellation diagram showing the correlations between the various species. Solid lines represent statistically significant correlations (see 7.A); the length of each line is inversely proportional to the value of the correlation coefficient between pairs of species.

Figure 8

Real and reconstructed events.
I: model of a 'real' event in which a habitat changes rapidly from one state to another; this basic model is the 'real' basis of the following models.
II: model in which, as a result of an event, Species B replaces Species A very rapidly; this rate of replacement is reflected in the sequence of sub-fossil assemblages, leading to the 'correct' interpretation.
III: model in which species A declines rapidly because its favoured habitat has diminished but Species B responds slowly because it has to colonise the new habitat; the interpretation of a series of sub-fossil assemblages showing these trends might be correct (not illustrated) or might suggest alternatives such as a time-transgressive event (III.a) or a series of seral changes (III.b).
IV: in this model, Species B responds slowly to the opportunities presented, allowing Species A to persist because levels of competition are low; the interpretation of a series of sub-fossil assemblages reflecting these changes might be correct (not shown), could be similar to those for model III, or could invoke a second event with a third type of environment (IV.c).
V: here the two species behave initially as in model IV but Species A rapidly declines once some threshold frequency of Species B has been reached; again, the interpretation of sub-fossil assemblages might be as for model IV, above, or could involve a further sophistication as shown in V.d, where the timing of the inferred second event is changed and the progression to Environment 2 is successional and not 'instantaneous'.

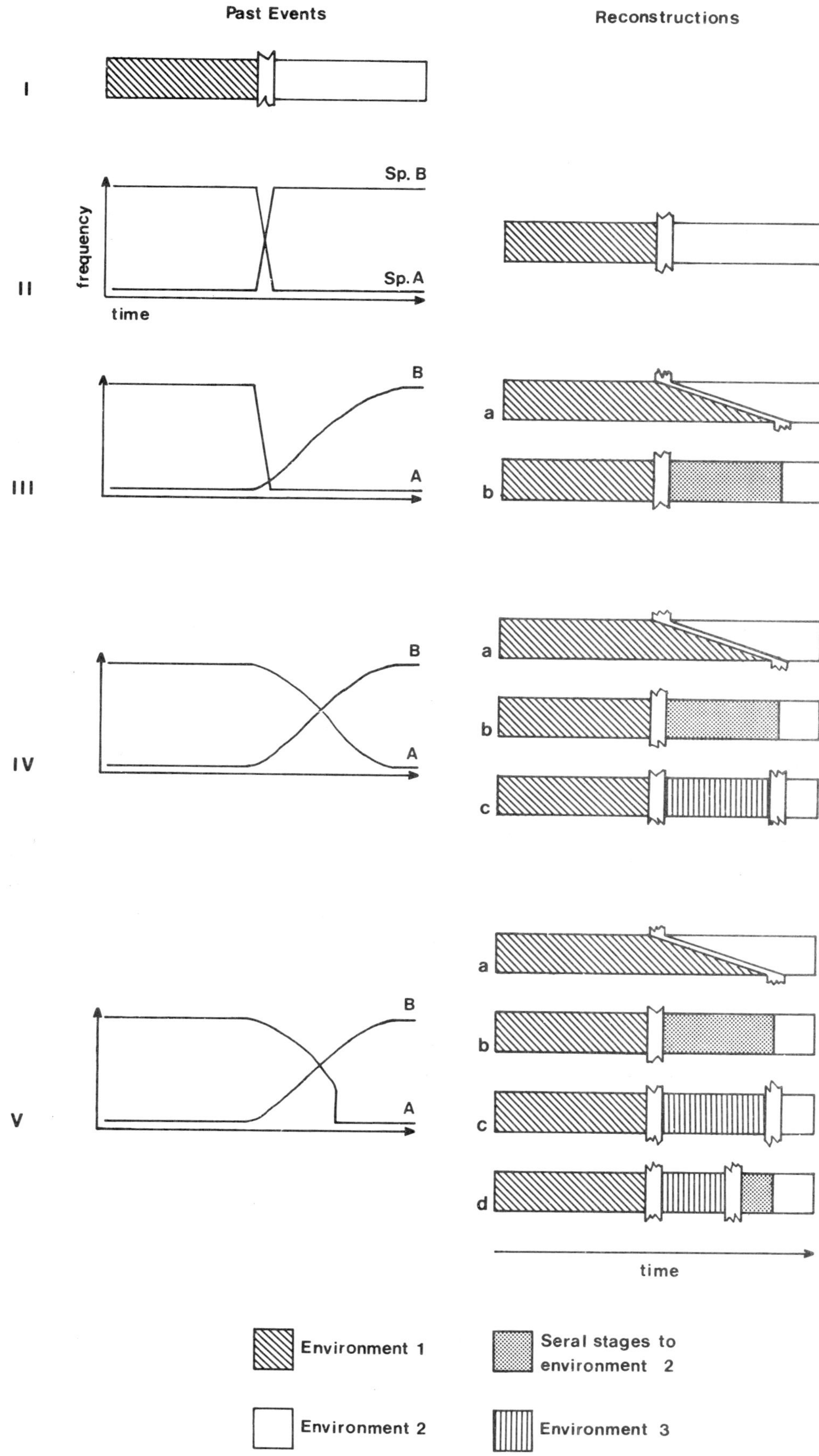
Past Events
Reconstructions
I
II
III
IV
V
frequency
time
Sp. B
Sp. A
B
A
a
b
c
d
time
Environment 1
Environment 2
Seral stages to environment 2
Environment 3

Figure 9

Possible trajectories of change in land-snail communities resulting from an extension of woodland clearance. Sampling location I is in a cleared area, location II is in an area which becomes cleared. The living snail communities at each site just prior to the clearance phase are taken as time-zero; subsequent changes are reflected in the age-stratified assemblages which are preserved in the soils. At site I it is assumed that there is no change in the molluscs over time. The possible changes at site II are numerous; they may be rapid or slow. The various blocks within the diagram show various phases of development of various trajectories of change; each block could represent the mollusc sequence which could be preserved in a soil buried at various times after clearance. Obviously, many other intermediate situations could have been illustrated, but some of the fundamental alternatives are presented here. Not all trajectories are equally probable, but it is not possible to choose, a priori, between them. Arrows connect those histograms for the living snails at any one time; the trajectories stop when the equilibrium communities (as represented at site I) have become established.

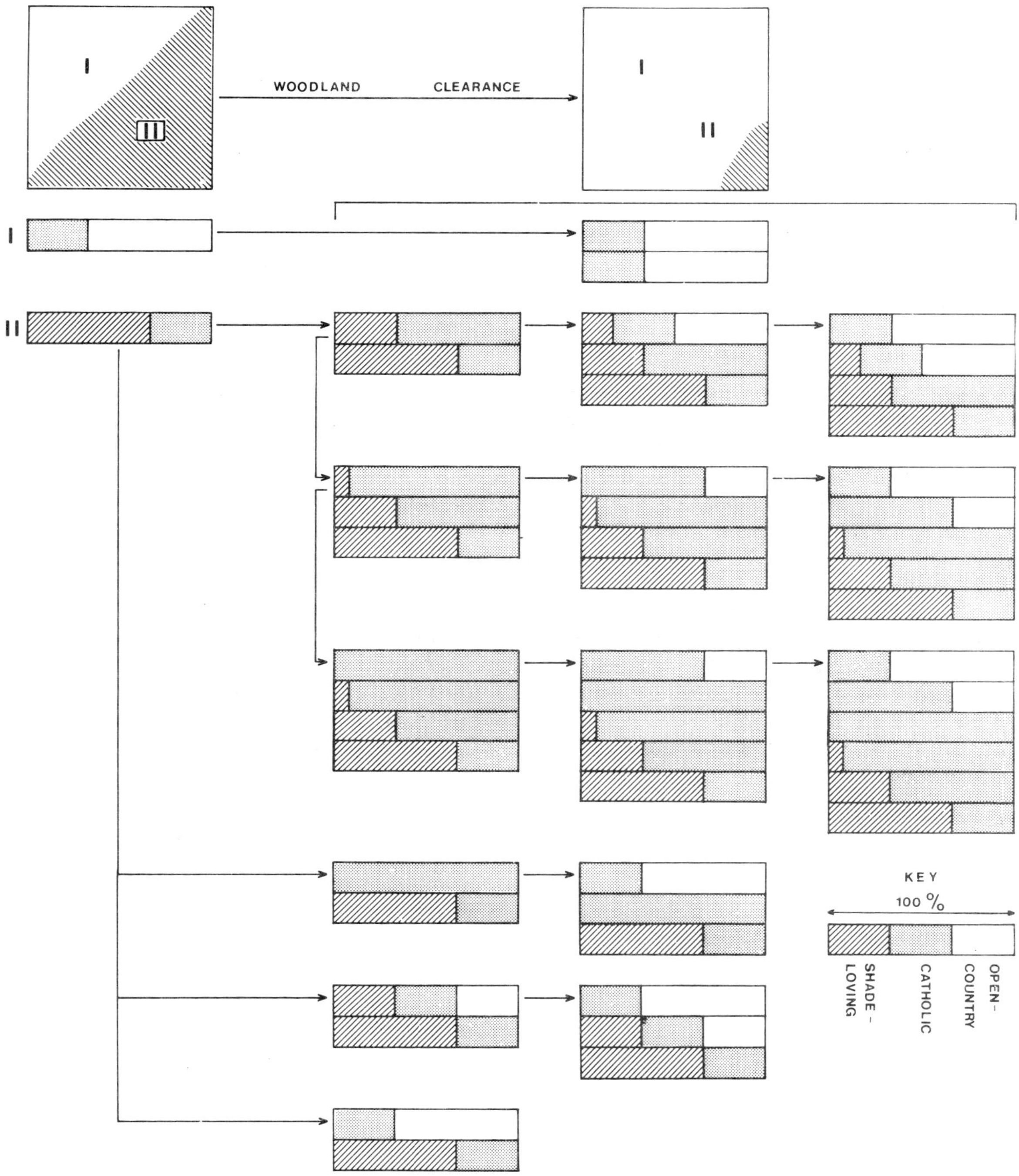

I
III
WOODLAND
CLEARANCE
I
II
I
II
KEY
100 %
SHADE-LOVING
CATHOLIC
OPEN-COUNTRY

Figure 10

Possible trajectories of change in land-snail communities resulting from a temporary localised clearing in woodland. 'S' is the sampling location. The living land-snail community in the woodland just before the creation of the clearance is taken to be time-zero; this and subsequent assemblages become incorporated into the soil as sub-fossil assemblages. The arrows link histograms showing changes over time in the living snail communities. Various trajectories are possible but not all are equally probable (e.g. it is not likely that the assemblages will show absolutely no ecological changes as a result of the environmental shifts; equally, it is improbable that the assemblages would reflect changes from fully shaded to fully open conditions, and then back again, over a short time-span). Models of this sort were used by Thomas (1982) in reconstructing the probable environmental setting of neolithic enclosures in Sussex, England.

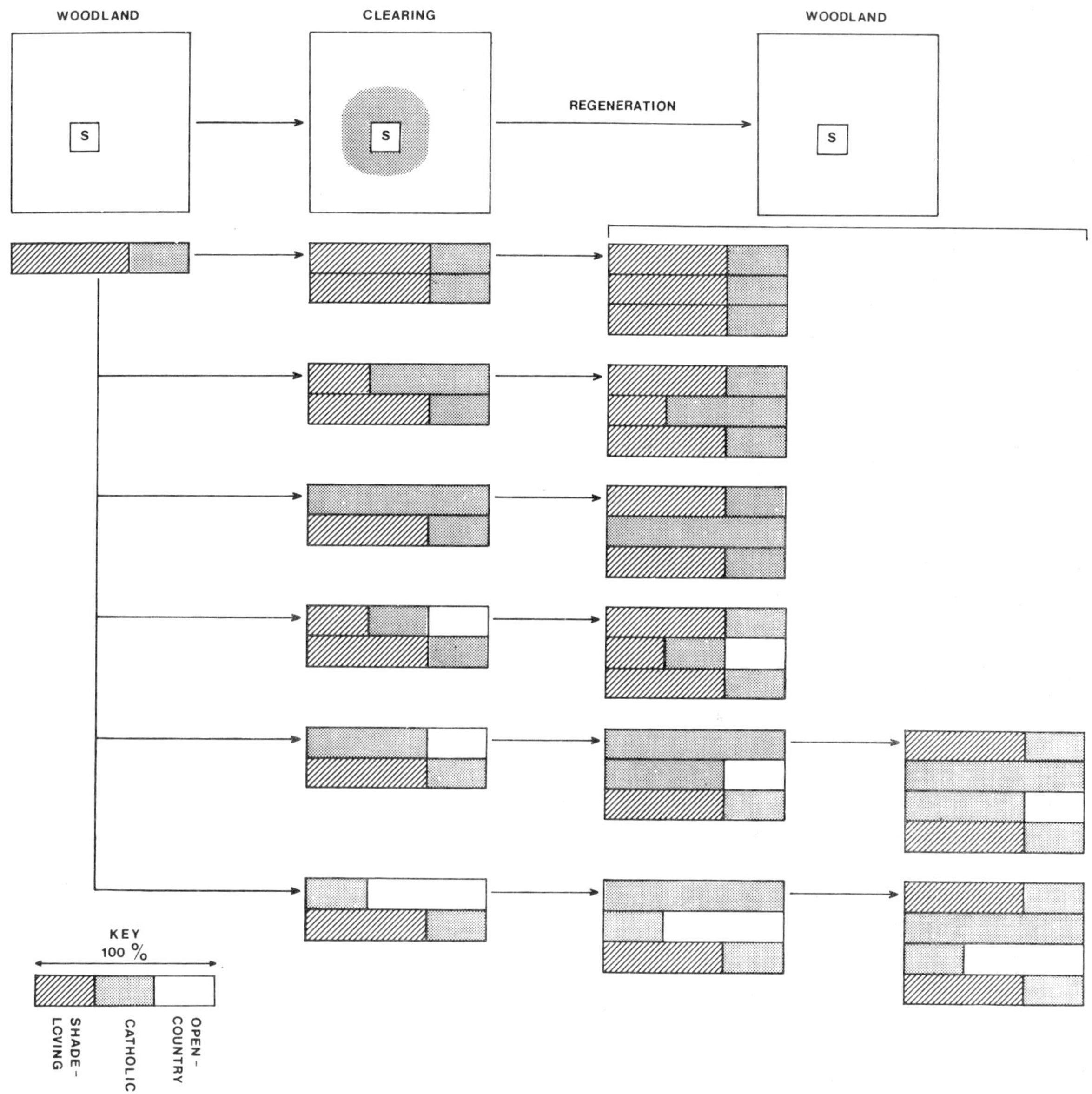
WOODLAND
CLEARING
WOODLAND
S
S
S
REGENERATION
KEY
100 %
SHADE-
LOVING
CATHOLIC
OPEN-
COUNTRY

Figure 11

Age (time)-depth relationships in land-snail assemblages from soils. Various models are illustrated (see text for discussion).

I: Linear time-depth function for the incorporation of stratified assemblages of shells.

II: A more realistic decreasing power-function.

III. The most realistic model, in which snail shells are lowered to a maximum limit, with age-compression of assemblages with increased depth, and probable age-mixed (and possibly ecologically-mixed) assemblages at the limiting depth.

IV: A model in which some older shells are moved upwards in the profile by worm action.

V: A development of IV in which younger shells are moved down by biological and other pedological processes.

VI: An extreme model of a thin soil with intense soil mixing.

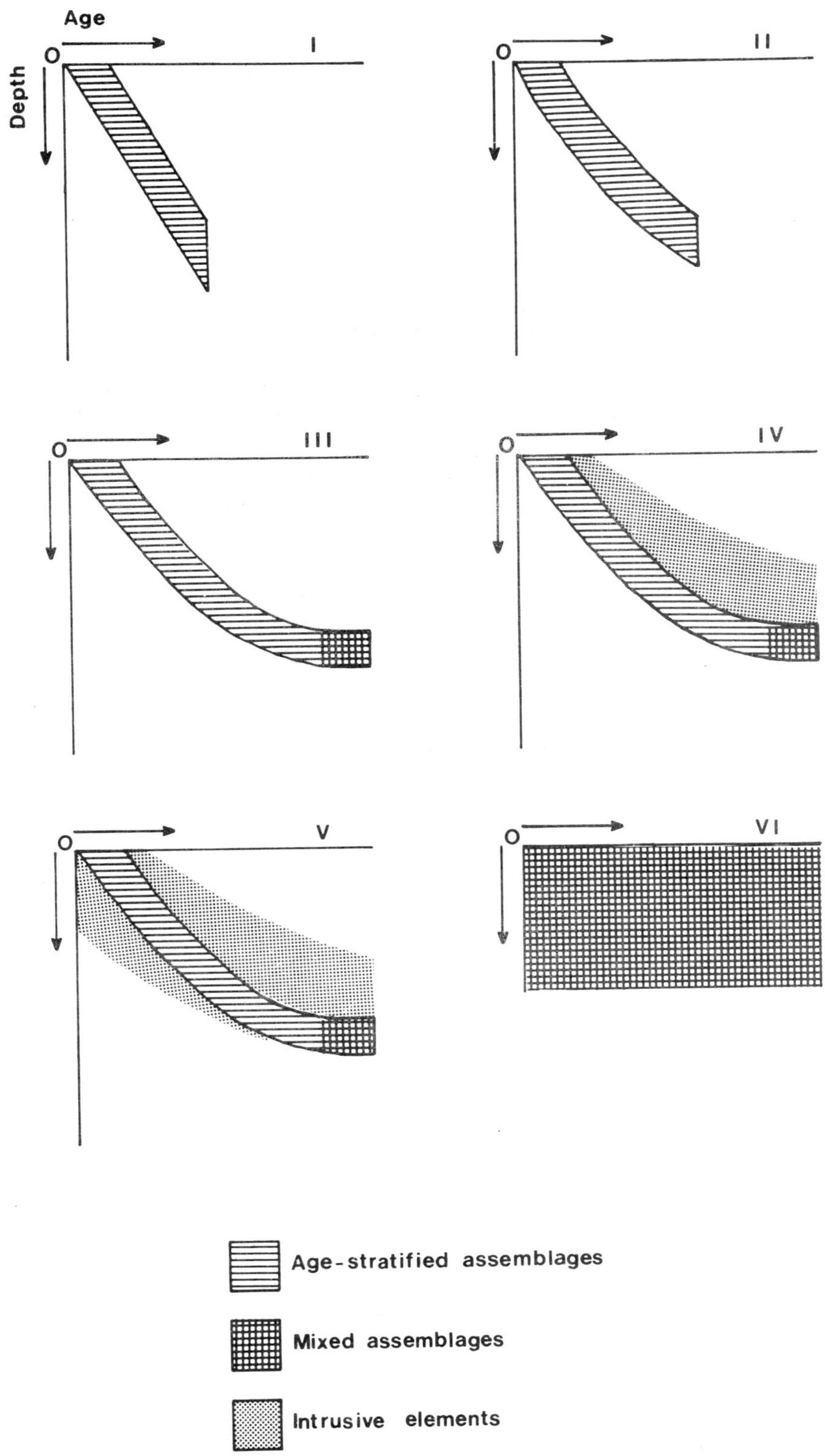
Age
Depth
O
I
II
III
IV
V
VI
Age-stratified assemblages
Mixed assemblages
Intrusive elements

Figure 12

Scales of environmental resolution.
A hypothetical model of a landscape with three major habitat units and a sampling location in a context which has accumulated and preserved assemblages of land snails. The context of the samples is located in a relatively small area of 'local environment'.

Catchment I: here, the molluscs would only be accumulated from species living within the context itself.

Catchment II: here, snail shells would be derived from those in the context and those of the local environment.

Etc. for Catchments III and IV.
A major problem lies in interpreting the true nature of the landscape even if, as in the case of catchment IV, the context is receiving information from all available habitats. Assemblages or communities living near to the sampling context would be over-represented in the final assemblage. It is also difficult to reconstruct the spatial arrangement of the habitat mosaic - even if it is detected.
The model applies to various scales of analysis:

Context	Local Envt.	Habitat 1	Habitat 2	Habitat 3
Arable field	Area of mixed farming	Open grassland	Woodland	Dense woods in valley
Pit	Settlement	Dry-valley floor	Valley sides	Upland pasture
Damp hollow	Lush vegetation	Ditch	Coarse vegetation	Arable field

The mollusc bar-charts, with the exception for that of catchment I, indicate only a few of the huge variety of possible combinations.

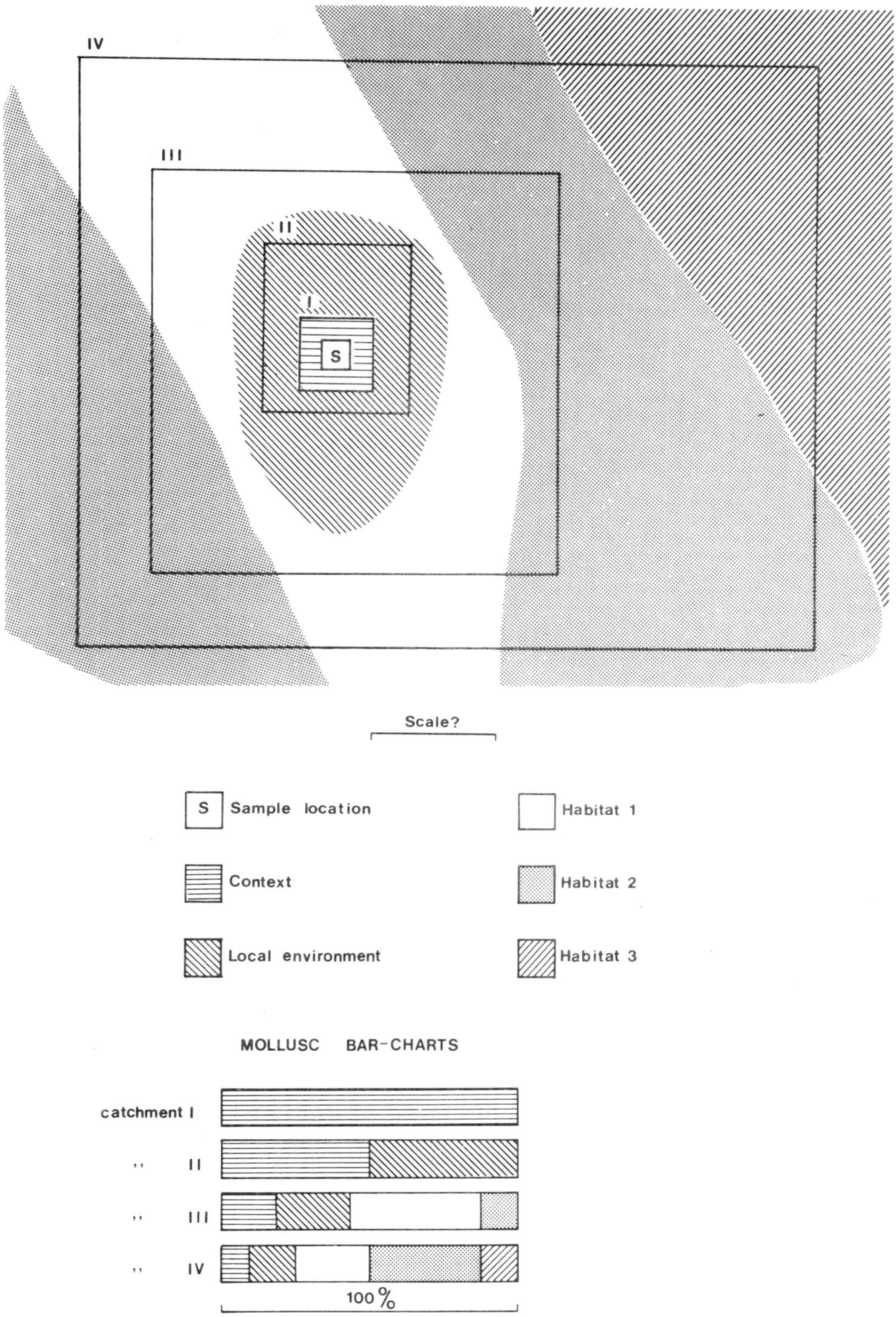
IV
III
II
I
S
Scale?
S Sample location
Context
Local environment
Habitat 1
Habitat 2
Habitat 3
MOLLUSC BAR-CHARTS
catchment I
,, II
,, III
,, IV
100%

FLANDRIAN MOLLUSCAN BIOSTRATIGRAPHY AND ITS APPLICATION TO DRY VALLEY INFILL DEPOSITS IN EAST SUSSEX.

Caroline Ellis
Department of Geology
Royal School of Mines
Imperial College of Science and Technology
University of London
London SW7 2AZ

ABSTRACT

Flandrian molluscan biozones have been established by work carried out in Kent (Kerney 1977; Kerney, Preece and Turner 1980) and have been correlated with the pollen biozones by radiocarbon dating. These molluscan zones have enabled biostratigraphical correlations between deposits in different parts of southern England and have been applied to dry valley infill deposits in the South Downs at Asham Quarry near Lewes and at the Devil's Dyke near Brighton. Sections through these infill deposits reveal Late-glacial deposits of shattered chalk fragments and chalky silts of solifluxion origin, overlain by a Postglacial soil that has been buried by colluvial (hillwash) deposits. The Postglacial deposits can all be assigned to molluscan biozones d, e and f, representing a time period of approximately 6000 B.P. to the present day.

INTRODUCTION: PREVIOUS WORK

Most Quaternary palaeoenvironmental studies have been carried out in non-calcareous areas on fossil pollen, plant macrofossils and coleopteran remains, thereby excluding the chalk and limestone areas of Britain. In calcareous areas it is only under exceptional conditions of waterlogging or in subaqueous environments that pollen is preserved. These sediments are usually totally oxidized, suitable only for the preservation of mollusc shells, charcoal and bone.

Some palaeoecological work has been carried out in these areas. Evans (1972) has studied stratified molluscan remains in an archaeological context on the Chilterns and on Salisbury Plain. On the North Downs in Kent, Kerney (1963, 1977), Kerney, Brown and Chandler (1964), and Kerney, Preece and Turner (1980) have studied the molluscan succession in dry valleys from the Late-glacial to the present day. On the South Downs, Thomas (1982) has studied neolithic enclosures and associated fossil molluscs and Bell (1981) has used molluscs to reconstruct the prehistoric land use around the settlement areas at Itford Hill and at Bullock Down. The molluscan succession in dry valleys has been studied by Ellis (1983) to compare with the biozones established in the North Downs by Kerney (1977). These biozones span the Late-glacial and the Flandrian and are based on the occurrence of certain indicator species and their changing dominance. In zones y to d inclusive, the changes are due to migration reflecting climatic and natural vegetation change whereas the faunas of biozones e and f result from the influence of man on the environment: clearing areas of woodland, creating open-country conditions and introducing alien species. The biozones are shown in an idealised diagram with modifications from work in Sussex (Fig. 1).

In the Late-glacial period (molluscan biozones y and z) there is a restricted open-country assemblage. Biozone z differs from y by the appearance of the thermophile, Abida secale (Draparnaud). The early Postglacial biozones a and b are characterised by the decline of the open-country genera and the expansion of woodland species including Discus ruderatus (Ferussac), a species now extinct in Britain. In biozones c and d the woodland genera predominate until forest clearance which marks the base of biozone e. These anthropogenic biozones (e and f) are characterised by the re-expansion of the open-country genera and the appearance and subsequent expansion of many of the introduced species. Within zone e, Monacha cartusiana (Müller) appears, this species is thought to be a molluscan "weed" (Kerney 1970) that colonises areas after they have been cleared for cultivation. Zone f is defined by the appearance of Helix aspersa Muller (the common garden snail) which is a Mediterranean species that is thought to have been introduced into lowland Britain during the Romano-British period. Throughout this zone many other alien molluscan species appear. Work in Sussex has established that Cernuella virgata (da Costa) and Candidula gigaxii (Pfeiffer) arrive after H. aspersa, and that Candidula intersecta (Poiret) and Monacha cantiana (Montagu) are very recent immigrants as they are found only in the topsoil (Ellis 1983). Other very late zone f introductions into Britain include Cochicella acuta (Müller) which is found very occasionally in fossil contexts, the age of which are uncertain, and Hygromia limbata (Draparnaud) and H. cinctella (Draparnaud), southern European species that are known to be twentieth century introductions to the Torquay area (Kerney 1966).

The establishment of these molluscan biozones has allowed biostratigraphical correlations between deposits in different parts of southern England. They have been applied in East Sussex to sections at Asham Quarry near Lewes and at the Devil's Dyke near Brighton. Both of these sites are scarp face dry valleys that have been infilled first by Late-glacial chalky solifluxion deposits and slope wash, and later by Postglacial colluvial deposits, often burying a soil that formed during the woodland biozones prior to clearance. The earliest Postglacial sediments found at these sites contain a zone d molluscan assemblage. The absence of the earlier biozones (a - c inclusive) is attributed to physical-chemical processes within the soil, soil loss through erosion and the gradual destruction of the shells by corrosion and solution. The shells found in the buried soils are usually of mixed preservation due to solution which results in an over-representation of the robust shelled species such as Clausiliidae and Pomatias elegans (Müller) and an under-representation of thin shelled species such as the Zonitidae. A fragmentary record of the earlier biozones may however occasionally be found in these dry valleys in chance subsoil hollows, preserved stratigraphically beneath the buried soils underlying anthropogenic colluvium.

Two sites will be discussed in detail to illustrate the sequence of the molluscan succession on the South Downs.

ASHAM QUARRY (TQ 440061)

Holocene deposits were first studied at this Quarry in the late 1960's by Evans and Williams (Williams 1971), who studied Late-glacial solifluxion deposits which were found to contain molluscan assemblages of biozones y and z. This particular section is no longer accessible but quarrying in the late 1970's provided new sections through Late-glacial and Postglacial dry valley infill deposits. These consist of Late-glacial solifluxion deposits of unstratified chalk rubbles interbedded with pockets of silt that are overlain by a brown Postglacial soil in turn buried by colluviation. Also present are subsoil hollows that are overlain by the buried soil (Fig. 2).

From the south section two columns of samples were taken for molluscan analysis and the results from column A are presented in a percentage histogram (Fig. 3). The species are plotted in an order of ecological preference, woodland species on the left, intermediate in the centre and open-country species on the right (a plus sign indicates less than 1% of the total).

The Late-glacial deposits were found to contain a zone z assemblage of *Trichia hispida* (L.), *A. secale*, *Pupilla muscorum* (L.), *Punctum pygmaeum* (Draparnaud) and *Nesovitrea hammonis* (Ström). Between the Late-glacial deposits and the earliest record of the Postglacial in the subsoil hollow, a hiatus of several thousand years occurs.

The subsoil hollow (column B) contains a woodland fauna (biozone d) characterised by *Discus rotundatus* (Müller) and *Oxychilus cellarius* (Müller). Towards the top of the hollow *Acicula fusca* (Montagu) occurs, a species that is considered to be a late zone d immigrant. Subsoil hollows are very common in chalk areas; those studied by Evans (1972) at Ascott-under-Wychwood and at Avebury contain woodland assemblages similar to those at Asham. The origin of these hollows is still a matter of debate. Evans (1972) considers them to form when a tree is blown over leaving a cavity which is gradually filled by the inwashing of the surrounding soil. Alternatively they could have formed by any one of a number of processes including solution, subsidence, or excavation by frost or by man. The hollow at Asham was probably filled rather rapidly as its contents show only a weak lithological and biological stratification.

Overlying the Late-glacial deposits and the subsoil hollow is a well developed dark soil that represents the ancient soil that covered the valley sides and bottom. This soil can be divided into a lower stone-free horizon and an upper stony horizon. The stone-free horizon represents the *in situ* soil and contains a woodland fauna of *Carychium tridentatum* (Risso), *D. rotundatus*, *Aegopinella pura* (Alder), *Aegopinella nitidula* (Draparnaud) and *O. cellarius*. The upper stony horizon contains large numbers of small chalk fragments that have been removed from the chalk bedrock on the valley sides and mixed and transported with the soil covering to the valley bottom. At the junction of these horizons there is a marked change in the Mollusca from a zone d woodland assemblage to a zone e open-country assemblage, reflecting clearance and the onset of colluviation. This change in environmental conditions is shown clearly by the marked expansion of the *Vallonia* species and of *Helicella itala* (L.), and by a peak in the abundance of *P. elegans*, a species that requires a loose rubbly substrate in which to burrow.

The colluvial deposits (including the transported soil) represent the removal of the surface material under the influence of gravity, a process which has been accelerated by the destruction of the protective vegetation cover. All the Mollusca found in these sediments are open-country species and can be assigned to biozone e.

Sections on the north side of the road-cutting show a more complete stratigraphy, more recent colluvial deposits (biozone f) being preserved here. Molluscan analysis shows a similar sequence of events, with the oldest Postglacial deposits being found in a subsoil hollow and with a marked change from woodland to open country species in the overlying buried soil. Within the soil there was a charcoal-rich horizon, stratigraphically above the clearance level. Carbonised seeds from this layer have been identified by Dr D.T. Holyoak as species that are secondary colonisers of bare cultivated ground (e.g. *Plantago lanceolata* L. and *Veronica hederifolia* L.)

In the upper colluvial deposits a zone f molluscan assemblage occurs; in these sediments the more recently introduced helicellids (*C. virgata*, *C. gigaxii* and *C. intersecta*) appear and expand in numbers, with a corresponding decline of *H. itala* and *M. cartusiana*.

The evidence at Asham demonstrates that the valley and the surrounding downland were once wooded. This supports the pollen work of Thorley (1981) at the Vale of Brooks who showed that this area of the Downs was once wooded and that clearance can be dated to 3190 ± 125 B.P. (I - 4454). Subsequent clearance of the woodland for cultivation accelerated soil erosion and as a result the soil on the valley floor was buried and preserved. These colluvial sediments become increasingly chalky upwards reflecting an increase in the bedrock component of the soil. The upper colluvial deposits are lithologically similar to the present day soil cover on the downland slopes and contain a similar open-country fauna.

DEVIL'S DYKE (TQ 266201)

This is a well known Sussex landmark cut into the face of the escarpment, south of Poynings. Its form is essentially similar to the many other chalkland scarp face dry valleys such as Rake Bottom, Hampshire (Small 1968, Gordon and Shakesby 1973, Shakesby 1975) and Pegsdon, Hertfordshire (Sparks and Lewis 1957).

In 1982 and 1983 seven pits were excavated in the floor of the Dyke. They all showed a similar stratigraphy of chalky, colluvial postglacial sediments 2-3m thick overlying solifluxion deposits of probable Devensian age. In the pit furthest down the valley (TQ 26641053) a buried soil 30cm thick was discovered at a depth of 205cm lying directly on top of Late-glacial deposits.

The pit highest up the valley (TQ 26641027) was sampled the most fully as it showed the most detailed stratigraphy, including a charcoal-rich layer with associated pottery fragments. (Fig. 4). A molluscan histogram (Fig. 5) has been constructed from the samples from Pit 1. The basal Postglacial sediments (170-200cm) contained a woodland fauna of C. tridentatum, D. rotundatus, Vertigo pusilla Muller, Helicigona lapicida (L)., Cochlodina laminata (Montagu), Clausilia bidentata (Ström) and Zonitidae. At 170cm there is a change to a more open-country fauna with V. costata, Vertigo pygmaea (Draparnaud) and H. itala expanding rapidly. V. costata becomes the dominant species between 140 and 70cm comprising up to 55% of the fauna; this is because it is a pioneering species able to colonise poorly vegetated habitats such as land that has been cleared of its vegetation cover and undergoing soil erosion. V. costata tends to prefer drier habitats than Vallonia excentrica Sterki (Evans 1972) and the change in relative dominance that occurs above 70cm can be attributed to successional processes towards a mature ecological community. As the ground cover increases after clearance, the soil becomes more stable and so the microhabitats become more humid. At the same time molluscan competition for these habitats is greatly increased. The occurrence at 30-40cm of Vallonia pulchella (Müller), a hygrophilous species, also indicates a change to damper conditions. The re-expansion of the shade-loving genera at the top of the section is a reflection of the tall grassland vegetation covering the site today - a grassland plagioclimax community.

In the colluvial deposits at the Devil's Dyke the molluscan succession is similar to that at Asham Quarry. Within biozone e M. cartusiana arrives and expands. This level at the Dyke has been radiocarbon dated to 2315±35 B.P. (BM 2137). This postdates clearance as the charcoal layer is 30cm above the base of biozone e. Zone f is also represented at the Dyke in the upper colluvial deposits and in the topsoil samples where Monacha cantiana (Montagu) and C. intersecta, very late introductions, occur. The earliest dated record of M. cantiana is from a 4th century occupation level at Lullingstone Villa, Kent (Kerney unpublished).

Samples from the other pits revealed a similar molluscan succession. Detailed sampling was also carried out at the pit furthest down the valley, that showing a buried soil. The molluscs preserved in this soil were in a very poor condition but formed a woodland assemblage similar to that present in the basal colluvial sediments in the other trenches.

To summarise, at the Devil's Dyke the Late-glacial solifluxion deposits contained no snails, any recovered from the Late-glacial samples being contamination from the overlying Postglacial silts. Only in the pit furthest down the valley was an in situ buried soil found; contained Mollusca were all woodland species indicating that the Dyke and surrounding downland were formerly covered by forest. This woodland persisted until permanent clearance which probably took place during the Late Bronze Age or Early Iron Age. It is possible that the woodland may have been cleared in order to build the Iron Age Fort on the crest of the Downs.

The Postglacial deposits found in the other pits were entirely of colluvial origin although the basal deposits contained a woodland fauna; this material is therefore interpreted as having accumulated rapidly at the onset of clearance and represents the reworking of the forest soil by slope wash. As at Asham Quarry, the record of the Postglacial is incomplete, the first half being apparently unrepresented. This is in contrast to the sections studied by Kerney et al. (1964, 1980) in dry valleys on the North Downs where molluscan assemblages of the early part of the Postglacial are preserved in tufaceous deposits that are overlain by colluvial deposits.

DISCUSSION

A general picture can be built up from the studies at Asham Quarry and the Devil's Dyke of the environmental changes that the chalklands in East Sussex have undergone since the mid Postglacial. It is now well established that the South Downs were wooded although the date of permanent clearance is known to have varied from valley to valley. The lateness of clearance at the Devil's Dyke is attributed to the steepness of the valley sides and hence its low value for agriculture or settlement.

An ecological model can be constructed based on the changes that the Downs have undergone. As a result of woodland clearance in these valleys, bare ground was exposed and this is reflected at both sites by a peak in abundance of P. elegans at the zone d/e boundary. Colonisation of the bare ground surface then took place by pioneering weed species of plants and associated molluscs such as T. hispida and V. costata. This is seen most clearly at the Devil's Dyke by the dominance of V. costata (55% of the total). Following the early plant colonisers are late colonisers such as grasses and thistles; these create more humid, less exposed microhabitats, enabling molluscs such as V. excentrica (which as said previously prefers slightly moister conditions than V. costata), V. pygmaea, H. itala and later M. cartusiana to become established. When the late coloniser plants have become firmly established a grassland plagioclimax ecosystem is reached; this sees the re-expansion of some of the shade-loving molluscs such as C. tridentatum, A. nitidula and the hygrophilous V. pulchella whose characteristic late expansion may be noted in dry valley deposits on both the North and South Downs. The establishment of a grassland plagioclimax is dependent on the cessation of arable activity and stabilisation of the cultivated ground by pioneering species of weeds. At these dry valley infill sites in Sussex the thickness of the colluvium suggests that cultivation has occurred for long periods and that it is only comparatively recently that grassland has become established. These changes are summarised in the table (Fig. 6).

A further amplification of the molluscan biostratigraphy is based on work at Asham, the Devil's Dyke and at Cow Gap near Beachy Head. At all these sites the earliest Postglacial deposits contain a zone d assemblage characterised by the late arrival of *Acicula fusca* towards the end of biozone d. After initial clearance, zone e is marked by the expansion of the open-country genera including *M. cartusiana*. The base of zone f is defined by the arrival of *H. aspersa* which at all these sites clearly predates the arrival of *C. virgata*, *C. gigaxii* and *C. intersecta*. *C. virgata* and *C. gigaxii* appear approximately together in Sussex, whereas *C. intersecta*, a species that abounds at all the sites today, occurs much later, and it is only found in the topsoil. Associated with the expansion of *C. gigaxii* and *C. virgata* is a decline in the numbers of *H. itala*; this may be due to the removal of the grassland and its replacement by a cleared land surface or alternatively could possibly result from competition from the introduced helicellid species.

CONCLUSIONS

In conclusion, the detailed sampling of dry valley deposits in East Sussex has enable the reconstruction of the palaeoecological history of the Downs during the later half of the Postglacial and a direct comparison with the biozones established on the North Downs. The sequence of biozone d to f is essentially similar to that of the North Downs although the arrival and expansion of the introduced helicellids has not as yet been well documented in chalk areas outside Sussex.

ACKNOWLEDGEMENTS

I would like to thank the Fourth International Flint Symposium for providing the money for the excavations and Dr R.B.G. Williams for arranging and helping with them. Also I would like to thank Brighton Corporation, Blue Circle Cement Company and Mr Lee for permission to dig on their land and Mr M. Wingman and Mr D. Gordon for helping with the excavations. Also Mr R. Burleigh at the British Museum for providing the radiocarbon date at the Devil's Dyke, Dr D.T. Holyoak for identifying the carbonised seeds from Asham Quarry and Mr I. Kinnes for examining the pottery from the Devil's Dyke. Finally, I would like to thank my supervisor Dr M.P. Kerney without whose help this work would never have been done.

I acknowledge NERC who are supporting me in a research studentship.

REFERENCES

Bell, M. (1981) *Valley sediments as evidence of Prehistoric land use: A study based on dry valleys in southeast England*. Unpublished PhD thesis, University of London.

Ellis, C.S. (1983) Dry valley fills at Asham and the Devil's Dyke, Excursion Guide Fourth International Flint Symposium, Brighton Polytechnic 10-15 April 1983, pp 9-23.

Evans, J.G. (1972) Land Snails in Archaeology. London: Seminar Press.

Gordon, P. and Shakesby, R.A. (1973) Study of the Valley Infill, Rakes Bottom, Butser Hill, South Hampshire Geographer 6, 10-20.

Kerney, M.P. (1963) Late-glacial deposits on the chalk of South-east England, Philosophical Transactions of the Royal Society of London B 246, 203-254.

Kerney, M.P. (1966) Snails and Man in Britain, Journal of Conchology 26, 3-14.

Kerney, M.P. (1970) The British distribution of Monacha cantiana (Montagu) and Monacha cartusiana (Müller), Journal of Conchology 27, 145-148.

Kerney, M.P. (1977) A proposed zonation scheme for Late-glacial and Postglacial deposits using land Mollusca, Journal of Archaeological Science 4, 387-390.

Kerney, M.P., Brown, E.H. and Chandler, T.J. (1964) The Late-glacial and Postglacial history of the chalk escarpment near Brook, Kent, Philosophical Transactions of the Royal Society of London B 248, 135-204.

Kerney, M.P., Preece, R.C. and Turner, C. (1980) Molluscan and plant biostratigraphy of some Late Devensian and Flandrian deposits in Kent, Philosophical Transactions of the Royal Society of London B 291, 1-43.

Shakesby, R.A. (1975) An investigation into the origin of the deposits in a chalk dry valley of the South Downs, Southern England, University of Edinburgh, Department of Geography Research Discussion paper 5.

Small, R.J. (1958) The origin of Rake Bottom, Butser Hill, Proceedings of the Hampshire Field Club 21, 22-30.

Sparks, B.W. and Lewis, W.V. (1957) Escarpment dry valleys near Pegsdon, Hertfordshire, Proceedings of the Geologists Association, London 68, 26-38.

Thomas, K.D. (1982) Neolithic enclosures and woodland habitats on the South Downs in Sussex, England. In Archaeological Aspects of Woodland Ecology S. Limbrey and M. Bell (eds) pp. 147-170. British Archaeological Reports International Series 146.

Thorley, A. (1981) Pollen analytical evidence relating to the vegetational history of the Chalk, Journal of Biogeography

8, 93-106.

Williams, R.B.G. (1971) Aspects of the Geomorphology of the South Downs. In Guide to Sussex Excursions R.B.G. Williams (ed), Sussex, Institute of British Geographers.

SUCCESSION ON CHALK IN SUSSEX

ENVIRONMENT	FLORA	MOLLUSCA
Late-glacial open ground solifluxion	grasses, sedges willow, birch	T. hispida A. secale V. costata P. muscorum P. pygmaeum
Postglacial closed forest Soil formation and slope stability	mixed deciduous woodland climax vegetation	D. rotundatus Clausiliidae Zonitidae C. tridentatum
--------------------------CLEARANCE--------------------------		
Open ground soil erosion	pioneering species: ragworts, dandelions nettles	V. costata P. muscorum T. hispida
Open ground Increasing stability	late colonisers: grasses, plaintains	V. excentrica V. pygmaea H. itala M. cartusiana
Slope stability	grassland plagioclimax: maintained by grazing	Decline of M. cartusiana Re-expansion of C. tridentatum and Zonitidae expansion of V. pulchella

Table 1: Succession on Chalk in Sussex.

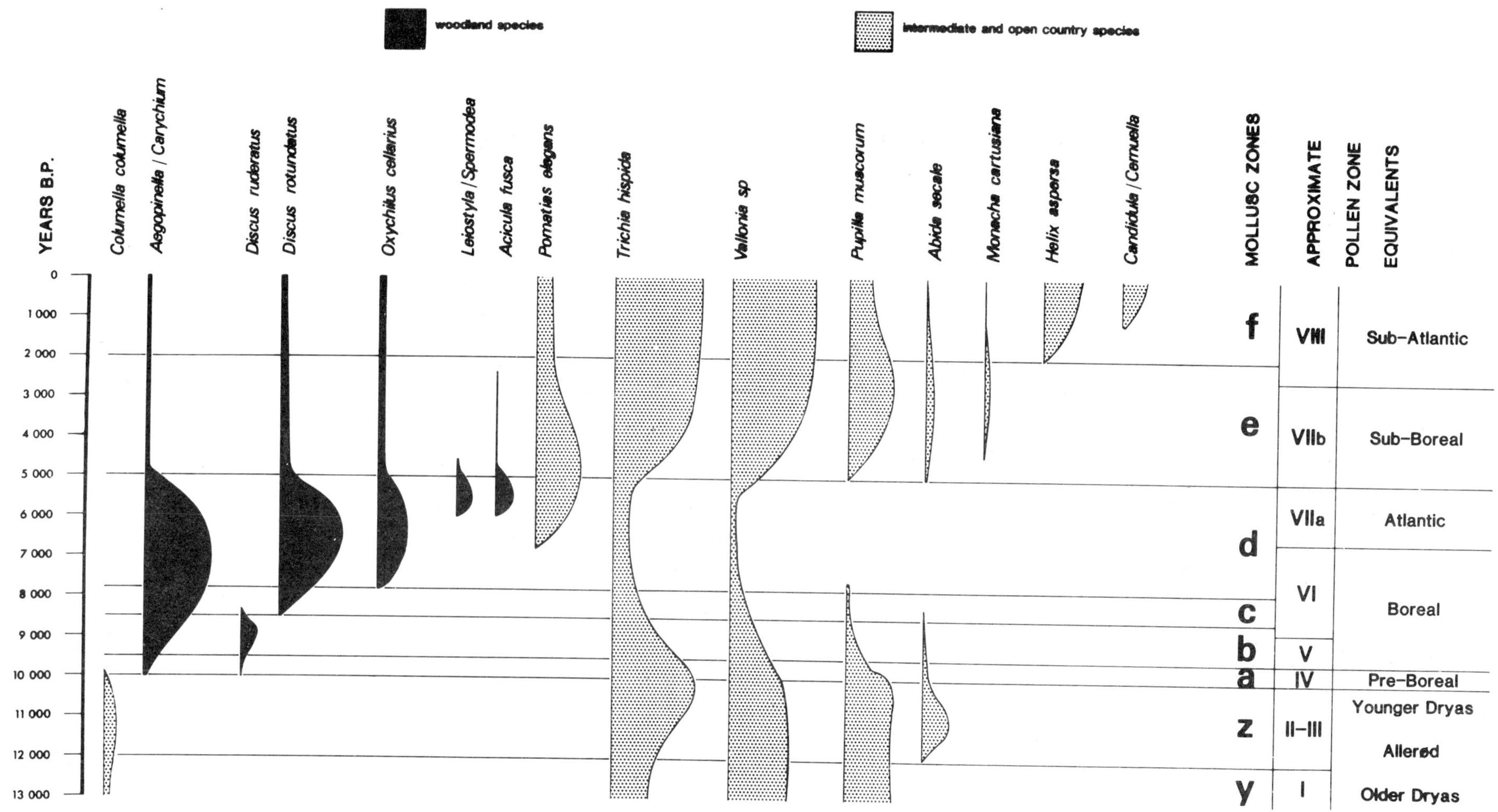

Fig. 1: Idealised diagram of Late-glacial and Postglacial molluscan biozones based on work in Kent and Sussex (only diagnostic species are shown).

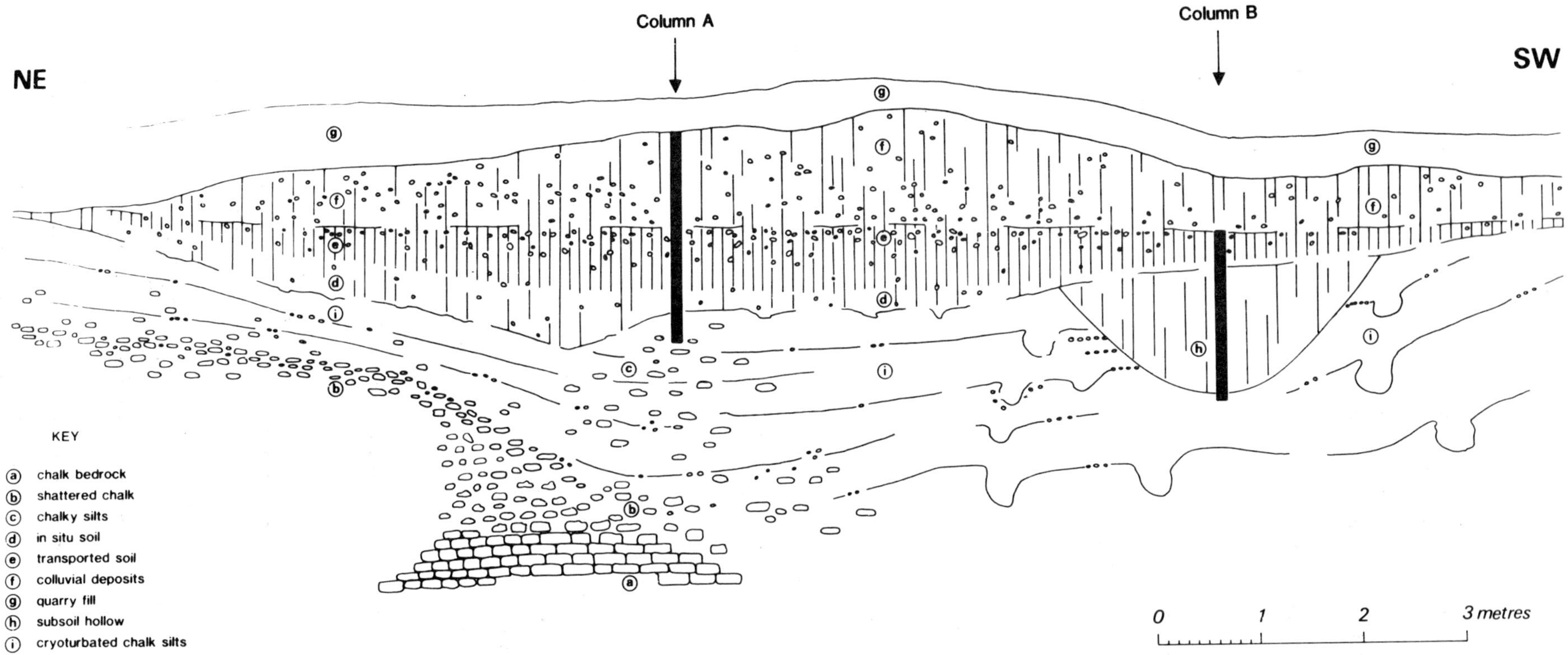

Fig. 2: Schematic diagram of the South section, Asham Quarry.

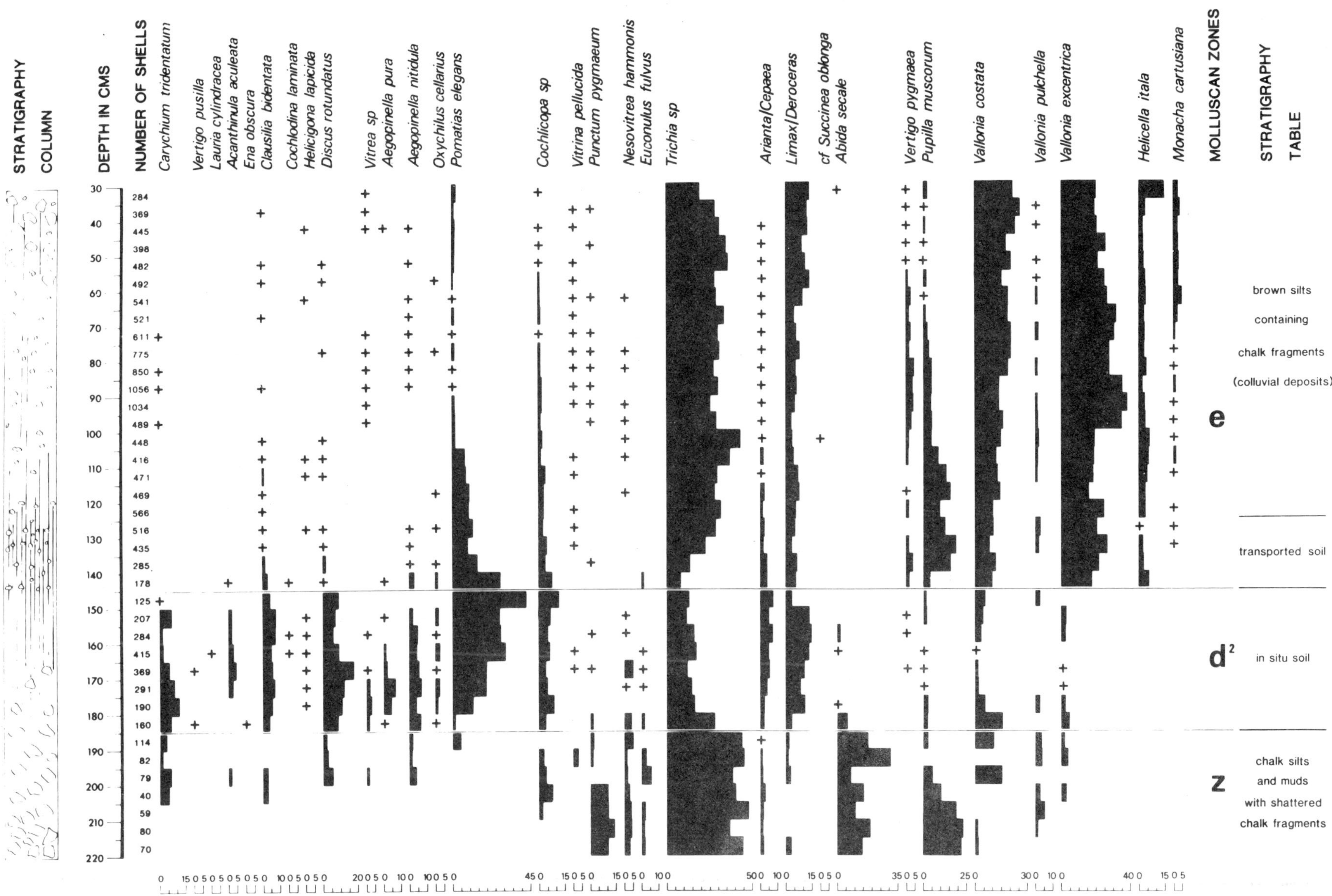

Fig. 2. Percentage histogram of the results from Column A

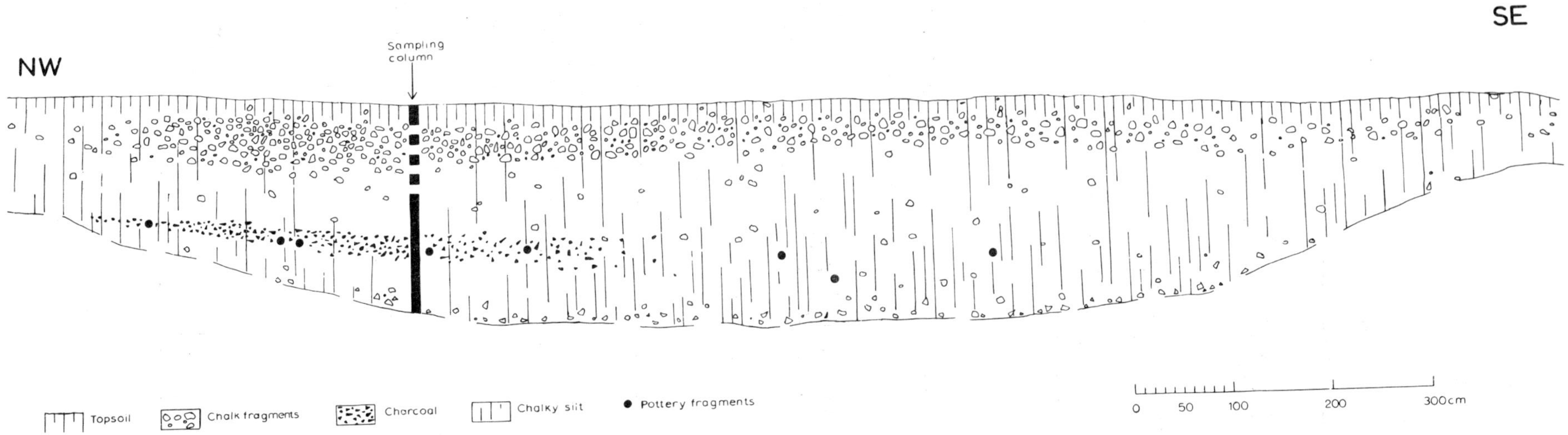

Fig. 4: Schematic diagram of Pit 1, the Devil's Dyke.

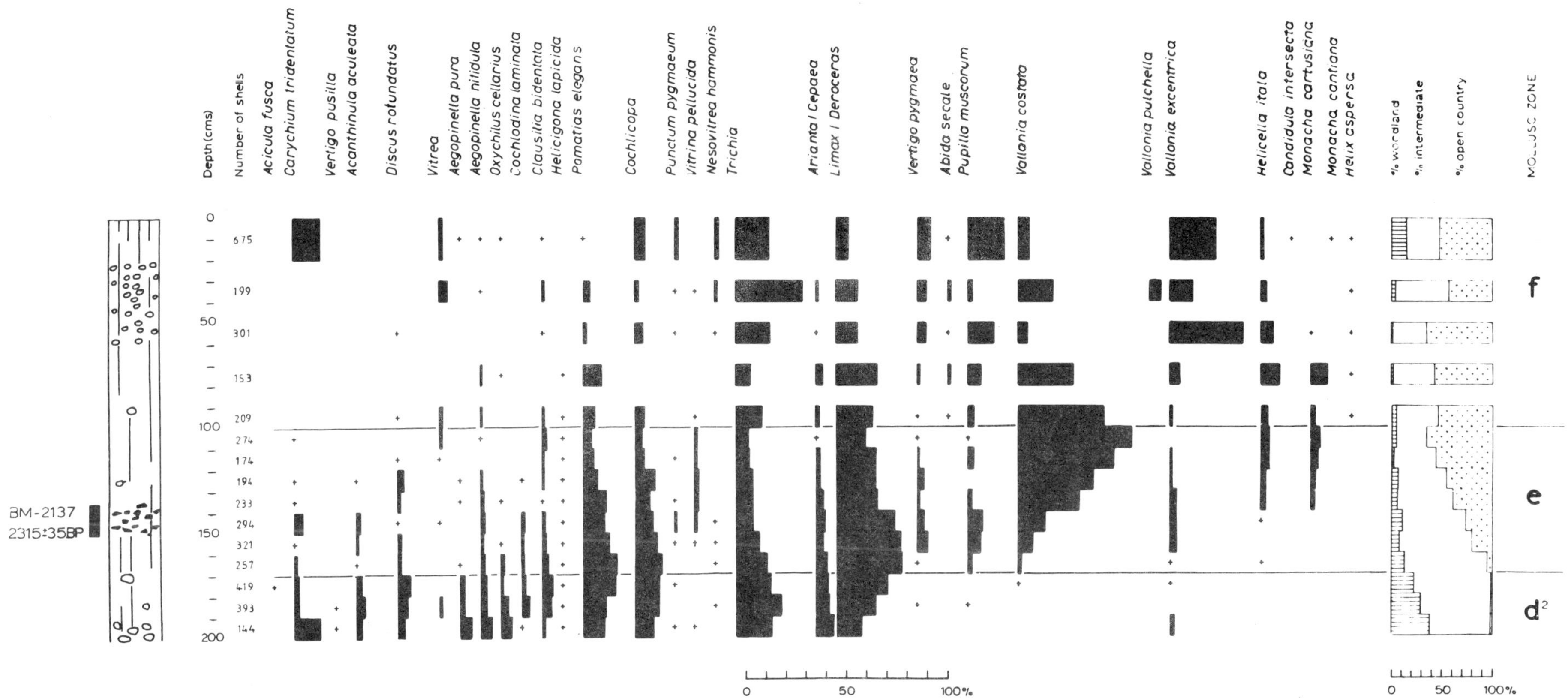

Fig. 5: Percentage histogram of the results from Pit 1, the Devil's Dyke.

PAST AND PRESENT DISTRIBUTIONS OF HYDROBIA ULVAE (PENNANT), GASTROPODA, AND MACOMA BALTHICA (L.), BIVALVIA, ALONG THE SOUTHERN SHORES OF THE SEVERN ESTUARY IN RELATION TO WATER POLLUTION, FIELD SURVEY AND LABORATORY METHODS.

D.D. Gilbertson
Department of Prehistory and Archaeology
University of Sheffield
Sheffield S10 2TN

and

A.B. Hawkins
Department of Geology
University of Bristol
Bristol BS8 1TR

ABSTRACT

A comparison is made between the recorded modern and Upper Quaternary distributions of the inter-tidal molluscs Hydrobia ulvae and Macoma balthica from sites found on the southern shores of the Severn Estuary. Difficulties of comparison and interpretation are discussed in detail; and modifications to field and laboratory methods are suggested. Qualified support is given for the proposal that the modern absence or rarity of Macoma at the mouth of the River Avon is due to a recent environmental change, possibly water pollution.

INTRODUCTION

Boyden and Little (1973) when undertaking a study of the present fauna along the southern shores of the Severn Estuary found the molluscs Hydrobia ulvae (Pennant) and Macoma balthica (L) to be either rare or absent in the modern soft, inter-tidal sediments at Avonmouth and downstream from the mouth of the River Avon. Sewage pollution transported into the estuary by the Bristol Avon was suggested as a possible cause. Their data have been re-drawn as Figure 1. Subsequent surveys indicated a slight increase in the numbers found around the mouth of the Avon (Little and Boyden 1976). These data may be compared with that from Late Quaternary beach/inter-tidal deposits in the region which differ from the modern estuarine sediments only by virtue of the fact that the former pre-date the impact of the nineteenth and twentieth century industrialisation, urban growth and associated pollution.

Sub-fossil Hydrobia ulvae and Macoma balthica occur at many sites in the area at which Quaternary inter-tidal soft sediments have been found (Figure 2). These Quaternary sediments represent deposition which took place either in the present Estuary between 4,000 to 7,000 years ago, i.e. in the middle Flandrian, or approximately 80,000 to 100,000 years ago in the Ipswichian interglacial along earlier shorelines of the Estuary (see Mitchell *et al.*, 1973 for explanations of terms). The more important Quaternary sites are summarised in Table 1. The two most critical exposures were the excavations for the inner gates of Avonmouth Dock at the south east of what was once Dumball Island (Bolton 1908) and at Kenn, Somerset, during the re-routing of the drainage system (Gilbertson and Hawkins 1978).

Wright (1973) emphasised the importance of the longer time perspective provided by Quaternary deposits. With appropriate analyses, these could offer the opportunity for evaluating recent human impacts by facilitating the reconstruction of the 'base-line' fauna of inter-tidal sediments in the Estuary before urbanisation and industrialisation. In addition interpretive problems caused by seasonal or short-term population fluctuations are likely to be reduced in view of the many (if not precisely quantifiable) years which the deposits took to accumulate. Often the potential of this comparative method is not realised for various reasons; some of these are discussed below and a solution offered for one major type of problem.

PROBLEMS OF COMPARISON

Geological age differences

The current views of the geological age(s) of the principal sub-fossil assemblages from Quaternary inter-tidal deposits are given in Table 1. Examination of the scientific literature recording the full sedimentary context and faunal associations of both Hydrobia ulvae and Macoma balthica, indicate that there are no reasons to suspect that the ecological tolerances and/or habitat preferences of the species have changed from the Ipswichian interglacial stage, through the Flandrian to the present. Hence differences between present and past distribution patterns are unlikely to be the result of the different ages of the inter-tidal deposits investigated.

Climatic differences

The available palaeoecological evidence currently obtainable from the Quaternary sub-fossil sites is incapable of precisely identifying the sea or air temperatures prevailing at the time of deposition. The valves of the thermophilous, freshwater bivalve Corbicula fluminalis (Müller), which during the Ipswichian interglacial periods were washed into more brackish water depositional environments at Kenn (Gilbertson and Hawkins 1978) and in the Burtle Beds (Kidson, Gilbertson, Haynes, Heyworth, Hughes and Whaley (1978) are important in this context. This species is now extinct in Europe; its closest location being northern Egypt (Ellis 1962). There is general agreement that its presence suggests mean annual air temperatures approximately 2°C higher than at present (Sparks and West 1970, Gilbertson and Hawkins 1978). There is as yet no evidence from the other local Ipswichian sites that air temperatures were significantly different from those prevailing today.

The general consensus of evidence summarised in Stuart (1979) and West (1977) suggests that mid-Flandrian mean annual air temperatures may have been 1 - 2°C higher than those at present. No species indicating sea or air temperatures higher than those prevailing today have been found in local deposits of this age. Examination of Kerney's (1976) distribution map of Hydrobia ulvae and the published maps of the marine molluscan survey of the British Isles (Seaward 1982) suggest that differences of this magnitude are unlikely to affect the distribution of either species along the Severn Estuary in any significant manner.

Facies variation

Seven principal types of estuarine bio-sedimentary facies are represented by the sub-fossil faunas listed in Table 1. The middle and lower zones of the Quaternary inter-tidal mud flats are represented at Greylake (Figure 2 this paper; and Kidson et al. 1978) and (Penzoy Farm Figure 2, this paper; and Kidson, Beck, and Gilbertson, 1981). The middle and upper levels of Quaternary salt marshes are represented at Gravel Banks and the ICI reservoir sites (Figure 2) where Hydrobia ventrosa agg., dominates (Gilbertson and Hawkins 1980 and unpublished data).

Sites representing locations where Quaternary streams entered either the upper parts of salt marshes or sandy beach zones are also relatively common in the local fossil record. They occur at Gravel Banks, Kenn (Gilbertson and Hawkins 1978), Greylake (Kidson et al. 1978; Gilbertson 1979), and Penzoy (Kidson et al. 1981). Hydrobia ventrosa agg., usually dominates, with Hydrobia ulvae present in low numbers. The latter species may have floated in with the tide (Newell 1962; Chatfield 1972; Little and Nix 1976). Alternatively H. ulvae may have been living in the slightly deeper water of a small tidal creek, possibly with Macoma balthica. A further possibility is that both H. ulvae and M. balthica shells may have been introduced by waves or tides.

The precise character of the Quaternary depositional environment at the important site of Dumball Island (Bolton 1908) is difficult to assess with confidence from the species list provided. Quite possibly a number of different habitats were represented through the sequence, but were not recognised in the original study of the 1.7m thick shelly layer. The "box sample", of unknown dimensions, which was studied by Bolton (1908) included shells from several different habitats. The recorded fauna suggests the following environments were represented: woodland or shrub; marsh; freshwater pools; streams and rivers; upper salt marsh levels; sandy, gravelly or rocky beaches and estuarine tidal and mudflats. The presence of species from such a wide range of habitats, and the reported frequency of separated valves indicates extensive re-working and transportations by river, waves or tides. Macoma balthica was obviously abundant; Hydrobia ulvae was not recorded, but there are records of the predominantly upper salt marsh species - Leuconia bidentata (G. Montague) = Leucophytia bidentata (Montagu), and of Paludestrina stagnalise (Baster) = Hydrobia ventrosa (G. Montagu), (see Ellis 1969 for synonomy).

Quaternary beach environments are represented at the other sites (Figure 2). At Weston-in-Gordano (ApSimon and Donovan 1956) only the largest, most strongly shelled taxa appear to be represented. This probably reflects the lack of sieving and detailed searching of sands, and the higher energy character of the depositional environment leading to greater fragmentation. The storm beach deposits typified by Swallow Cliff (Prestwich 1892; Gilbertson and Hawkins 1977) are dominated by very strong shelled species such as Macoma and Patella spp. These have been derived from adjacent mud-flats, sand-flats and rocky shores by waves and tides, and accumulated in very densely packed, cobbly, storm beaches. It is reasonable to assume that the absence of the smaller hydrobiids in such deposits is because the small thin shells would be preferentially abraded or removed by the very strong backwash associated with storm beach formation.

It is clear that nearly all the sub-fossil accumulations reported are death assemblages, rather than the living communities studied by Boyden and Little (1973), Little and Boyden (1976). Penecontemporaneous re-working, derivation, transportation, and preferential destruction have all played a part in the assemblage formation; the significance of each factor varying according to the particular circumstances of the local depositional environment.

Methodolocal problems

Several types of methodological problems occur when using this comparative approach. The first concerns curation and species identification. The earlier finds from Dumball Island (Bolton 1908), Swallow Cliff (Prestwich 1892), and some of the Burtle Beds sites (Bulleid and Jackson 1937, 1941) could not be checked as losses occurred through war damage or other unknown reasons. The small hydrobiids sometimes present difficulties of identification, principally Hydrobia ventrosa, but not usually H. ulvae.

Potentially more significant are difficulties resulting from past differences in field and laboratory extraction and counting methods. Before the mid-1960s the majority of studies of sub-fossil Mollusca from the region yielded lists of taxa noted by eye, often after only limited or no washing, and then with a coarser sieve than the 0.5mm mesh used today. Inevitably, smaller taxa such as the hydrobiids were missed. Exceptions are the study by Bolton (1908) and in the Carnelly Collection from the Burtle Sand Beds near Othery (Table 1, see Gilbertson 1979). Here smaller taxa such as the genus Pisidium and hydrobiids are recorded or preserved.

Assessments of frequency are either non-existent, or usually at best relative (for example - rare, abundant). Field and laboratory procedures were designed to search sections for particularly 'good' specimens or unusual taxa, with little or no attempt being made to sample the fauna systematically, or to make faunal counts per unit weight, or per volume of sediment.

In most circumstances it is difficult or impossible to make quantitative comparisons between the species lists of earlier palaeoecological studies and the quantitative frequency data of modern palaeoecological surveys. Where only species lists are available for comparison, all species records tend to have the same significance. Consequently it is very difficult to make comparisons, other than in terms of presence/absence measurements. This causes difficulties. For example, at this nominal scale of measurement it is extremely difficult from the published literature to identify species or groups of species whose presence in small numbers would be so anomalous that they would be usually interpreted as the result of derivation. This reworking might be from adjacent contemporary faunas and environments, or as a result of re-cycling from much older deposits. This latter process can be seen operating on the contemporary Gravel Banks, 1.5 km seaward of ICI Severnside (Figure 2) where small numbers of mid-Flandrian terrestrial and freshwater molluscs are being mixed with large numbers of modern estuarine species as a result of erosion of old sediments and subsequent re-deposition of the fossils.

In the last decade palaeoecological studies have concentrated on identifying and counting the number of individuals per unit weight of sediment: usually 1 or 2 kg of material oven dried at 100°C. Dry unit weights rather than unit volumes are used as the latter are subject to significant, but often not precisely quantifiable volume changes, due to de-watering or compaction.

In contrast the faunal survey along the southern Estuary shore by Boyden and Little (1973) was conducted by 'normally' extracting the fauna from a 0.25m^2 quadrant which was sampled to a depth of approximately 0.1m and then sieved through a 0.5mm sieve; the sample was supplemented by cores when the modern inter-tidal sediments were either too hard or shallow to yield the standard volume. Consequently it is impossible to compare directly the molluscan frequency data from the quantitative weight-based palaeoecological studies with the modern volume-based ecological studies, other than in terms of species presence/absence.

DISCUSSION

These observations indicate that useful comparisons between present and past distributions can be made as long as problems of derivation, transportation and fragmentation are borne in mind. However, problems resulting from differences in past study methods are more intractable. In the context of the shoreline deposits discussed here, problems due to differences in geological age, climate, facies variation, are seen not to rule out the use of the comparative method. Indeed faunal transportation and mixing enables the palaeoecologist to identify species and environments at present nearby, when the actual original habitat may not be exposed in the section studied.

Macoma balthica

It is clear from the data (Figure 2) that sub-fossil *Macoma balthica* was abundant to the south of the River Avon during the middle Flandrian and the Ipswichian interglacial (Figure 2) in contrast to its 'modern' scarcity in the region between Anchor Head and Avonmouth (Figures 1 and 2). At the critical Dumball Island site it was sufficiently common for Bolton (1908:241) to write:

> "The whole surface of this layer could be seen dotted over with the separated valves of *Tellina balthica* (= *Macoma balthica*). An examination of the material showed that the shells were distributed throughout the mass, and were not confined to the surface layers only."

The present palaeoecological data (Figure 2) indicates *Macoma* is either rare or absent in the sub-fossil assemblages north of the mouth of the Avon. However, the biofacies found in this area to date are all of insufficient water depth or salinity to expect to find the species in quantity (Table 1). In the critical area further south, between Avonmouth and Anchor Head, the species is common in the low energy inter-tidal silt, sand bank, and beach deposits south of the Avon at Kenn, Weston-in-Gordano, and in the Burtle Sand Beds (Figure 2, Table 1). Unlike the hydrobiids, it is abundant in the storm beach environments at Swallow Cliff, North Bay and Anchor Head.

Therefore it is concluded that the palaeoecological evidence indicates Macoma balthica should be present nowadays in significant quantities at the mouth of the River Avon and further south on the coast between Avonmouth and Anchor Head. Its contemporary rarity or absence as reported by Boyden and Little (1973, 1976) must either be the result of chance, short term population fluctuations, disease, or an effect induced by human activity; quite possibly pollution introduced by the River Avon (Boyden and Little 1973).

Hydrobia ulvae

Hydrobia ulvae is common in low energy early- and mid-Flandrian deposits north of Avonmouth at Gravel Banks, ICI Severnside, further south at Kenn, and in the Burtle Beds (Figure 2, Table 1). At Weston-in-Gordano its Quaternary rarity is associated with the absence of suitable habitat; sandy and cobbly beaches occurring rather than inter-tidal mudflats. However, in the critical Dumball Island deposits it is absent, even though similarly small and fragile upper salt marsh species such as Leucophytia and Hydrobia ventrosa are recorded along with slightly deeper water estuarine species such as Macoma balthica and Scrobicularia.

The presence of all the above inter-tidal species suggests the past absence at Dumball Island of Hydrobia ulvae is not the result of lack of suitable habitat, differential destruction, transportation, its floating habit, or inadequate study methods. Short-term population fluctuations are similarly unlikely to be the explanation, since although we can only guess at the period of time represented by the sediment in the box sample, it is likely it would encompass several such fluctuations. Consequently, we conclude that the evidence presently available suggests that either the species was genuinely absent at the mouth of the River Avon for reasons currently unknown (clearly not involving human interference with the environment), or that through chance of misidentification, the species was missed in Bolton's (1908) study. Given the number of species he did record, the latter is unlikely.

IMPROVEMENTS TO FIELD AND LABORATORY PROCEDURES

This study once again demonstrates that precise quantitative comparisons of population numbers in present and Quaternary deposits are prohibited by the impossibility of comparing mollusc numbers per unit volume of sample with mollusc numbers per dry unit weight of sample. Both sampling units are likely to continue to be used since biologists require data on individual spacing and density in an area (volume) while Quaternary deposits are prone to compaction and therefore are usually represented in terms of dry weight measurements. Consequently it is advocated that in modern faunal surveys in addition to the area or volume sampling procedures a small sample of known volume be also taken using a sealable container; the oven-dried weight of the sample be noted and the Bulk Density recorded (B.D.) - oven dried weight per unit volume. Conversion to molluscan frequencies per unit weight is then a simple calculation. Similarly, palaeoecologists may determine the volume of the sample under study in the field by measurement, or noting the volume of sand poured from a graduated container into the sampled hole. Both data sets may then be similarly recorded in terms of bulk density, permitting conversion of the frequency data to numbers per unit volume or weight.

CONCLUSIONS

With certain provisions it is possible to compare successfully the modern and Upper Quaternary molluscan faunas of present and past inter-tidal sediments in the Severn Estuary with the objective of assessing whether an apparent contemporary absence or scarcity is the result of (a) fairly recent human activity: and/or (b) short term population fluctuations: and/or (c) whether the absence/scarcity is due to other factors.

Whereas Macoma balthica is currently rare or absent in the vicinity of the mouth of the River Avon and downstream from it, this does not appear to have been the case in the middle Flandrian or the Ipswichian interglacial when it was abundant and widely distributed along the southern shores of the Severn Estuary between Avonmouth and Anchor Head. Its contemporary scarcity is therefore anomalous, and possibly related to pollution introduced by the Bristol Avon as discussed by Boyden and Little (1973).

Hydrobia ulvae was similarly widely distributed in the past, except in the area of the mouth of the River Avon, when for unknown reasons probably not related to Bolton's (1908) study methods, it was not found in a mid-Flandrian site at or near where suitable habitats should have occurred. Consequently its modern scarcity/absence may be due to unknown but natural factors as well as possible pollution effects.

Further comparative studies would be greatly aided by ecologists and palaeoecologists organising field and laboratory studies of this type of deposit so that the bulk density data is obtained for the samples.

ACKNOWLEDGEMENTS

The authors are grateful for the award of grants from the Natural Environment Research Council and the University of Sheffield (URG no.419) which helped finance this research; and to Dr Colin Little of the Department of Zoology, University of Bristol, for his comments on the manuscript.

REFERENCES

ApSimon, A.M. and Donovan, D.T. (1956) Marine Pleistocene deposits in the Vale of Gordano, Somerset, Proceedings of the Spelaeological Society 7 (3), 130-136.

Bolton, H. (1908) On the occurrence of a shell-bearing gravel at Dumball Island, Proceedings of the Bristol Naturalists' Society 10, 241-244.

Boydon, C.R. and Little, C. (1973) Faunal distributions in soft sediments of the Severn Estuary, Estuarine and Coastal Marine Science 1, 203-223.

Bulleid, A. and Jackson, J.W. (1937) The Burtle Sand Beds of Somerset, Proceedings of the Somerset Archaeological and Natural History Society 83, 171-195.

Bulleid, A. and Jackson, J.W. (1941) Further notes on the Burtle Sand Beds of Somerset, Proceedings of the Somerset Archaeological and Natural History Society 87, 111-116.

Chatfield, J.E. (1972) Studies on the variation and life history in the proso-branch Hydrobia ulvae (Pennant), Journal of Conchology, London 7, 463-473.

Ellis, A.E. (1962) A synopsis of British freshwater bivalve Molluscs. Synopses of the British Fauna No. 13, London: Linnaean Society.

Ellis, A.E. (1969) British Snails. Oxford: Clarendon Press.

Gilbertson, D.D. (1979) The Burtle Sand Beds of Somerset: Significance of the Freshwater Molluscan Faunas, Transactions of the Somerset Archaeological and Natural History Society 129, 115-117.

Gilbertson, D.D. and Hawkins, A.B. (1977) The Quaternary Deposits at Swallow Cliff, Middlehope, County of Avon, Proceedings of the Geologists' Association 88 (4), 255-266.

Gilbertson, D.D. and Hawkins, A.B. (1978) The Pleistocene Succession at Kenn, Somerset London, H.M.S.O. Bulletin of the Geological Survey of Great Britain 66, 44pp.

Gilbertson, D.D. and Hawkins, A.B. (1980) An early Flandrian sea level in the Severn Estuary, Proceedings of the Ussher Society 5 (1), 92. (Abstract).

Kerney, M.P. (1976) Atlas of the Non-Marine Mollusca of the British Isles. Conchological Society of Great Britain and Ireland/ Institute of Terrestrial Ecology-NERC, 68 Hills Road, Cambridge, UK.

Kidson, C., Gilbertson, D.D., Haynes, J.R., Heyworth, A., Hughes, C.E. and Whaley, R.C. (1978) Interglacial marine deposits of the Somerset Levels, South West England, Boreas 7, 215-228.

Kidson, C., Beck, R.B. and Gilbertson, D.D. (1981) The Burtle Beds of Somerset: temporary sections at Penzoy Farm, near Bridgwater. Proceedings of the Geologists' Association 92 (1), 39-45.

Little, C. and Boyden, C.R. (1976) Variations in the fauna of particular shores in the Severn Estuary, Estuarine and Coastal Marine Science 4, 545-554.

Little, C. and Nix, (1976) The burrowing and floating behaviour of the gastropod Hydrobia ulvae, Estuarine and Coastal Marine Science 4, 537-544.

MacFayden, A. (1970) Geological Highlights of the West Country. Butterworths:London.

Mackintosh, D. (1868) On the mode and extent of encroachment of the sea on some parts of the shores of the Bristol Channel, Quarterly Journal of the Geological Society of London 24, 279-283.

Mitchell, G.F., Penny, L.F., Shotton, F.W. and West, R.G. (1973) A correlation of Quaternary deposits in the British Isles, Geological Society of London Special Report No. 4, 99p.

Newell, R. (1962) Behavioural aspects of the ecology of Peringia (= Hydrobia) ulvae (Pennant) (Gastropoda, Prosobranchia), Proceedings of the Zoological Society of London 138, 49-75.

Palmer, L.S. (1931) On the Pleistocene succession of the Bristol district, Proceedings of the Geologists' Association 42, 345-361.

Prestwich, J. (1892) The raised beaches and 'head' or rubble drift of the south of England, Quarterly Journal of the Geological Society London 48, 263-343.

Seaward, D.R. (1982) Sea Atlas of the Marine Molluscs of Britain and Ireland. Conchological Society of Great Britain and Ireland, Nature Conservancy Council: Cambridge.

Sparks, B.W. and West, R.G. (1970) Late Pleistocene deposits at Wretton, Norfolk. 1. Ipswichian Interglacial deposits, Philosophical Transactions of the Royal Society of London B258, 1-30.

Stuart, A.J. (1979) Pleistocene occurrences of the European pond tortoise (Emys orbicularis L.) in Britain. Boreas 8, 359-371.

Welch, F.B.A. (1956) Notes on gravels at Kenn, Somerset, Proceedings of the Spelaeological Society 7 (3), 137.

West, R.G. (1977) Pleistocene Geology and Biology with Special Reference to the British Isles (3rd edition). Longmans: London.

Wright, H.E. (1973) Quaternary Science and Public Service, Quaternary Research 3, 515-519.

Site	Deposits	Age
Gravel Banks, Severnside (ST 525833) (Gilbertson and Hawkins, unpublished)	inter-tidal silts and perimarine (terrestrial freshwater, brackish) peats	Flandrian 7030 ± 115 yrs bp (I4903)
ICI Reservoir, Severnside (ST 546833) (Gilbertson and Hawkins, unpublished)	salt marsh peats and inter-tidal silts	Flandrian 3110 ± 100 yrs bp (IGC C/14-27) 3905 ± 100 yrs bp (IGS C/14-28) 5100 ± 100 yrs bp (IGS C/14-29)
Dumball Island, Avonmouth (ST 506787) (Bolton 1908); Gilbertson and Hawkins, unpublished)	estuarine riverine gravel bank and marsh	early-mid Flandrian
Weston-in-Gordano (ST 4474) (ApSimon and Donovan 1956)	beach conglomerate, gravel and marine sands	Ipswichian interglacial?
Kenn (ST 41/69 & ST 42/69) (Gilbertson and Hawkins 1978, Welch 1956)	beach and inter-tidal sands	Ipswichian interglacial?
Swallow Cliff, Middlehope (ST 325662) (Gilbertson and Hawkins 1977, Palmer 1931, Prestwich 1892)	storm beach conglomerate	Ipswichian interglacial?
North Bay, Middlehope (ST 337664) (Gilbertson and Hawkins 1977, ApSimon and Donovan 1956, Macfayden 1970)	storm beach conglomerate	Ipswichian interglacial?
Anchor Head, Weston-super-Mare (ST 308626 approx) (Mackintosh 1868, Prestwich 1892)	storm beach conglomerate	Ipswichian interglacial?
Shapwich, Nr Glastonbury (ST 420404) (Kennard in Bulleid and Jackson 1937, 1941, Gilbertson 1979, Kidson *et al* 1979)	inter-tidal, salt marsh, stream silts	Ipswichian interglacial?
Greylake pits, Othery-Sedgemoor (ST 385336 390335) (Bulleid and Jackson 1937, 1941, Gilbertson 1979, Kidson *et al* 1979)	inter-tidal silts beach deposits	Ipswichian interglacial?
Penzoy Farm, Westonzoyland (ST 337348) (Kidson *et al* 1981)	inter-tidal silts	Flandrian 3900 ± 80 yrs bp (HAR 545)

TABLE 1: Sites with (sub)fossil *Hydrobia ulvae* and/or *Macoma balthica* along the southern shores of the Severn Estuary: for locations see Figure 2.

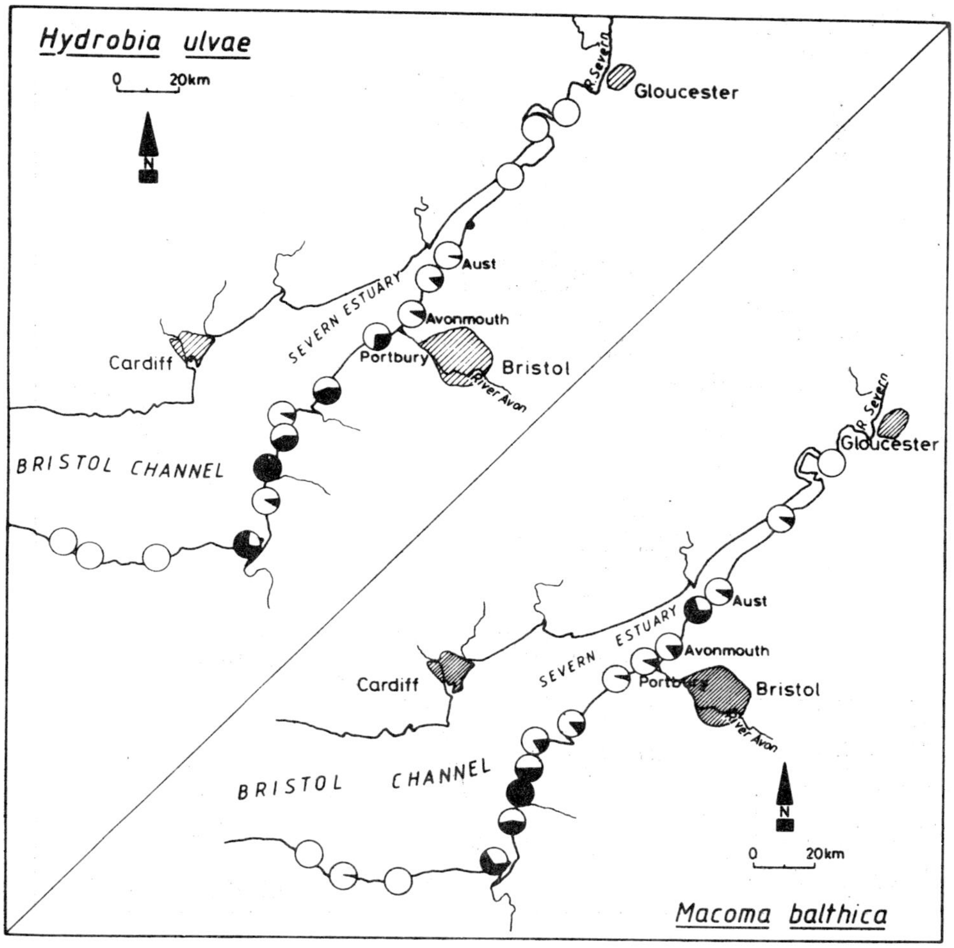

Figure 1. Maximum densities of Hydrobia ulvae and Macoma balthica found along the soft particulate southern shores of the Severn Estuary, expressed as a percentage of the greatest overall densities; 9600/quadrat for H.ulvae; 880/quadrat for Macoma balthica (redrawn from Boyden and Little, 1973).

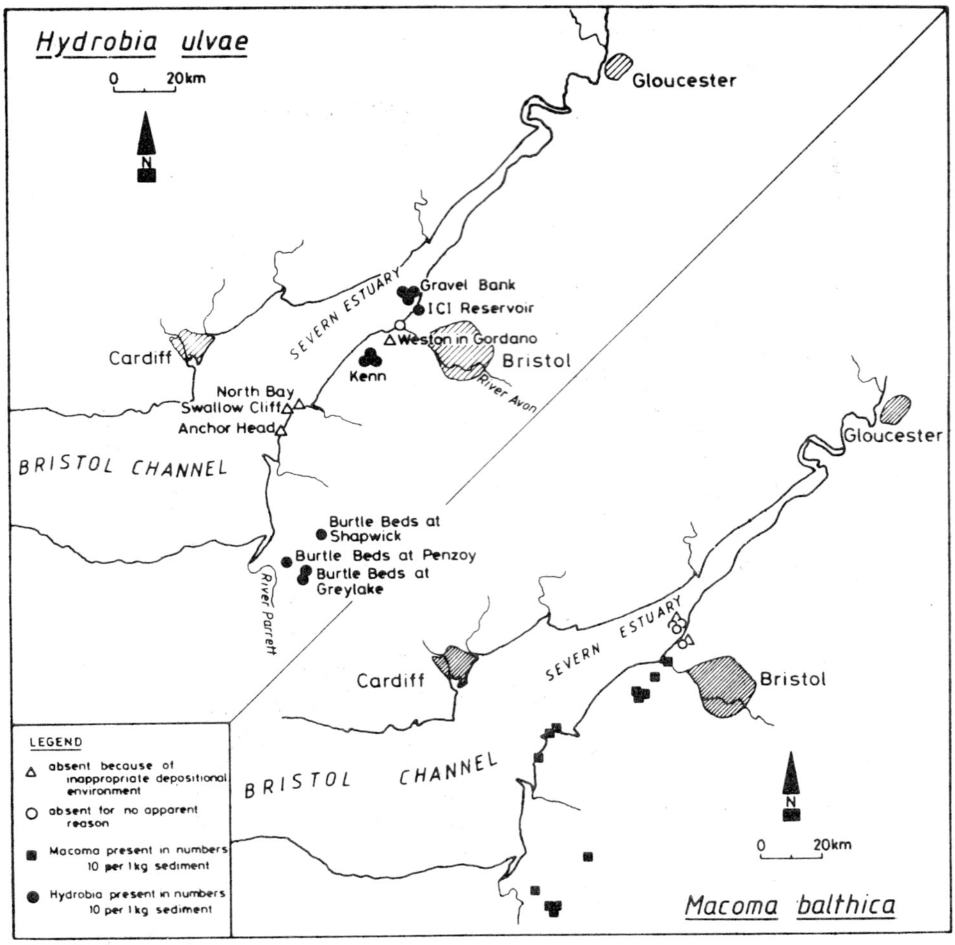

Figure 2. Occurrence of (sub)fossil remains of Hydrobia ulvae and Macoma balthica in Upper Quaternary inter-tidal deposits of the former southern shore of the Severn Estuary.

Botanical approaches

RECENT ADVANCES IN POLLEN EXTRACTION TECHNIQUES: A BRIEF REVIEW.

C.O. Hunt
Department of Prehistory and Archaeology
The University of Sheffield
Sheffield S10 2TN.

ABSTRACT

Environmental archaeologists and Quaternary geologists would find substantial improvements in pollen preparation quality and savings in time if they adopted some of the pollen extraction techniques which have recently been developed, mostly by palynologists working on pre-Quaternary sediments. These techniques include deflocculation in Sodium pyrophosphate, micropore filtration, 'swirling', ultrasonic deflocculation and filtration, and froth filtration. Micropore filtration and 'swirling' are particularly suitable for teaching purposes, noxious chemicals being far less frequently used than in conventional techniques. Alternatively, these new techniques offer supplements or alternatives to more widely-known preparation techniques for research palynology.

INTRODUCTION

Pollen analysis has great potential as a palaeoecological tool for archaeologists, for the pollen record gives an impression of vegetational changes in the past. Since these vegetational changes are almost synchronous over limited areas, pollen analysis may also be useful as a dating tool, if a well-dated pollen diagram is available nearby as comparison.

Three problems confront the prospective archaeological pollen analyst. The first is to select sediment that will contain pollen that can usefully be interpreted. The second is to obtain pollen from the sediment. The third is to identify the pollen and other microfossils present.

The problems connected with sampling have been dealt with by, among others, Dimbleby (1976) and Van der Veen (for plant macrofossils, this volume): pollen identification is less of a problem than it was, since pollen keys such as Moore and Webb (1977) are now widely available (although a type collection is still essential for precise work). Methods of pollen extraction have also improved considerably in recent years, mainly because of advances of pre-Quaternary palynology connected with the oil industry, but several of these methods are now widely known to environmental archaeologists and Quaternary palynologists. This paper summarises some of these recent advances.

ACETOLYSIS AND HYDROFLUORIC ACID MACERATION.

Most pollen analytical work is carried out on organic sediments, such as peat, using the acetolysis technique of Erdtmann (1934). Mineral sediments are generally processed using HF (Hydrofluoric acid) maceration (Assarsson and Granlund 1924). These procedures concentrate pollen by solution of the organic or mineral matrix of the sample. As an alternative or supplement to HF maceration, inorganic samples are disaggregated and the pollen separated by centrifuging in heavy liquids (Knox 1942). All these techniques are extensively reviewed in Brown (1960), and in most text-books on pollen analysis. They share the disadvantages of requiring expensive equipment (fume cupboards, centrifuges) to handle the toxic and very unpleasant chemicals involved. A number of workers have also noted that acetolysis is selectively destructive of the spores and pollen of certain taxa (Erdtmann 1952, Wenner 1947). Despite the 'affection' of many workers for these methods, it is clear that, from the point of view of laboratory safety, expense of chemicals, expense of equipment, time, teaching, and preparation quality, other, 'simpler' techniques are highly desirable.

RECENT ADVANCES

DEFLOCCULATION. Bates, Coxon and Gibbard (1978) have described the use of Sodium pyrophosphate to disaggregate clay-rich samples. Their procedure is (i) to boil the sample in Na OH, centrifuge, and decant, (ii) boil the sample in Sodium pyrophosphate for 20 minutes, centrifuge, and decant. The clay minerals are deflocculated and remain in suspension when the pollen is sedimented by centrifuging. They can be poured off, leaving a sample relatively enriched in pollen and large mineral grains.

This technique may be used in place of HF maceration for clay-rich samples, since silt and sand remaining after pyrophosphate treatment are easily removed by other techniques. Thus Hunt (1984) obtained excellent collections from tills (boulder clays) by boiling the samples in Sodium pyrophosphate, sieving through a 10 micron nylon micropore filter, and swirling.

MICROPORE FILTRATION. Neves and Dale (1963) pioneered the first micropore filtration technique, using a sintered glass filter-funnel with an aperture of 2 micron, linked to a reversible pump, to remove fine organic particles after HF maceration (and neutralisation). Sintered glass funnels have subsequently been found to be useful for washing pollen with acids, alkalis, stains and solvents, and dewatering samples. They can thus be used as an alternative to centrifuging, and are so used extensively by the oil industry. Sintered glass funnels are, however, prone to clogging, even if used with a reversible pump.

The introduction of nylon boulting-cloth has solved the problem of clogging. These filters are made with apertures of 5, 10, 20, 40 and

60 microns. The 10 micron aperture is sufficiently small to retain Quaternary pollen, even tiny grains such as Chrysosplenium while letting through a maximum of fine particles. It is, however, too coarse to retain all algal microfossils; if these are to be studied the 5 um aperture is more suitable. Surface tension will prevent water and fine particles passing through the boulting cloth; a strong, but fine jet of filtered water is generally used to break the surface effect.

Micropore filtration will rapidly remove clay and fine organic particles from pollen preparations previously disaggregated in Sodium pyrophosphate or Potassium hydroxide (KOH), giving very 'clean' preparations. Silt and sand-size particles, however, remain in the sample, and are best removed by 'swirling'.

'SWIRLING'. This procedure is extensively used by pre-Quaternary palynologists. An aqueous suspension of sediment is 'panned' in a clock-glass. If the clock-glass is very gently agitated, silt and sand will fall out of suspension, while pollen remains in suspension and may be poured off. Samples treated in this way, and subjected to micropore filtration, are comparable in quality to those subjected to heavy-liquid separation.

ULTRASOUND. Ultrasound has been used by a number of workers (Marceau 1969, Faegri and Iversen 1974) to disaggregate flocculated organic matter. Recently Carratini (1980) has described the use of an ultrasonic sieving device, where the ultrasound generator is linked to a micropore filter to aid the passage of water and fine detritus. Excessive exposure to low-frequency ultrasound may damage pollen grains, so ultrasound generators emitting more than 80 kHz are preference (Marceau 1969).

FROTH FLOTATION. Dumait, Marceau, Devin and van Campo (1963) and Campbell (1976) have described using the evolution of froth to carry pollen away from a mineral matrix.

DIELECTRIC SEPARATION. Wang Yu-xian (1981) has described a technique which makes use of the different dielectric properties of sediment and microfossils to separate them. This technique has only been applied to calcareous microfossils, but may have applications in palynology.

DISOUSSION

The method used to obtain pollen from any sample must depend on the lithology of that sample. Some generalisations may, however, be made (Figure 1).

Most peats, silts, sands and clays will give excellent preparations with 10 minutes boiling in 5% KOH; followed by sieving through 150 micron mesh to remove large particles; sieving on a 10 micron micropore filter to remove fines; and swirling to remove sand and silt.

Samples full of lignin (opaque organic matter) are improved by a 10 second oxidation in a mixture of concentrated bleach and HCl (Erdtmann and Erdtmann 1933). Samples full of cellulose (for example cell walls) may, however, need acetolysis. The combination of oxidation and acetolysis tend to destroy certain pollen types (Hafsten 1959) and should not be used.

Tenacious clays may not be disaggregated by boiling in KOH. They should be boiled in 5% sodium pyrophosphate for 20 minutes, and vigorously stirred, after which they can be sieved and swirled.

Samples containing a high proportion of silt or mica will usually disaggregate easily; but it may prove difficult to sieve them because the silt particles block the pores of the filter. 'Swirling' before micropore filtration will remove the worst of the silt, and following filtration repeated swirling will remove most of the rest. Should the pollen still be unacceptably diluted it may be necessary to centrifuge in heavy liquids or to macerate in HCl and HF.

EXAMPLES OF THE USE OF THESE 'NEW' METHODS

These techniques have served well in several palynological investigations of Quaternary archaeological and geological sites:

(i) sediment from an interglacial site at Kenn, Avon, (Hunt 1981) where conventional pollen extraction techniques had failed to yield satisfactory assemblages (Gilbertson and Hawkins 1978);

(ii) in investigations of tills (boulder-clays) (Hunt, 1984);

(iii) in investigations of fluvial and estuarine sediments from Somerset (Bradley, Macklin and Hunt, 1984);

(iv) in investigations of archaeological sequences from caves at Creswell (Jenkinson and Gilbertson 1984) and Lancashire (Gale, Hunt and Southgate 1984, in press);

(v) sediment of Medieval age from fluvial sequences in Tuscany (Gilbertson, Holyoak, Hunt and Paget, in press, Hunt in prep.);

(vi) lacustrine and marsh sequences of saline and carbonate-rich deposits at Azraq Oasis in the Jordanian Desert (Gilbertson, Hunt and Bradley in press);

(vii) Late-glacial lacustrine deposits from Skipsea on the Holderness coast (Gilbertson 1984);

(ix) the fills of Iron Age wells and pits from Aulnat, France (Hunt, in Collis and others, in prep.).

In all cases except at Kenn, where HF maceration was necessary to remove excess silt after micropore filtration and 'swirling' acceptable pollen concentration was achieved after less than 10 minutes 'hands-on' time per sample. The rapidity, safety and cheapness of these techniques make them ideal for teaching purposes as well as research, while the 'cleanness' of the preparations makes identification of pollen easier and the counting of pollen a much less tiring procedure.

These techniques have been, and continue in use at the Department of Geology at the University of Sheffield, whence some of the methods ultimately derive (for instance, Neves and Dale 1963); and now are also being successfully used and taught as standard procedures at the Department of Prehistory and Archaeology at the same University.

CONCLUSION

The techniques outlined above offer alternatives or supplements to acetolysis and HF maceration. They make pollen analysis less time-consuming and much cheaper and safer: they reduce or eliminate the need for specialist laboratory facilities, and commonly produce better-quality palynological preparations. An added advantage is that other microfossil groups - diatoms, amoebid and dinoflagellate cysts, desmids, Chrysostomaceae, other algal remains and fungal spores are not automatically destroyed in the quest for pollen and spores.

ACKNOWLEDGEMENTS

The author would like to thank M. Macklin, S. Bradley and D.D. Gilbertson, who read early drafts of this paper, and Ken Dorning for facilities.

REFERENCES

Assarson, G. and Granlund, E. (1924) En metod for pollen analys av minerogena jordarter, Geologiska Foreningen i Stockholm, Forhandlingar 46 (1-2), 76-82.

Bates, C.D., Coxon, P. and Gibbard, P.L. (1978) A new method for the preparation of clay-rich sediment samples for palynological investigation, New Phytologist 81, 455-465.

Brown, C.A. (1960) Palynological Techniques. Baton Rouge, Louisiana, privately published.

Campbell, J.B. (1977) The Upper Palaeolithic in Britain. Oxford: Clarendon Press, 2 vols.

Caratini, C. (1980) Ultrasonic sieving to improve palynological processing of sediments, International Commission for Palynology Newsletter 3 (1), 4.

Dimbleby, G.W. (1976) A review of pollen analysis of archaeological deposits. In Geoarchaeology, Earth Science and the Past, D.A. Davidson and M.L. Shackley (eds) pp. 347-354. London: Duckworth.

Dumait, P., Marceau, L., Devin, C. and van Campo, M. (1963) Nouvelle methode de concentration des pollens dan les sediments pauvres par microflottation, Comptes Rendus (Paris) 265, 231-233.

Erdtmann, G. (1934) Uber die Verwendung von Essigsaureanhydrid bei Pollenuntersuchungen, Svensk Botanissker Tidskrift 28 (2), 354-358.

Erdtmann, G. (1952) Pollen morphology and plant taxonomy. Angiosperms: an introduction to palynology. Stockholm: Almqvist and Wiksell, Vol. 1.

Erdtmann, G. and Erdtmann, H. (1933) The improvement of pollen analysis technique, Svensk Botanissker Tidskrift 27, 347.

Faegri, K. and Iversen, J. (1974) Textbook of pollen analysis. 3rd edition. Oxford: Blackwells.

Frey, G.D. (1951) Pollen succession in the sediments of Singletary Lake, North Carolina, Ecology 31 (3), 518-533.

Gale, S.J., Hunt, C.O. and Southgate, G. (in press) Kirkhead Cave: new evidence of the colonisation of Britain after the last glaciation. Naturwissenschaft.

Gilbertson, D.D. (1984) Late Quaternary Environments and Man in Holderness. Oxford. B.A.R.

Gilbertson, D.D. and Hawkins, A.B. (1978) The Pleistocene succession at Kenn, Somerset, Bulletin of the Geological Society of Great Britain. HMSO No. 66, 44pp.

Gilbertson, D.D., Holyoak, D.A., Hunt, C.O. and Paget, N. (1984 in press) Palaeoecology of Late-Quaternary floodplain deposits in Tuscany: the Feccdia Valley, Archaeologia Medievale, X, 340-350.

Gilbertson, D.D., Hunt, C.O. and Bradley, S.B. (1985 in press) Micropalaeontological and stratigraphic investigations of recent sediments from the Azraq Marshes of the Jordanian Desert. In Hadini, A (ed.) 1985. Studies in the History and Archaeology of Jordan London, Pitmans and Amman; Department of Antiquities, Royal Hashemite Kingdom of Jordan.

Hafsten, U. (1959) Bleaching + HF + acetolysis, a hazardous process. Pollen et Spores 1 (1), 77-79.

Hunt, C.O. (1981) Pollen and organic-walled microfossils from interglacial deposits at Kenn, Avon, Proceedings of the Somerset Archaeological and Natural History Society 125, 73-76.

Hunt, C.O. (1984) Erratic palynomorphs and Palaeolithic stratigraphy. In Fieller, N.R.J., Gilbertson, D.D. and Ralph, N.G.A. Palaeoenvironmental Investigations: Research Design, Methods and Interpretations. Oxford: B.A.R.

Jenkinson, R.D.S. and Gilbertson, D.D. (1984) In the Shadow of Extinction: A Quaternary Geology and Palaeoecology of the Lakes, Fissures and Smaller Caves at Creswell Crags. Sheffield. University of Sheffield Monographs in Prehistory and Archaeology. Chapter 2.

Knox, A.S. (1942) The use of Bromoform in the separation of non-calcareous sediments, Science 95 (2464), 307-308.

Marceau, L. (1969) Effets, sur le pollen, des ultrasons de basse frequence, Pollen et Spores 11, 147-164.

Moore, P.D. and Webb, J.A. (1978) An Illustrated Guide to Pollen Analysis. London: Hodder and Stoughton.

Neves, R. and Dale, B. (1963) A modified filtration system for palynological preparation, Nature (London) 198, 775-776.

Veen, M. van der, (1984) Carbonised seeds, sample size and on site sampling In Fieller, N.R.J., Gilbertson, D.D. and Ralph, N.G.A. Palaeoenvironmental Investigations: Research Design, Methods and Interpretations. Oxford: B.A.R.

Wang Yu-Xian. (1981) High frequency dielectrical separation of microfossils, Lethaia 14, 261-268.

Wenner, C-G. (1947) Pollen diagrams from Labrador. Thesis, reprinted from Geografisker Annaler 1-241.

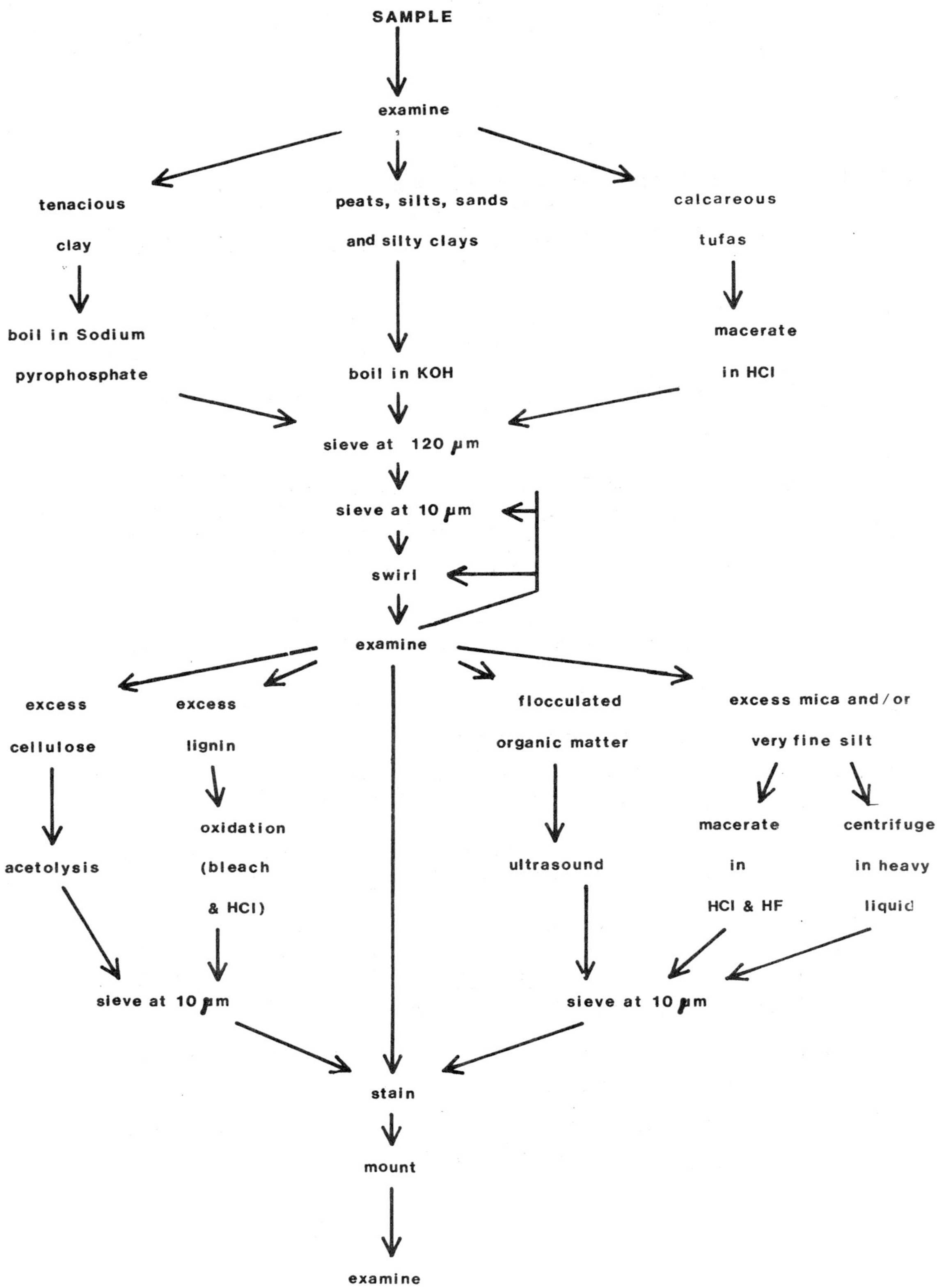

Figure 1. Sequence of pollen preparation techniques.

THE PROBLEM OF THE TIME SPAN REPRESENTED BY POLLEN SPECTRA IN PODZOL TURVES, WITH EXAMPLES FROM THE ROMAN SITES AT BAR HILL AND MOLLINS, CENTRAL SCOTLAND.

W.E. Boyd
Department of Botany
University of Glasgow
Glasgow G12 8QQ

ABSTRACT

Ditch infill associated with the Roman forts at Bar Hill and Mollins, in central Scotland, includes well-preserved fossil turves. On the basis of the archaeological context, they represent the land surface present on the arrival of the Romans at these sites, and were analysed to provide details regarding the immediately pre-Roman (Iron Age) environment and its development. Interpretation of this data is difficult, and discussion in this paper concentrates on one aspect of the interpretation; that is the time-span represented by the pollen content of the turves. At these sites, the recorded Iron Age vegetational changes are shown to have occurred over 150 to 350 years. The uncertainty in this conclusion, however, emphasises the need for a coordinated archaeological-cum-environmental research programme.

INTRODUCTION

This paper examines one of the problems encountered in the interpretation of the pollen spectra contained in fossil podzol soils. Following a brief description of the principal forms of interpretation for soil pollen spectra, the problem, that of assessing the time span represented in the soil pollen spectra, is discussed using data from the Roman Antonine fort at Bar Hill and the Agricolan marching camp at Mollins.

POLLEN ANALYSIS OF SOILS: AN INTRODUCTION

Sediment recovered from archaeological contexts is often analysed for its pollen content, providing data about the vegetation at the site before or as the sediment was incorporated into the structure concerned.

Pollen analysis is also frequently used to provide dating information or control of archaeological sites. Frequently the sediment used comprises fossil turves. The interpetation of pollen data from soils contrasts with that from peat, lake deposits and other accumulated sediments. The processes by which pollen is incorporated into these latter are relatively straightforward, and consequently stratigraphic changes can usually be interpreted in terms of temporal changes in the vegetation and, to a lesser extent, pollen transport processes. However, pollen entrainment into soils is more complex, and account must first be taken of the soil processes responsible for the movement and destruction of pollen and other material within the soil.

In base-rich and/or well-drained soils, such as those typically found at lowland archaeological sites, it is generally considered that pollen is incorporated from the surface, organic-rich layers into the lower, inorganic layers by faunal mixing and, to a lesser extent, down-washing. In such soil, the processes controlling pollen distribution are dominated by those causing rapid destruction of pollen; not only is the chemical environment itself highly unsuitable for the preservation of pollen, but also under such conditions a highly active microbial fauna is present, contribution to the pollen degradation. Consequently, the contained pollen at any time is relatively recent, and what there is, is mixed throughout the profile, providing a homogeneous pollen spectrum down to the lower limit of faunal mixing (Dimbleby and Evans 1974). If acidity increases (for whatever reason), the soil mixing decreases and ultimately ceases as faunal activity declines, and pollen preservation improves. The pollen spectra, while still being homogeneous throughout the soil, then represents an earlier period, dating from a time prior to the increase in acidity (e.g. Dimbleby 1963, Tubbs and Dimbleby 1965). Several phases of increasing acidity may be recognised by the presence of discrete (pollen) biostratigraphic units within a soil, representing the periods prior to critical pH and faunal changes (e.g. Godwin 1958, Keef, Wymer and Dimbleby 1965, Keatinge 1983).

The situation is different with acid soils (pH less than c. 5.5), such as podzols, which are now widespread in upland Britain. In many cases, turves of podzol soil have been incorporated and preserved into archaeological structures (see references in Table 1), and where pollen analysis is conducted through such turves, the patterns of pollen stratigraphy are often seen to be different to those described above. Rather than discrete biostratigraphic units, the changes tend to be gradual, forming what may be termed a "graduated stratigraphy". It is not possible to explain such a stratigraphy in terms of faunal mixing or preservation processes. In this case downwashing and pollen preservation dominate the processes, the pollen haveing been slowly washed down through the soil, probably in aggregates bound together by organic matter. Consequently deeper pollen represents pollen deposited on the soil surface earlier than pollen which occurs nearer the top of the profile (Dimbleby 1961). In certain cases, it can be demonstrated that pollen introduced into the soil in this manner was also mixed by the soil fauna, where the soil pH was slightly higher (Andersen 1979).

A graduated pollen stratigraphy may be interpreted as a time-sequence, and offers important opportunities to gather palaeoenvironmental data at archaeological sites, not only regarding the vegetation present when the feature was constructed, but also the most recent history of the vegetation at the site. This contrasts with the situation in which faunal mixing and pollen destruction are dominant, and which therefore provide data regarding the vegetation at isolated periods only. One important consequence of the presence of graduated pollen stratigraphy in certain soils is that the analysis of single samples taken from fossil turves (e.g. Casparie 1976, Jones 1981) may not provide data regarding the vegetation when the turf was incorporated into the archaeological structure.

INTRODUCTION TO AN EXAMPLE

If the pollen spectra from podzol soils do represent time-sequences, there remains the problem of estimating the length of time represented. This problem arose during the interpretation of pollen data from two Roman sites in central Scotland (Figure 1), where fossil podzol turves were recovered from infilled ditches. At Bar Hill (NS 707 759), a native Iron Age ditch system was infilled, mostly with turf fragments, immediately prior to the construction of the Antonine Wall fort in AD 142 (Keppie, in prep.), and about 4km south at Mollins (NS 714 719), the ditch surrounding an Agricolan marching camp was deliberately backfilled with turves and other soil debris when the camp was abandoned after several years of occupation around AD 80-90 (Hanson and Maxwell 1980, Hanson, pers. comm.). At both sites the turves (Table 1) and their contained pollen are remarkably well-preserved. The pollen analyses (Figures 2 and 3) indicate that in both cases, graduated pollen stratigraphies are present, and they can, with minor exceptions, be interpreted as time-sequences.

The main questions to be answered concerned the vegetation at or immediately prior to the Roman occupation at these sites, and the history of that vegetation. As part of the interpretation of the vegetational history, it was important to establish the time span represented by the pollen. The problems regarding the vegetational history can be answered in a standard manner by reference to contemporary vegetation, and this aspect of the interpretation will not be discussed here, but will be presented elsewhere; briefly, the vegetational history of the area is that after some earlier period(s) of probable clearance not recorded in these diagrams, the local vegetation consisted of open or scattered woodland, primarily of *Alnus* (alder), *Quercus* (oak) and possibly *Corylus* (hazel), followed by further forest clearance associated with pastoral agriculture and later partial expansion of *Calluna* (heather). A difficult problem to solve is the time scale over which such changes occurred: do they represent only immediately pre-Roman Iron Age activity (say over tens of years) or longer term activity attributable to Iron Age and possibly even Bronze Age communities (that is, over hundreds or thousands of years)?

DISCUSSION The pollen data used in this discussion are presented on Figures 2 and 3. Three independent approaches are taken to estimate the time during which the vegetational changes took place. The data presented here is compared with (i) modern rates of vegetation changes, and (ii) past regional vegetational changes recorded at sites elsewhere. Thirdly, the data sets from the two sites are compared with each other.

Plant and vegetation ecology are widely discussed in the literature, from which it may be possible to estimate rates of past vegetational change. However, all the major changes recorded at Bar Hill and Mollins, as probably is the case in most archaeological contexts, were influenced by human related activity, and changes probably did not progress at natural rates. In the example discussed here, one trend which may be significant, in turf 3 at Bar Hill, is the *Calluna* curve. The sharp rise in values may represent the period between the establishment of seedlings as *Calluna* colonised the area, and the period in the growth cycle in which profuse flowering occurs about 15 to 20 years later (Gimingham 1972). Since the growth of young *Calluna* plants may have been retarded by grazing (at Bar Hill), this time span is probably a minimum value, and suggests that the turves contain pollen representing at least 1 year per mm depth of soil, and thus that the vegetation changes described above occurred during, at least, 100 years, and possibly during a longer period.

If the local vegetational changes, especially woodland clearance, recorded at these sites occurred widely in central Scotland, they should be recognisable in regional diagrams, and indeed at several regional sites, trends around or before *c*. 2,000 b.p. are similar to those at Bar Hill and Mollins (Durno 1956, Turner 1965, 1975, Walker 1975, Stewart 1979, Boyd 1982). Mostly these trends are not dated accurately. However, at Bloak Moss, in North Ayrshire (Godwin *et al*. 1975, Turner 1975), vegetational changes representing extensive clearances are similar to the changes recorded at Mollins and, especially, Bar Hill, and two radiocarbon dates [2,375 ± 90 b.p. (Q723); 1,535 ± 90 b.p. (Q722); difference = 840 ± 130 years] bracket an associated *Alnus* decline which follows a decline in *Quercus*. It is difficult to apply dates from one site to another about 40km away. However, as a first working hypothesis, the vegetational events recorded at these sites are assumed to represent similar human activity which occurred at about the same time. The date difference is best taken as an order of magnitude value, and suggests that at both Bar Hill and Mollins, vegetational changes, similar in form and timing to those at a distant site, may have occurred over several hundred years and possibly over 1,000 years. This suggests that the pollen within the turves could represent up to approximately 9 years per mm depth of soil.

Since Bar Hill and Mollins are only about 4km apart, there is good reason to compare the data from these two sites. Examination of the pollen curves at Bar Hill and Mollins reveals that parts of the curves at the two sites probably represent similar vegetational changes; the upper part of the Mollins diagram is equivalent to the lower part of the Bar Hill diagrams. Since the turves at Bar Hill represent the soil surface at about AD 142 and that at Mollins represents the soil just prior to AD 80-90, the vegetational developments represented at Bar Hill but not at Mollins presumably occurred between these two dates, i.e. over approximately 50-60 years. This period is represented at Bar Hill by pollen in approximately 20 to (mostly) 40mm of soil depth, and thus that rate at which pollen represents time is about 1.25 to approximately 3 years per mm depth of soil. At Bar Hill, this range of rates indicate that the recorded vegetational changes occurred over a period of approximately 150-350 years, probably nearer the former.

CONCLUSIONS

These three approaches provide results which are slightly at variance with each other. The first and third methods used local data, and both indicated that the rates of time with soil depth represented by the pollen are of the same order (about or slightly greater than 1 year/mm).

In terms of the vegetational history recorded at Bar Hill, this represented a time span of the order of 100-200 years. In terms of the original questions posed, this represents moderately long-term Iron Age interference in the local vegetation, but precludes records of earlier Bronze Age activity.

Of the two methods it is considered that comparison between the two sites provides the most reliable results, since there is greater control on the data used. The second method, applying radiocarbon dates from a regional site, is fraught with difficulties. Apart from the obvious problems regarding the contrasting nature of the sites and pollen evidence being used (regional versus local), problems regarding the transposition of radiocarbon dates from one site to another and uncertainties regarding the dates themselves (there are some spurious data regarding other radiocarbon dates from the Bloak Moss cores since these dates - nine in all - indicate very irregular peat-growth rates, and two of the dates are inverted) introduce a lack of confidence into this method of dating from Bar Hill and Mollins, and consequently, this method is regarded as the least reliable of the three.

To put this discussion into perspective, it is interesting to compare the results discussed here with results of similar pollen analysis undertaken on soil profiles, mostly fossil podzols, at sites elsewhere. Some of the available data is summarised on Table 2. The values must be treated with caution, and are, at best, order of magnitude values. The values represent pollen analyses of various soil thicknesses, the sediment analysed always including the uppermost, organic-rich horizons and, in many cases, part or all of the leached and lower, inorganic rich horizons. Whether the rates of time represented vary between horizons is unclear, although this remains a strong possibility. Certainly pollen concentration may be variable, especially being greater than average in the humus layer, where that layer may be compacted. The rate values presented in Table 1 represent averages through the entire analysed parts of the soil described, and potential differences between horizons are masked. Most of the rates lie between 1 and 10 years/mm, and it is encouraging to note that the rates estimated for the data at Bar Hill and Mollins fall within this range. Although there are certain cases where the rate is very much higher than these values, it appears that where there is a recognisable graduated pollen stratigraphy which can reasonably be interpreted as a time-sequence, the rate at which time is represented is moderately low.

CONSEQUENCES FOR RESEARCH DESIGN

There are certain weaknesses in this analysis which could be rectified by changes in and/or development of a research design. There are problems often specific to work such as this which was undertaken as two separate specialist items on contract to archaeological investigations. As such, the soils which were analysed were essentially chance finds, and the opportunity of analysing and comparing similar types of evidence from two nearby sites also happened by chance. Since the comparison of data from similar material from nearby dated sites provided the most reliable results in the example discussed above, more active involvement by the specialist both at the sampling stage and, perhaps more important, at the excavation planning stage, should reduce the element of chance from this type of research, and increase the reliability and value of the results provided by the specialist.

Although the comparison of the soil pollen data with the regional data is regarded as being the least reliable method in the case discussed here, it is felt that this is till a potentially valuable method. If a regional site was chosen with care, it would provide a long and continuous record of the vegetational history in the vicinity of the archaeological site(s) being investigated, a record which would not only be directly relevant to the archaeology of the area, but also be free of some of the problems discussed here. It would provide a strong framework for discussion of samples taken at a site, and increase the value on return of the investment in employing specialist investigators within the archaeological investigation.

ACKNOWLEDGEMENTS

This work was funded by the Ancient Monuments Branch of the Scottish Development Department, and carried out in the Department of Botany, University of Glasgow.

REFERENCES

Andersen, S. Tho. (1979) Brown earth and podzol: soil genesis illuminated by microfossil analysis, Boreas 8, 59-73.

Ashbee, P. and Dimbleby, G.W. (1976) The Moor Green Barrow, West End, Hampshire: excavations 1961. Proceedings of the Hampshire Field Club Archaeological Society 31, 5-18.

Boyd, W.E (1982) The stratigraphy and chronology of Late Quaternary raised coastal deposits in Renfrewshire and Ayrshire, Western Scotland. Unpublished PhD Thesis, University of Glasgow.

Case, H.J., Dimbleby, G.W., Mitchell, G.F., Morrison, M.E.S. and Proudfoot, V.B., (1969) Land use in Goodland Townland, Co. Antrim from Neolithic times until today, Journal of the Royal Society of Antiquaries of Ireland 99, 39-53.

Casparie, W.A. (1976) Palynological investigations of the celtic field near Vaassen, the Netherlands. In Air Photography and Celtic Field Research in the Netherlands. J.A. Brongers (ed). Netherlandse Oudheden 6, 105-119.

Crampton, C.B. (1963) The development and morphology of iron pan podzols in mid and south Wales, Journal of Soil Science 14, 282-302.

Dimbleby, G.W. (1961) Soil pollen analysis, Journal of Soil Science 12, 1-11.

Dimbleby, G.W. (1963) Pollen analysis of a mesolithic site at Addington, Kent, Grana Palynologica 4:1, 140-148.

Dimbleby, G.W. and Bradley, R.J. (1975) Evidence of pedogenesis from a Neolithic site at Rackham, Sussex, Journal of Archaeological Science 2, 179-186.

Dimbleby, G.W. and Evans, J.G. (1974) Pollen and land-snail analysis of calcareous soils, Journal of Archaeological Science 1, 117-133.

Durno, S.E. (1956) Pollen analysis of peat deposits in Scotland, Scottish Geographical Magazine 72, 177-187.

Florin, M-B. (1975) Microfossil contents of two soil profiles from western Kolmarden, southern central Sweden, Geologiska Foreningensi Stockholm Forhandingar 97, 135-141.

Gimingham, C.M. (1972) Ecology of Heathlands. London: Chapman and Hall.

Godwin, H. (1958) Pollen-analysis in mineral soil. An interpretation of a podzol pollen-analysis by Dr. G.W. Dimbleby, Flora 146, 321-327.

Godwin, H., Willis, E.H. and Switsur, V.R. (1975) Cambridge University natural radiocarbon measurements VII, Radiocarbon 7, 205-212.

Hanson, W.S. and Maxwell, G.S. (1980) An Agricolan praesidium on the Forth-Clyde isthmus (Mollins, Strathclyde), Brittania 11, 43-49.

Jones, R.L. (1981) Report on the pollen samples: Appendix 6 in Neil, M.R.J., A Bronze Age burial mound of Holland, St. Ola, Orkney, Glasgow Archaeological Journal 8, 32-45.

Keatinge, T.H. (1983) Development of pollen assemblage zone in soil profiles in southeastern England, Boreas 12, 1-12.

Keef, P.A.M., Wymer, J.J. and Dimbleby, G.W. (1965) A Mesolithic site on Iping Common, Sussex, England, Proceedings of the Prehistoric Society 31, 85-92.

Munaut, A.V. (1966) Un type do sol rarement observe en Belgique: le podzol humique hydromorphe developpe sous foret, Agricultura 1, 105-109.

Stewart, D.A. (1979) The Flandrian Vegetational History of the Loch Lomond Area. Unpublished PhD Thesis, Department of Botany, University of Glasgow.

Tubbs, C.R. and Dimbleby, G.W. (1965) Early agriculture in the New Forest, Advancement of Science 22 (96), 88-97.

Turner, J. (1965) A contribution to he history of forest clearance, Proceedings of the Royal Society of London B161, 343-353.

Turner, J. (1975) The evidence for land use by prehistoric farming communities: the use of three-dimensional pollen diagrams. In The effects of man on the landscape: the Highland zone. J.G. Evans, S. Limbrey and H. Cleere (eds) Council for British Archaeology Research Report 11, 86-95.

Vuorela, I. (1976) An instance of slash and burn cultivation in Southern Finland investigated by pollen analysis of a mineral soil, Fauna Flora Fennica 52, 24-46.

Walker, A. (1975) A Study of the Vegetation History of Dubh Lochan, a small loch near Loch Lomond. Unpublished M.Sc. Thesis, Department of Botany, University of Glasgow.

Table 1: Summary stratigraphies of the turves. Layer thicknesses in mm.

Description		Interpretation
Bar Hill turves 1 to 4		podzols
2 - 5	compressed moss fragments, mostly *Polytrichum commune*; charcoal abundant; *Calluna* twigs & leaves common; *Juncus conglomeratus/effusus* seeds present to abundant; *Carex* cf. *panicea* nutlets and *Potentilla erecta* achenes occasional.	A horizons (humus-rich)
20 - 30	dark brown humus-rich sandy silt; charcoal abundant; *Juncus conglomeratus/effusus* seeds present to abundant; moss fragments and *Calluna* twigs & leaves common; *Carex* cf. *pilulifera* nutlets, *Potentilla erecta* achenes and *Polygala* sp. seeds present.	
min. 90	light-coloured sandy silt, becoming less organic and increasingly leached towards the base; charcoal absent to common; *Juncus conglomeratus/effusus* common to abundant; other plant remains scarce.	E horizon (leached)
Mollins turf 1		podzol
5 - 10	brown humus-rich sand; plant remains scarce; occasional moss fragments and *Juncus* spp. seeds.	A horizon (humus-rich)
min. 65	very light yellow-brown sand with occasional orange patches and sandstone gravel; no plant macro-remains.	E horizon (leached)

	AUTHOR	PLACE/SITE	SOIL TYPE	THICKNESS OF ANALYSED PART OF OF SOIL PROFILE (mm)	TIME SPAN (YEARS)	INTERPOLATED RATE (YEARS/MM)
1.	Andersen 1979	E. Jutland, Denmark	podzol	300	c 200	c 0.7
2.	Dimbleby 1961	Matley Wood	acid brown forest soil	300	200	0.7
3.	Andersen 1979	E. Jutland, Denmark	podzol	220	200	0.9
4.	Crampton 1963	Wales	gleyed & peaty gleyed podzol	510	500	1.0
5.	Vuorela 1976	S. Finland	podzol	90	100	1.1
6.	Dimbleby 1963	Addington, Kent	partly disturbed cultivated soil	250	300	1.2
7.	Tubbs & Dimbleby 1965	Hincheslea	podzol	300	c 500	c 1.7
8.	Tubbs & Dimbleby 1965	Holmsley, New Forest	humus podzol	300	c 1,000	c 3.3
9.	Godwin 1958	Cock Head, Glaisdale Moor N. Riding, Yorks	humic (?) podzol	230	c 1,000	c 4.4
10.	Case *et al*. 1969	Goodland Townhead, Co. Antrim	podzol	c 200	1,000	5.0
11.	Ashbee & Dimbleby 1976	Moor Green Barrow, Hamps.	humic (?) podzol	280	1,500	5.4
12.	Florin 1975	S. Sweden	podzolized till	450	c 2,500	c 5.5
13.	Dimbleby & Bradley 1975	Rackham, Sussex	well-leached podzol	275	2-2,500	c 8.2
14.	Munaut 1966	Belgium	humic podzol	300	c 5,000	c 16.7
15.	Keef *et al*. 1965	Iping Common, Sussex	humus-iron podzol	130	c 5,000	c 38.5

TABLE 2: Estimates of rate of pollen accumulation (years per mm thickness of soil) for fossil podzols. In most cases the rates are estimated from data provided by the authors, and rarely is the time span, let alone the rate, expressly or accurately stated in the original papers. NOTE: these values do not necessarily provide an age for the soils, only the contained pollen.

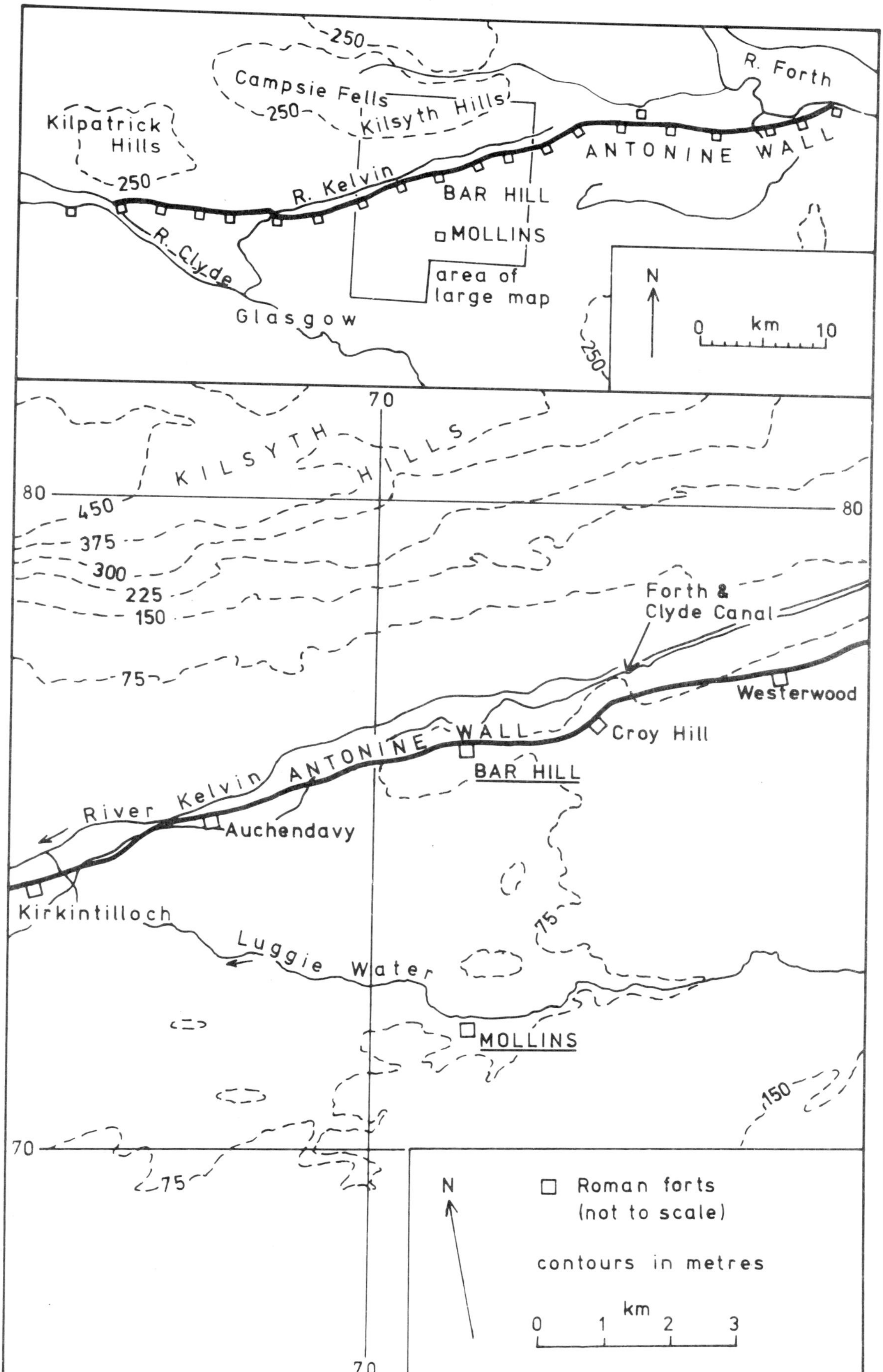

Figure 1: Location map.

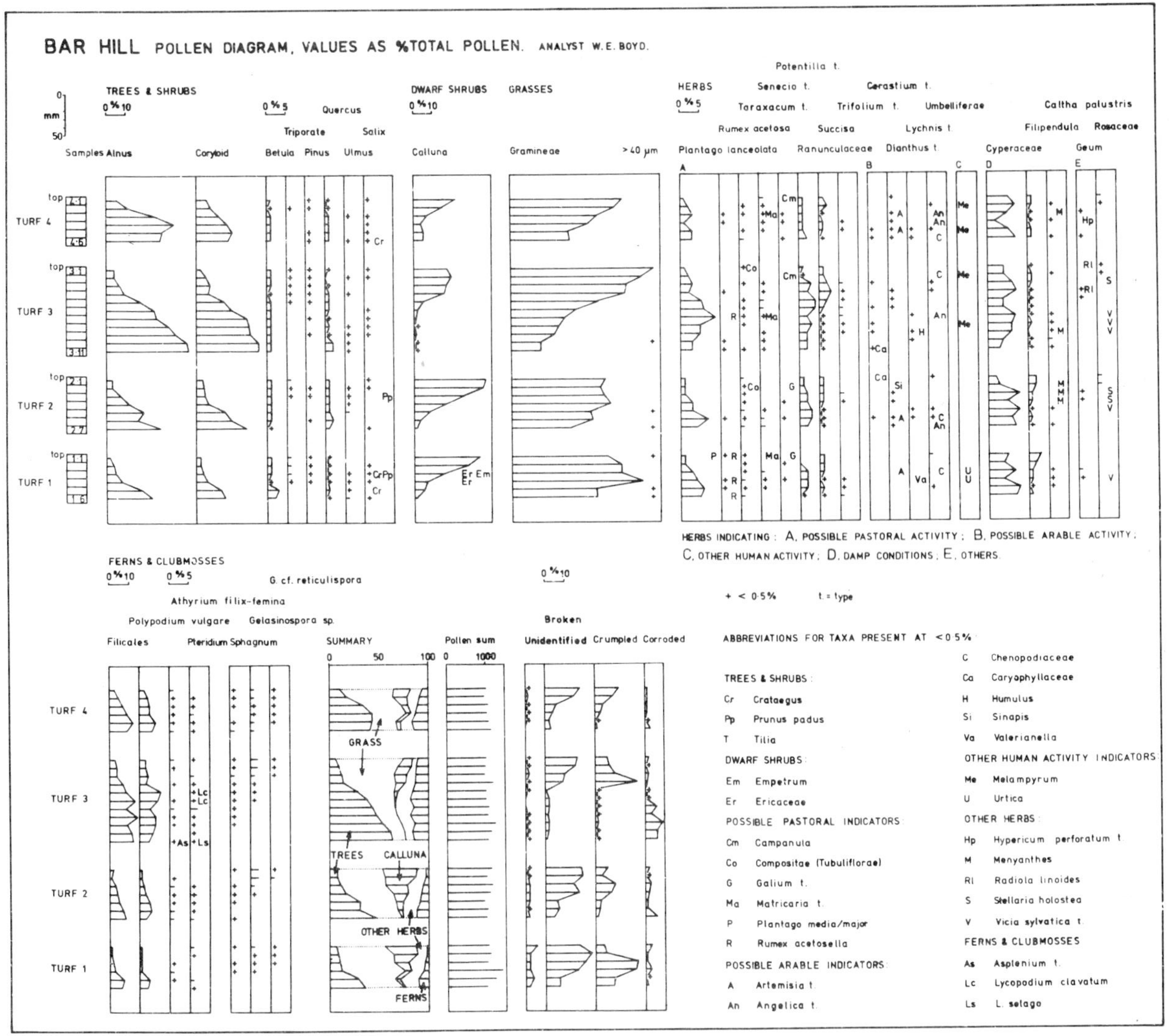

Figure 2: Pollen diagram for Bar Hill, Scotland.

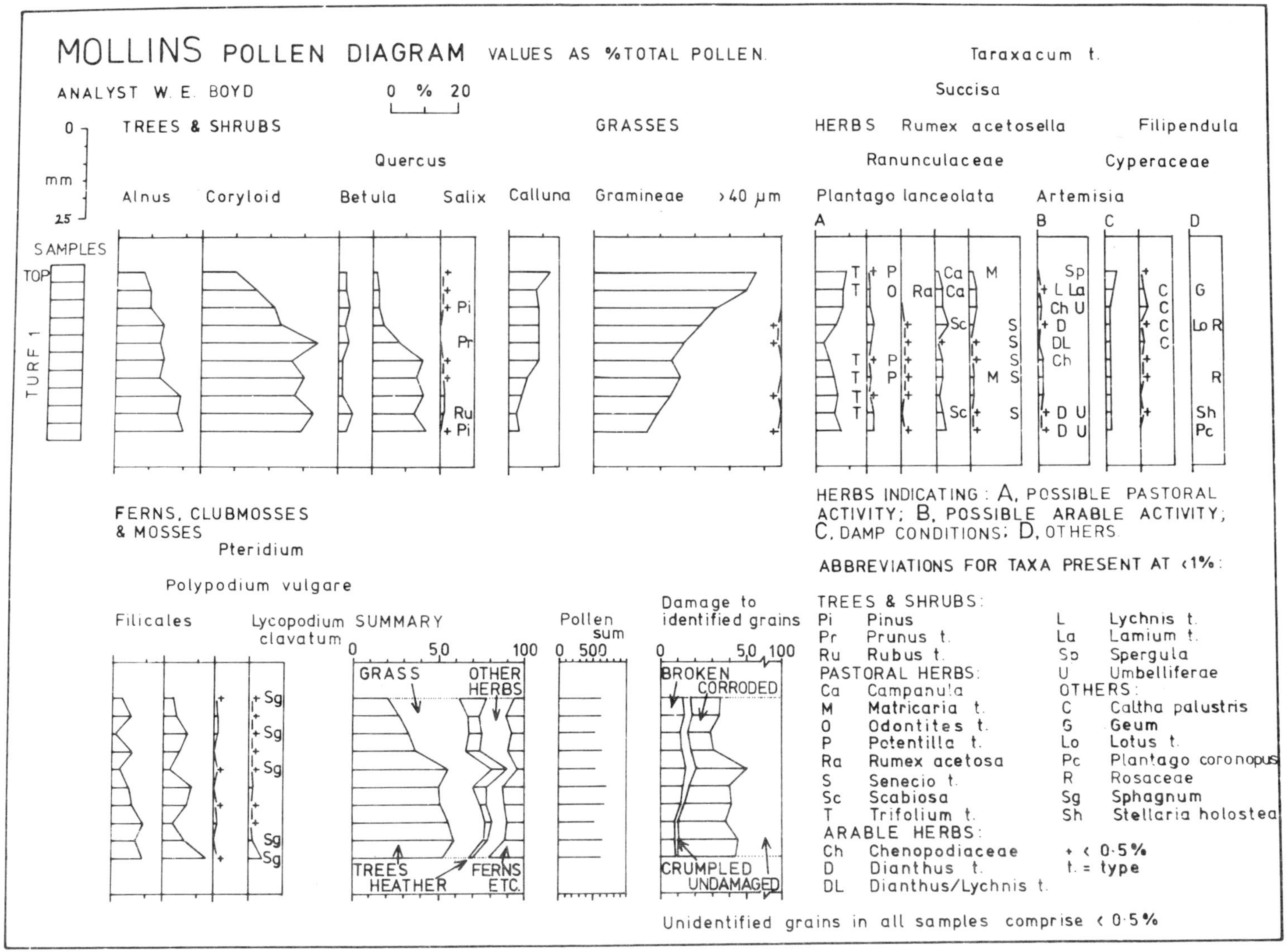

Figure 3: Pollen diagram for Mollins, Scotland.

EVIDENCE FOR CROP PLANTS FROM NORTH-EAST ENGLAND: AN INTERIM OVERVIEW WITH DISCUSSION OF NEW RESULTS.

Marijke van der Veen
Department of Archaeology
University of Durham
England.

ABSTRACT.

In this article the evidence for arable agriculture in the North-East of England during the prehistoric period is discussed. The results of the pollen diagrams from this region are summarised and the results from the analysis of carbonised plant remains from five prehistoric sites are presented. It is suggested that the changes in landscape and agriculture known to have occurred around 500 BC. in the south of England, did not take place in the North-East until much later.

In diesem Artikel wird die Möglichkeit eine agrarische Landwirtschaft im Nord-Osten von England in der Vorgeschichte diskutiert. Die Ergebnisse der Pollendiagrammen aus dieser Gegend werden zusammengefasst, und das Resultat der Analyse der verkohlten Pflanzenreste von fünf vorgeschichtlichen Ausgrabungen wird vorgelegt. Es wird vorgeschlagen, dass die Veränderungen der Landschaft und der Landwirtschaft, die, unserem Wissen nach, um 500 v. Chr. in Süd-England stattfanden, im Nord-Osten erst später stattfanden.

INTRODUCTION.

Many readers will be familiar with Piggott's description of the people living north of the Humber: "Celtic cow-boys and shepherds, footloose and unpredictable, moving with their animals over rough pasture and moorland"(Piggott 1958: 25). The absence in the North of England of storage pits for grain, Celtic-fields and finds of carbonised grain of prehistoric date, reinforced by Caesar's comments that "the people of the interior do not for the most part cultivate grain, but live on milk and meat..." (Piggott 1958: 16) led Piggott to divide the Iron Age economy into two types: the Woodbury type, prevalent in the south-east of England, consisting of stable farming communities, growing cereal crops, using well ordered field systems and capable of producing a surplus, and the Stanwick type, characteristic of the North and West of the country, an economy based on pastoralism and with a probable element of limited nomadism. Agriculture in this region was limited to hoe-cultivation in the Bronze Age tradition. The area north of the Tyne - Solway line was assumed to have remained in a retarded Bronze Age cultural tradition, until refugees from southern Britain arrived in the first century BC (Piggott 1958).

Piggott's view of the north of England as a sparsely inhabited region with a subsistence base of semi-nomadic pastoralism no longer finds general favour, but, so far, very little has been put in its place. Recent surveys of the archaeology of the North-East of England (Clack 1982, Gates 1982b, Gates 1983, Haselgrove 1982 and Jobey 1982) are all very cautious in expressing their ideas about the prevailing economy. It is generally accepted that the population in the North-East was larger than originally postulated and that some form of crop production was present, especially in the coastal areas. However, notions about the role of agriculture in the overall subsistence base remain vague. Extensive fieldwork and aerial photography have greatly improved our knowledge of the distribution of settlements and field systems in the North-East (Gates 1982b and 1983, Harding 1979). Storage pits for cereal grains, however, are still absent from the region, but this might be related to differences in soil types. In the last two years finds of cereal grains from prehistoric contexts in the North-East have come to light, so that by now at least two of Piggott's three criteria have been proved incorrect.

In addition to the extensive fieldwork and aerial photography, the results of pollen analysis have greatly improved our knowledge of settlement and land use of the region. For the North-East of England alone, the results of some 25 pollen diagrams have been published in the last 10 years, showing the vegetational changes through time in the various parts of the region and especially showing the impact of man on the vegetation. Various phases of forest clearance and regeneration could be detected, and pastoral and agricultural indicators offer some suggestion for the type of activities carried out in the vicinity of the pollen site. The occurrence of cereal pollen in the diagrams has often been the first indication of the practice of agriculture in the region. However, the technique of pollen analysis can rarely go further than indicating the presence of crop production in a particular region.

If we want to improve our understanding of the subsistence patterns in the region it is now the time to start looking at the primary products of the activities we want to study: i.e. the crop plants and domesticated animals themselves. Very little is known about the animal husbandry, because the frequent acid soils in the region do not preserve animal bones. However, the survival of carbonised plant remains is not influenced by soil conditions and the material should be present on most sites. Unfortunately, carbonised plant remains were not systematically collected on any site prior to October 1981. Since that date I have tried to make the collection of carbonised plant remains an integrated part of all excavations and now, almost two years later, some fruits can be reaped.

In this article a summary is given of the results of pollen analysis in the region, focussing on the evidence for arable agriculture. In the second half, new evidence from the analysis of the carbonised plant remains from five excavations, spanning the period of the Late Neolithic to the Late Iron Age, will be presented. The evidence from both the pollen analysis and the carbonised plant remains will then be set against the information from the south of the country. The scope of the article is geographically restricted to the counties Northumberland, Tyne and Wear, Durham and Cleveland north of the River Tees. The locations of all pollen sites and excavations mentioned in the text are given in Figure 1.

THE POLLEN EVIDENCE FOR ARABLE AGRICULTURE.

In this section the evidence for crop production as reflected in the pollen diagrams will be discussed. For convenience the region is here divided into three areas: 1. Northumberland, 2. West-Durham, i.e. upper Weardale and upper Teesdale, and 3. East-Durham, i.e. the coastal zone.

1. Northumberland.

A total of 6 pollen diagrams have been published for this region, three from central and northern Northumberland: Camphill Moss, Broad Moss and Steng Moss (Davies & Turner 1979), and three from the south western part: Fellend Moss (Davies & Turner 1979), Coom Rigg Moss (Chapman 1964) and Muckle Moss (Pearson 1960). Only three of these diagrams have been radiocarbon dated (Camphill Moss, Steng Moss and Fellend Moss), and some of the dates from Camphill Moss are, in fact, not reliable, because of suspected contamination. As a result the dating evidence for crop production relies heavily on the diagrams from Steng Moss and Fellend Moss.

Of the diagrams from central and northern Northumberland only one covers the Neolithic period (Steng Moss). During this period the area was still heavily wooded and no substantial human interference with the vegetation could be detected. In all three diagrams small clearances start appearing in the Bronze Age, each clearance normally followed by a phase of forest regeneration. At Steng Moss these clearances are dated to 1644±45 bc (only pastoral indicators present), 1065±45 bc and 636±45 bc (in the last two phases cereal pollen present). From 578±35 bc onwards a period of limited, but continuous forest clearance starts. At Broad Moss three clearance phases are present at supposed Bronze Age levels, with cereal pollen and pollen of arable weeds present in the last two phases. At Camphill Moss an increase in the frequencies of pastoral indicators (Gramineae, *Plantago lanceolata*, *Rumex* spp.) is dated between 1560±70 bc and 1160±80 bc. Another clearance phase starts at 720±70 bc, containing cereal pollen and pollen of arable weeds. The date for the end of this phase (ad 1325±80) is assumed to be incorrect, possibly due to modern contamination.

In all three diagrams these clearance episodes are followed by widespread and large-scale forest clearance, during which tree pollen frequencies are reduced to very low values, often similar to those of today. The beginning of this phase is dated to 20±60 bc at Steng Moss. Here cereal cultivation seems to have obtained some importance. At Broad Moss the supposed Late Iron Age/Romano-British levels contain cereal pollen, though not in large quantities. It is assumed that the pastoral economy here continued to be important, associated with only small-scale cultivation. At Camphill Moss a clearance phase containing very low values for trees and high ones for cereals, Cannabis and Juglans (with cereal pollen rising to a maximum of 22% A.P.) could possibly be dated to the Roman period, but unfortunately no radiocarbon dates are available. The end of this large scale forest clearance phase is dated to ad 460 ± 60 in the Steng Moss diagram.

Of the three diagrams from south-western Northumberland only Fellend Moss is radiocarbon dated. In this diagram the first sign of man's impact on the forest is dated to the Bronze Age, when a small forest clearance is associated with substantial amounts of pollen of Gramineae, Plantago lanceolata, Rumex etc. No cereal pollen were found, but some pollen of arable weeds were recorded. This phase is dated around 1735±60 bc. After this phase the forest regenerates and not until ad 2±45 is a new clearance detected. Pollen of pastoral indicators are prevalent and only a few cereal grains are recorded from one level. This phase ends c. ad 620±40.

At Coom Rigg Moss evidence for forest clearance is recorded for levels assumed to pre-date the Roman occupation (Chapman 1964). At Muckle Moss pollen of PlantagO lanceolata, Compositae and cereals were found in levels dated to AD. 100 - 200 (Pearson 1960). At both sites, however, the dating is based entirely on Godwin's pollen zones and estimated rates of peat growth. The boundary of zone VIIb and VIII is set at 500 BC. by both authors. This method of dating pollen diagrams is nowadays regarded as inaccurate (Godwin 1975).

The picture emerging from the diagrams of this region is fairly uniform. The first small clearances occur in the Bronze Age (dates ranging from 1738±60 bc to 636± 45 bc), each followed by forest regeneration. Most phases, except the earlier ones, contain cereal pollen and/or pollen of arable weeds, but pollen of pastoral indicators tend to dominate all clearances. In all diagrams the first large scale clearance occurs in the Late Iron Age/early Romano-British period, with dates for the beginning of this phase only available from Steng Moss: 20±60 bc and Fellend Moss ad 2±45 , i.e. starting 30 - 150 years before the Roman army arrives in the region. During this phase the cultivation of crops appears to become more important. In addition to wheat and barley, some rye is recorded at Steng Moss, and Cannabis and Juglans pollen are recorded in the possible Roman levels at Camphill Moss. The end of this phase is dated to ad 460+60 at Steng Moss and to ad 620+40 at Fellend Moss, that is, after the withdrawal of the Roman army at the end of the fourth century. This suggests that a degree of political and economic stability in the region was maintained after the Romans had left (Davies & Turner 1979: 802).

2. West - Durham.

This region can be divided into upper Weardale and upper Teesdale. Two pollen diagrams are available from upper Weardale: Steward Shield Meadow and Bollihope Shield (Roberts, Turner & Ward 1973). The diagrams of both these sites start at c. 1200 bc and show a reasonable amount of woodland, but little oak. At Steward Shield Meadow a sharp decrease in the tree pollen frequencies occurs at 110+120 bc, associated with a rise in the pollen of pastoral indicators and some cereal pollen. The end of this phase is dated to ad 1110+100, but there is the possibility of a hiatus in the peat growth. At Bollihope Shield the clearance begins at ad 200+100 but cereals do not occur until much later. The end of this phase is dated to ad 1700, but again the possibility of a hiatus in the peat growth makes this date a little unreliable.

A large number of pollen diagrams from the upper Teesdale area have been published : Red Sike Moss, Widdybank Moss, Slapestone Sike Moss, Foolmire Sike Moss, Weelfoot Moss, Dead Crook, Dead Crook Moss, Furness Moss, Dubby Moss and Weelhead Moss (all in Turner *et al*. 1973). The localities of most of these diagrams are now submerged by the Cow Green Reservoir (Fig. 1). The evidence for crop production is not as easily extracted from these diagrams as in some other areas of the region, mainly because of the lack of radiocarbon dates for most of the clearances. For this reason the diagram from Valley Bog, though just in Cumbria, is incorporated here, as it provides some radiocarbon dates for land use in the region (Chambers 1978).

The upper Teesdale diagrams are all divided into local pollen assemblage zones, two of which concern us here: i.e. zone A and zone G. Zone A is an oak-alder assemblage and is roughly equivalent to Godwin's zone VIIb (Sub-Boreal), starting at c. 3000 bc. Zone G is a grass-plantain-heather assemblage and is similar to Godwin's zone VIII (Sub-Atlantic). Starting dates for zone G are: 1200+100 bc at Weelhead Moss WHI and 620+ 80 bc at Red Sike Moss TSI.

It is not possible here to discuss the diagrams individually, and only a very general impression can be offered. From c. 3000 bc onwards there are slight changes in the pollen frequencies of Gramineae and *Plantago lanceolata* and from 1200+100 bc at Weelhead Moss and 620+80 bc at Red Sike Moss these fluctuations become more marked , probably indicating the varying intensities of human occupation in the area (Turner *et al*. 1973: 403).

There is an early occurrence of cereal pollen in the Widdybank Moss diagram (i.e. below zone A), there are three diagrams with occurrences of cereal pollen in zone A, and seven with cereal pollen in zone G, besides two with pollen of arable weeds (Cirsium) without associated cereal pollen, also in zone G.

From c. 3000 bc onwards the existing open woodland was replaced by blanket peat except on the better drained soils where grassland developed. By 1000 - 500 bc most of the present day blanket peat was in existence. Only the better drained soils still carried some woodland.

Again, the picture of land use arising from these diagrams is one of small scale human interference during the Bronze Age with both arable and pastoral activities represented, though cultivation presumably only on a small scale, and an increase in the interference with the vegetation from the Iron Age onwards, with probably a slightly more important role for arable agriculture, a remarkable fact considering the altitude of the sites: 450+ m. In large parts of the area the trees were not able to regenerate, the development of blanket peat proved to be an irreversible process (Turner et al. 1973: 403).

Equally, the diagram at Valley Bog (Cumbria) shows woodland clearances during the Bronze Age, mainly of a pastoral nature, though some cereal pollen are present. These are, however, assumed to have been blown in from more lowland areas (Chambers 1978: 279). The major phase of deforestation at Valley Bog begins in the Late Iron Age at 262±55 bc to 225±45 bc.

3. East-Durham.

This area includes lower Weardale, lower Teesdale and the east-Durham plateau. Eight pollen diagrams from this area have been published: Hallowell Moss (Donaldson & Turner 1977), Cranberry Bog (Turner & Kershaw 1973), Hutton Henry, Thorpe Bulmer, Bishop Middleham, Nunstainton Carrs, Mordon Carr and Neasham Fen*1 (Bartley et al. 1976). The diagrams will be briefly discussed starting in the northern part of the area, with Cranberry Bog and Hallowell Moss, both in lower Weardale.

No peat was formed at Cranberry Bog during the Atlantic (VIIa) and Sub-Boreal(VIIb), but deposits dated to the Sub-Atlantic(VIII) are present. During this zone there is evidence for forest clearance, with high values for Gramineae and Plantago lanceolata. No cereals have been recorded. The diagram is not radiocarbon dated. At Hallowell Moss small clearances are recorded from 1695±60 bc to 482±60 bc (no cereal pollen present) and again between c. 400 bc and 6±70 bc. In both cases there is only evidence for slight grazing pressure (Donaldson & Turner 1977). At 6±70 bc a period of dramatic change starts. The landscape rapidly becomes as open as that of today and remains open until ad 600. Large areas of pasture and meadows are postulated; the evidence for crop production is limited.

The diagram from Bishop Middleham stands out in that it is the only one with large scale deforestation as early as the Bronze Age. Some minor fluctuations are recorded at levels dated around 3230±110 bc. The decline in tree pollen starts at 1710±80 bc. The first cereal pollen appear at 1410±80 bc. At this level *Plantago lanceolata* reaches 41% and remains around 10% for a considerable period. The intensity of the clearance is not matched anywhere in the North of England. There are no dates for the upper levels of the diagram, but after this large clearance in the middle Bronze Age there is very little recovery of the forest. The area around the site was never covered with trees again.

A similar development is recorded in the diagram from Hutton Henry, although the Bronze Age clearance here was not as dramatic as at Bishop Middleham. At 1594± 80 bc single occurrences of cereal and *Cannabis* pollen are recorded. After the major middle Bronze Age clearance the frequencies for trees and herbaceous pollen fluctuate. A large peak in the agricultural activity takes place around ad 108±70. At Thorpe Bulmer the major clearance starts at 114±60 bc. The high values for cereals and *Centaurus cyanus* and the extremely high values for *Cannabis* (up to 19% at ad 220) indicate the widespread practice of agriculture. The diagram shows evidence for continuous settlement and cultivation until ad 1098±60.

The highest levels of the Nunstainton Carrs diagram show low values for tree pollen and high values for herbaceous species. Cereal pollen are present in small amounts. The diagram is not dated. At Morden Carr small scale clearances are dated around 2986±85 bc and more clearly at 2593±70 bc. From this date onwards cereal pollen occur occasionally. The middle Bronze Age clearance is moderate in scale. There is an increase in cereals in the later part of the diagram, but no radiocarbon dates are available.

Neasham Fen is the southern most pollen site of the region. During the early Bronze Age a small forest clearance is recorded mainly for pasture, but the occurrence of a few arable weeds suggests the practice of some arable farming. At 1292±70 bc the first cereals appear , but the clearance is of moderate scale. There is another small scale clearance in the late Bronze Age - early Iron Age (900±60 bc to 538± 75 bc), after which the forest regenerates. No further clearances are recorded until ad 737±60 when a dramatic reduction in tree pollen is associated with an increase in arable weeds.

The information from these diagrams does not show a uniform picture of land use for the entire region. A distinction can be made between the pollen sites situated on the east Durham plateau (a Magnesian Limestone outcrop) with its more fertile, well-drained soils (Hutton Henry, Thorpe Bulmer and Bishop Middleham) and those in the Wear and Tees lowlands. On the east Durham plateau large scale deforestation and agriculture started as early as the middle Bronze Age (c. 1410 bc), and at Bishop Middleham the forest never regenerated. At Thorpe Bulmer the large scale deforestation appears to have started later, not until 114 bc, but agriculture is then practised on a very large scale. The other diagrams from the region, that is those located in the Wear and Tees lowlands, show the more familiar pattern of small-scale clearances during the Bronze Age and early Iron Age (though very minor at Hallowell Moss), and large-scale clearances during the Late Iron Age/early Romano-British period. Neasham Fen is the only exception. Possibly due to the very poorly drained soils this area is not cleared until the 8th century AD.

Discussion.

When we combine the evidence for crop production from all three areas (i.e. Northumberland, West- and East-Durham), we see very rare occurrences of cereal pollen during the Neolithic period (Widdybank Moss and Morden Carr). During the Bronze Age small scale clearances are recorded for all parts of the region, during which there is generally slightly more evidence for pastoral than arable agriculture. After these clearances the forest would normally regenerate, with the exception of the area around Bishop Middleham where the middle Bronze Age clearance initiated permanent deforestation. During most of the Iron Age, clearances occur on roughly the same scale as those of the Bronze Age, with the exception of Fellend Moss. The values for pastoral indicators are high, but pollen of cereal pollen and arable weeds are normally present as well. With the exception of the east Durham plateau and at least part of the Tees lowlands, all areas of the region become largely deforested at the end of the Iron Age or early Romano-British period. During this period the values for cereal pollen and arable weeds are higher than before. The various dates for the beginning of this phase convincingly suggest a start before the arrival of the Roman army in this region (c. AD 70), and the cleared land remained in use until after the Romans had withdrawn again (Turner 1979).

The evidence from the diagrams shows a general increase in the occurrence of crop production towards a climax in the Late Iron Age/early Romano-British period (with the exception of Bishop Middleham and Neasham Fen). The high intensity of agriculture is then maintained until at least the 6th or 7th century ad, and in some cases until the arrival of the Normans in the late 11th - early 12th century. However, it is not at all clear whether this means a gradual change in the role or importance of agriculture in the overall subsistence base, or whether it only reflects a general increase in the population. An increase in the population would almost automatically cause an increase in deforestation, and this would facilitate the recording of cereal pollen in the diagrams.

It is a well known fact that cereal pollen do not travel very far. Cereals are self-pollinated and only release small amounts of pollen, in contrast to Plantago lanceolata and Gramineae which both produce pollen in abundance. This fact, combined with the strong 'filtering' effect of heavily wooded areas might well cause an underrepresentation of the arable indicators, and a bias in favour of the pastoral indicators, for the period up to the late Iron Age. However, the application of a Principal Components Analysis to pollen diagrams (see Turner 1983) might well solve some of these problems. This technique recognizes the variation present in the data without making any assumptions about the ecology of the pollen taxa, and it gives equal weightings to the pollen taxa, irrespective of whether they have high or low pollen frequencies. The method is specifically designed to detect the nature of forest clearances and " even when a pollen or spore frequency is very low, as is often the case for some of the major crop plants, a principal component analysis is often sensitive enough to detect the way in which the taxon is varying " (Turner 1983: 11).

Nevertheless, there are certain questions that pollen analysis is not able to answer. It cannot tell us very much about the type of agriculture practised, about which species of cereals were cultivated, or when exactly new species were introduced. Nor can it give us information concerning inter - site differences and specialisations. This is the sort of detail that can only be acquired by the analysis of the crop plants themselves, as they are preserved on site. The results of the analysis of carbonised plant remains will be the subject of the next section.

EVIDENCE FROM THE ANALYSIS OF CARBONISED PLANT REMAINS.

None of the prehistoric settlement sites excavated in this region has contained waterlogged deposits, and thus the analysis of the crop plants is based on the study of carbonised plant remains only. A study of these remains can provide different kinds of information. In the first place it will show which crop plants were grown and when major changes occurred, i.e. the history of croplants. These changes can be related either to changes in dietary preferences, or to the introduction of new species by invading people, or to a new demand for crops specifically suited to new conditions, such as tolerance to acid soils or suitability for heavy soils. Changes in the crop record can refer to periods of population increase and pressure, during which an intensification of crop production is required (Jones 1981). Secondly, the specific character of the assemblage of carbonised plant remains can suggest the type of activities that it represents: not only information concerning the various crop-processing techniques that were carried out (e.g. which size of sieve was used for cleaning grain), but also more fundamental issues like whether we are dealing with activities of crop production or consumption (Jones forthcoming).

Both these aims of archaeobotanical research are served best by high quality data. However, while the second line of analysis, concerned with a characterisation of the agricultural activites, cannot be carried out unless adequate sampling techniques have been applied (Van der Veen 1984 and forthcoming a), the first line of analysis, concerned with the history of crop plants, largely requires well dated records, and can therefore use small and/or isolated samples, as long as they are accurately dated. As the information available from north-east England is still limited and patchy, only the first line of analysis is applied here.

The analysis of carbonised plant remains has only just started in the North of England, and consequently the information available is still very limited. However, in the last year and a half a small collection of data has been studied from five different sites in the North-East, two Late Neolithic, one late Bronze Age and two Late Iron Age. Except for the material from Coxhoe (Van der Veen & Haselgrove 1983) the material is still unpublished. In this section the plant remains from all five sites will be presented in chronological order and the general implications of the results will be discussed.

All samples were brought back to the laboratory in Durham, and were dried before flotation. Simple manual flotation was applied, mixing the sediment with water in a bucket and pouring the water off into a 0.5 mm mesh sieve. This guaranteed the recovery of both floating and suspended plant remains. It was felt necessary to dry samples from wet sediments before flotation to improve the recovery rate. However, samples with a high clay content were not allowed to dry out completely, to prevent rock-hard 'lumps' from developing. The sample volume is given in the Tables.

The Neolithic Period.

Whitton Hill, Northumberland (NT 933 347).

During the 1982 excavation of two ring-ditches at Whitton Hill (site 1 and 2)(Miket 1982) three samples were collected for flotation. Two of the samples came from the central cremations set within the ring-ditches, and one sample came from one of the ditch terminals of site 1. The results are set out in Table 1. Some plant remains other than charcoal were found in the radiocarbon samples and are listed under the column 'miscellaneous'.

The samples are dated to the Late Neolithic period. The ditch of site 1 produced sherds of a small bucket-shaped vessel of Grooved Ware tradition. The central cremation of site 1 contained pottery of a regional variation in the Peterborough style. Site 2 did not produce pottery, but the juxta-position of the two sites would seem to suggest that they may be contemporaneous (R. Miket, pers. comm.). The radiocarbon samples have not yet been examined. *2

Four grains of barley were found. They were all in a rather poor state of preservation. However, one grain clearly belonged to the naked variety of barley, as it had no ridges on the dorsal surface. Six-row barley is the only barley species present in prehistoric Britain, and both naked and hulled varieties occur. Here certainly Hordeum vulgare, var. nudum (naked, six-row barley) is present, but the presence of the hulled variety cannot be totally ruled out. One grain of Triticum dicoccum, emmer wheat, was found.

Seeds of various weed species were found, but unfortunately several were too badly preserved to allow detailed identification. The small vetches (Vicia sp.) are common weeds in arable fields. Ribwort plantain (Plantago lanceolata) favours grassland habitats and is one of the main pastoral indicators used in pollen analysis. Interesting is the presence of carbonised tubers of onion couch (Arrhenatherum elatius, var. bulbosum). These tubers have also been found in a late Bronze Age ditch on Rockley Down, Wilts. (Godwin 1975), and in Bronze Age cremation pits in Abingdon, Oxon. (Jones 1978). It is a strongly tuberised grass that can be a serious weed of cultivated ground (Godwin 1975). The fact that it occurred with barley in the Rockley Down sample and that it was found in cremation pits in Ashville, made both Godwin and Jones suggest that it might have been purposely gathered. Here in Whitton Hill it is also associated with cremation remains. Only a few fragments of hazelnut shell (Corylus avellana) were found in the samples; this is in contrast to the samples from Thirlings (see below), where hazelnut fragments occurred in very large numbers.

Thirlings, Northumberland (NT 956 324).

Excavations were carried out at Thirlings from 1973-1981 (O'Brien 1982, Miket & O'Brien forthcoming). The main concern was an Anglo-Saxon settlement. However, finds of Neolithic pottery associated with some scattered pits and postholes demonstrate the presence of possibly two phases of occupation during the Neolithic. Most of the Neolithic features were excavated during earlier seasons, but in 1981 four large pits containing Grooved Ware pottery were uncovered. They were c. 1 m in diameter and 25-30 cm deep. Each of these was sectioned, and half of each pit was taken for flotation. The results are set out in Table 2.

The samples are dated to the Late Neolithic. No radiocarbon dates are available. Pit 1858 contained two sherds of pottery with fingertip and incised decoration, both in Grooved Ware tradition.

Pit 1894 contained one sherd, not very diagnostic; and a flint scraper and knife, neither of which is unusual in a Grooved Ware context. Pit 1898 contained twenty sherds with fingertip decoration and one sherd with incised decoration in chevron pattern and horizontal lines. At least two vessels are present, both in Grooved Ware tradition. Pit 1901 contained four very small sherds of Neolithic appearance, but otherwise not very diagnostic (Colm O'Brien, pers. comm.).

Only two cereal grains were recovered in the Thirlings samples. The barley grain showed no ridges on the dorsal surface and can thus be identified as Hordeum vulgare, var. nudum (six-row naked barley). The other grain was too badly preserved to justify further identification. One grain of oat was found, Avena sp. It could not be established whether it was of cultivated or wild origin. In the absence of floret bases it is impossible to make reliable identifications to species level. The size ranges for wild and cultivated oats are not mutually exclusive, which prevents a distinction purely on size. The measurements for the grain from pit 1894 are: length 6.75 mm, width 2.25 mm and thickness 1.26 mm.

A large number of hazelnut shell fragments (Corylus avellana) are present in the samples, although one should remember that hazelnut shells can break up into many more fragments than any of the other species present. Their presence in large numbers points to the presence of nearby woodland, as does the presence of hawthorn (Crataegus) and bramble (Rubus) fruits. The small vetches (Vicia sp), bromegrass (Bromus sp.), fat hen (Chenopodium album) and chickweed (Stellaria media) are all weeds of arable fields or other disturbed ground. Blinks (Montia fontana, spp. chondrosperma) and sedge (Carex sp.) indicate the presence of damp ground nearby.

Discussion.

Carbonised plant remains from Late Neolithic sites are still a rare phenomenon in Britain, which made some authors suggest that the economy of the Grooved Ware cultures might have had a predominantly pastoral base (eg. Megaw & Simpson 1979). As Jones has rightly pointed out (Jones 1980) the absence of plant remains was purely a function of inadequate recovery techniques and a wrong interpretation of the absence of grain impressions. When adequate recovery techniques are applied, i.e. large scale flotation, cereal grains and other plant remains are found on every site, though in small numbers, as was shown for Mount Farm and Barton Court Farm, Oxon., Down Farm, Wilts. (Jones 1980), Bromfield, Shrops. (Colledge 1982), Trelystan, Powys (Hillman 1982), and at Whitton Hill and Thirlings. Only at Trelystan were cereal grains absent. The results from Whitton Hill and Thirlings are very similar to those from the above mentioned sites, i.e. very small quantities of cereals and weed seeds, some collected fruits and large quantities of hazelnut fragments.

The presence of cereals on Grooved Ware sites seems now sufficiently demonstrated to allow a revision of the original view of their economy. It seems certain that arable agriculture played a part in their society, although we are still uncertain about the actual role it played in the total subsistence base. Unfortunately, no chaff fragments have been recovered in any of the samples, but the association of the cereal grains with arable weeds, at least at the Whitton Hill and Thirlings samples, might suggest small scale local production. The consistent presence of large numbers of hazelnut fragments and other woodland food plants indicates that the exploitation of woodland resources still formed an important part of the overall economy.

The Bronze Age.

Hallshill Farm, East Woodburn, Northumberland (NY 906 886).

The excavations in 1981 revealed the wall line and postholes of an unenclosed timber round house, 9 m in diameter and of ring-grooved construction (Gates 1982a). The site was associated with a 0.6 hectare field and several small clearance cairns. In view of its rather insubstantial nature, the bank around the field was probably not intended to act as a barrier. The field was therefore not assumed to have acted as a stock pen, but rather as an arable field. The bank may have been formed as a product of field clearance (Gates 1982a).

A total of seven samples was collected for flotation. All but one contained plant remains. Sample no.8 came from the central hearth of the house. All the other samples are posthole fills. The results of the analysis are given in Table 3. The samples are dated to the late Bronze Age by three radiocarbon dates: 830±80 bc, 610± 60 bc and 570±70 bc (Gates 1983).

The samples are dominated by the remains of wheat: 59 grains and 146 chaff fragments. Only the reasonably well preserved grains were identified to species level, though bearing in mind the fact that identifications of wheat grains based on shape alone are never completely reliable. Of the 59 wheat grains 34 could be identified as *Triticum dicoccum*, emmer wheat. The chaff fragments consisted of one spikelet fork of *Triticum dicoccum*, six rachis internodes of a brittle rachis wheat and 139 glumebases of a glume wheat. Of the latter 121 were well enough preserved to be measured. Figure 2 gives a histogram of the glume widths, measured at the level of spikelet-articulation. The measurements indicate that we are dealing with one species only, as the histogram approaches the curve for a normal distribution very closely. The range of glume dimensions corresponds well with those for *Triticum dicoccum*, emmer, using figures given by Helbaek (1952). We can therefore conclude that both the grains and the chaff fragments indicate that *Triticum dicoccum*, emmer wheat, was the only wheat species present.

Nine barley grains and 8 barley rachis internodes were found. As the grains were all badly preserved, it could not be established whether both central and lateral grains were present. However, two of the rachis internodes show the bases of the glumes and belong clearly to the six-row variety. None of the few grains that still possessed their surface coat, had the characteristic features of hulled barley, i.e. ridges on the dorsal surface and a slightly angular cross-section, which means that we are probably dealing with the naked variety. The barley in the samples can therefore be identified with reasonable certainty as Hordeum vulgare, var. nudum, six-row, naked barley.

Two grains of oats, Avena sp., were found in the samples. Cultivated oat grains can only be reliably distinguished from wild ones when the flower bases have been preserved. Unfortunately this was not the case here.

One seed of flax or linseed was found. The seed was badly blistered, but the beaked shape and cellular surface structure were distinct enough to justify this identification. The length of the seed is 3.33 mm, the width 1.81 mm. This makes it about the same size as Linum bienne Mill. (or L. angustifolium Huds.), the seeds of which are c. 3 mm long. However, if we assume that the length of the seed is reduced through carbonisation by ca. 10%, the seed comes in fact in the lower range of Linum ussitatissimum, cultivated flax. Linum bienne occurs nowadays in dry grassland, especially near the sea, from the Isle of Man, Lancashire and Yorkshire southwards and is commonest in the south-west (Clapham et al. 1962). Hallshill Farm lies outside the present- day range of Linum bienne. We might therefore be dealing with cultivated flax, which is, in fact, thought to have derived from Linum bienne. Measurements of carbonised seeds of Linum ussitatissimum from the Late Neolithic settlement Aartswoud, the Netherlands, range from 2.71 - 3.43 mm (length) and 1.71 - 2.06 mm (width)(Buurman & Pals 1974). The Hallshill Farm seed fits nicely in this range. However, with only one rather badly preserved seed present, we have to confine ourselves to an identification of Linum sp.

Chenopodium album (fat hen) is common in waste places and cultivated ground. Rumex acetosella (sheep's sorrel) grows on heaths, grassland and cultivated land, commonly on acid soils. Plantago media (hoary plantain) occurs in grassy places, on neutral and basic soils.

Discussion.

The site of Hallshill Farm is c. seven km south- west of Steng Moss. As mentioned above, this diagram shows for the later Bronze Age and Iron Age a series of small-scale clearance and regeneration episodes. The part of the diagram that covers the occupation phase of the settlement (c. 830-570 bc) records only one occurrence of cereal pollen, i.e. at a level dated to 636±45 bc, and in fact this record only represents one pollen grain. This highlights the problems of using pollen analysis for the detection of arable agriculture, and illustrates the need for a close cooperation between pollen analysis and the analysis of carbonised plant remains.

The most striking aspect of the carbonised plant assemblage from Hallshill Farm is its 'conservative' character. At Black Patch (Sussex) a late Bronze Age site with radiocarbon dates ranging from 1130±70 bc - 830±80 bc, i.e. slightly earlier than Hallshill Farm, spelt, celtic beans and hulled barley are present in addition to emmer wheat (Hinton 1982). The absence of hulled barley in Hallshill Farm is remarkable. An increase in the occurrence of hulled barley has been recorded for the later Bronze Age in Britain. From the Iron Age onwards naked barley only plays a very minor role (Helbaek 1952).

The first half of the first millennium BC. is regarded as a period of innovation, witnessing the introduction of many new crop plants, such as spelt wheat (*Triticum spelta*) bread wheat (*Triticum aestivum*), celtic bean, (*Vicia faba*, var. *minor*) oat, (*Aven sativa*) and rye, (*Secale cereale*) (Helbaek 1952, Jones 1981). There are a few earlier records for bread wheat and celtic beans, and there is of course the much discussed Neolithic record for spelt wheat in Hembury (Field *et al*. 1964). Not until the first half of the first millennium BC., however, do these species occur with any regularity. By c. 500 BC. the new introductions seem firmly established, often at the expense of the previously cultivated species, like naked barley and emmer wheat. None of these new introductions are represented at Hallshill Farm, which assemblage is still firmly based in the earlier Bronze Age tradition.

The Iron Age.

Coxhoe, co. Durham (NZ 326 360).

This crop mark site was excavated in 1979 and 1980 (Haselgrove & Allon 1982). The rectilinear enclosure ditch, c. 0.4 hectare in area, encompassed a central circular feature, probably a house. An earlier phase was recognised, represented by a long linear gully. No radiocarbon dates are available from the site. The presence of a fragment of a rotary quern suggest a date not earlier than the 2nd century BC. and the absence of Roman imports suggests an abandonment by at least the end of the first century AD. The earlier phase can only be said to ante-date this period (Haselgrove and Allon 1982).

Unfortunately, no flotation samples were collected during the excavation, but three soil samples were still available for analysis. The results are published elsewhere (Van der Veen & Haselgrove 1983), but they will also be briefly presented here.

Sample 506 belongs to the earlier phase of the settlement (the linear gully), while sample 1013 and 504 belong to the enclosure ditch and circular house respectively. The results are given in Table 4.

The cereal grains were all badly preserved and only two of them could be identified to genus level: one as wheat (*Triticum* sp.) and one as barley (*Hordeum* sp.). However, a total of 11 glumebases was recovered, nine of them well enough preserved to be measured. The measurements of the glume-widths at the level of spikelet articulation are given in Figure 2. At least four fall outside the range of *Triticum dicoccum* (emmer wheat) the other five are at the top end of this range or at the bottom end of *Triticum spelta* (spelt wheat) using figures given by Helbaek (1952). In addition, the four widest glumebases, i.e. those which fall within the range of spelt, also show the venation pattern characteristic for spelt wheat. The other glumebases are either too damaged or show an indistinctive venation pattern. Thus *Triticum spelta* (spelt wheat), is definitely present in the Coxhoe samples. There is not sufficient evidence to exclude the presence of *Triticum dicoccum* (emmer wheat).

Thorpe Thewles, Cleveland (NZ 396 243)

Thorpe Thewles is a late Iron Age settlement site situated on the foothills of the east Durham plateau. Aerial photographs revealed a crop mark site consisting of a large (0.7 hectare) sub-rectangular enclosure with a central house. The site was excavated during 1980 - 1982 (Heslop 1983). The excavations have pointed to the existence of occupation phases both before and after the main occupation phase, which is represented by a sub-rectangular bank-and-ditch enclosure. This main enclosure phase dates from the late 3rd or 2nd century BC. During the first century BC. and the first century AD. the bank and ditch were levelled and the settlement expanded beyond the original enclosure. During this phase a small amount of imported Roman pottery (Samian) found its way to the site. The settlement is not thought to have continued after the end of the 1st century AD.

During the first season of excavations (1980) no samples for flotation were collected. During the second season (late 1981-1982), however, the collection of flotation samples became an integrated part of the excavation. Because of the large scale of the excavation it was not possible to collect samples from every feature, and a programme of random sampling was applied. A description of the random sampling programme can be found in Van der Veen (1984 and forthcoming a). A total of 130 samples was collected, of which 100 have been analysed so far, together containing c. 16000 seeds. It is not possible here to give a full table of the results, as this would require too much space. A detailed discussion of the material is forthcoming (Van der Veen forthcoming b). In Table 5. a summary of the results is given. The sample volume was standardised to ca. 28 litres of sediment (i.e. two buckets).

Two types of cereal grains were present: spelt wheat and six-row, hulled barley. The wheat grains were on the whole badly preserved and identification to species level was rarely possible. The best preserved caryopses could be identified as Triticum spelta (spelt wheat) though there might be a few grains of Triticum dicoccum (emmer wheat) present. This still needs further analysis. Large numbers of chaff fragments of wheat were present and the identification of spelt wheat is mainly based on the characteristics of the glumebases. A total of 1000 glumebases has been measured and a histogram of the measurements is given in Figure 2. The histogram approaches the curve for a normal distribution, indicating that it is likely that we are dealing with just one species; and the measurements fall clearly in the range for spelt wheat. Moreover, all the well preserved glumebases show clearly the venation pattern so characteristic for spelt wheat, i.e. one prominent vein and several secondary veins. There can therefore be no doubt that the wheat present in the samples belongs to spelt wheat. The evidence from the glumebases would suggest an absence of emmer wheat.

The barley grains were equally poorly preserved. In fact, the majority of the cereal grains had to be put in the class: Cerealia indet. However, of the few grains that were well enough preserved several were lateral grains and all showed the ridges on the dorsal surface and a slightly angular cross-section characteristic of hulled, six-row barley, Hordeum vulgare. The rachis internode fragments confirmed this identification.

No further crop plants have been found so far, with the exception of possibly oat. Small numbers of Avena sp. (oat) are present in the samples. Unfortunately, very few floret bases were found, and all of these are badly fragmented, but there is possibly one floret base of cultivated oat (Avena sativa) present.

In addition to the cereal grains and chaff fragments large numbers of grasses were present in all samples, especially Sieglingia decumbens (heath grass) and Bromus sp. (bromegrass). Furthermore, large numbers of arable weeds and other herbaceous species were present. They cannot be dealt with here.

Discussion.

The evidence from both Coxhoe and Thorpe Thewles shows clearly that by the 2nd century BC. spelt wheat and hulled, six-row barley are firmly established as crop plants, at least in the Durham lowlands. However, there is as yet no evidence for the cultivation of bread wheat, celtic beans or rye, and the evidence for oat is very slight. The introduction of these new crop plants in the first half of the first millennium BC. and their establishment by c. 500 BC. in the south of the country has been suggested to be the reaction to population pressure, as the new species allowed the sustenance of cultivation on deteriorating land (oat, rye, beans) and the expansion onto hitherto marginal land (spelt, bread wheat)(Jones 1981). The evidence from the sites studied so far would suggest that this population pressure was not felt in the North-East of England until much later.

SUMMARY AND SUGGESTIONS FOR FUTURE WORK.

While the region is reasonably well off for pollen diagrams, the information available from carbonised plant remains is still very limited. Nevertheless, some interesting patterns can already be distinguished. The Late Neolithic samples from Whitton Hill and Thirlings are important in that they confirm the evidence (admittedly still limited) from other parts of the country, that arable agriculture played some part in the economy of the Grooved Ware cultures. The exact nature of the role of crop production in their society is, however, still undetected and many more excavations are required. The activities of the Late Neolithic people are rarely recorded in the pollen diagrams, which is probably a function of the fact that the area was still only sparsely populated and of the fact that pollen analysis is inadequate in picking up arable activities in densely wooded areas. The most striking aspect of these two Late Neolithic plant assemblages, however, is that they are remarkably similar to those from southern Britain.

The one site of Hallshill Farm can obviously not be regarded as representative for the Bronze Age of North-East England, and it is hoped that many more sites will soon be excavated. Nevertheless, the rather conservative character of the assemblage from this site is remarkable and would suggest that the developments as we can trace them in southern Britain, take place in the North-East at a slightly later time.The Lowland - Highland Zone division is generally thought to have become important at the beginning of the first millennium BC. (Evans 1975), and it is possible that what we are seeing at Hallshill Farm is evidence for this statement.

Unfortunately, no information from early Iron Age sites is yet available, so we are unable to trace when exactly spelt wheat arrives in the North-East of England. The evidence from Coxhoe and Thorpe Thewles demonstrates that the cultivation of spelt wheat is firmly established by the Late Iron Age or 2nd century BC. However none of the other new introductions appear to be present then.

The beginning of the large-scale deforestation of the landscape in southern England is generally dated to c. 500 BC. (Turner 1981), although in parts of South-East England the landscape was already fairly open in the Bronze Age. This development coincides with the establishment of a whole series of new crop plants and the expansion of agriculture onto previously marginal land. Although no direct link has ever been suggested between these two developments, it is very likely that they are closely related. The forest clearances were, undoubtedly, largely the result of the intensification and expansion of agriculture. This large-scale deforestation of the landscape occurred progressively later in the north of the country, and as we have seen above, did not start in the North-East until the last century BC. or the first century AD. (with the area around Bishop Middleham on the east Durham plateau as the only exception). Although the information concerning the crop plants in the North-East is still scarce, the evidence available suggests that the introduction of new crop plants and the associated developments, as known from the south of England, took place here at a much later date. Again, the results from pollen analysis can be closely linked with those from the study of carbonised plant remains. Thus, the evidence available at the moment would suggest that the population pressure which is thought to have brought about these changes in landscape and agriculture in the south of the country, was not felt in the North- East until several centuries later.

As will be clear from the above discussion, there are still many blanks in our knowledge of the development of crop production in the North-East. It is hoped that soon many more samples for plant remains can be collected from many excavations to fill in some of the detail. Some suggestions for possible future work are given here:

1. Upland Northumberland.

Carbonised plant assemblages are required from more Bronze Age and Iron Age settlements to test when exactly the developments in crop production start lagging behind and when exactly, if ever, the new crop plants are introduced. The Lowland - Highland zone division and possible environmental constraints for the cultivation of crops should be incorporated in the analysis.

2. The Roman Wall area.

The Roman Wall forms one of the most remarkable features in the landscape of the region and the arrival of very large numbers of Roman troops must have made a profound impact on the economy of the area. Piggott argued that it was very unlikely that the North of England was capable of producing grain for a surplus and that the advance of the Roman army northwards would have had to rely upon elaborate arrangements for its corn supply (Piggott 1958). Manning, however, argued that for economic and military reasons the Roman army always tried to avoid moving supplies over long distances and argued that a considerable part of the requirements was locally produced (Manning 1975). It should be possible to answer this question by studying the plant remains from native sites and comparing them with those from vici and forts in the Wall zone.

3. Lowland Durham.

At least part of this area appears to have been deforested as early as parts of South-East England. Plant remains are required from Bronze Age and early Iron Age settlements to find out whether this early clearance was associated with an early introduction of the new crop plants, earlier than in the rest of the region. Excavations of Romano-British sites are required to study the reaction of local crop production to the arrival of the Roman garrisons.

What happened during the later period, after the Roman withdrawal from the region, is still very much an enigma. Pollen evidence suggests continuation and stability of cultivation, but the archaeological evidence strikes by its absence and consequently no carbonised plant assemblages are available. We hope that in a few years time more detailed evidence will be available for the periods discussed here, in addition to new information for the earlier Neolithic and the Saxon period.

Footnotes:

*1 The grid reference for Neasham Fen is given by Bartley et al. (1976) as NZ 332 166. I suggest that this is a typing error. From the location map given by the authors and from the Ordnance Survey Map (1 : 50 000) it is clear that the correct location is NZ 332 116. On Fig. 1 this corrected grid reference is used.

*2 The radio-carbon dates have since become available. They indicate that the central burial Site 2 is of much later date, probably first millenium BC (Nov. 1984).

Acknowledgements:

The work at the Biological Laboratory in Durham is financed by a grant from the Department of the Environment (Ancient Monuments Laboratory) to the University of Durham. I would like to thank Gordon Hillman and Martin Jones for their help with the identifications, and Julian Bennett, Martin Jones and Judy Turner for critically reading an earlier draft of this article. Finaly, I would like to express my gratitude to Colm O'Brien, David Heslop, Tim Gates and Roger Miket for allowing me to collect large bags of soil on their sites, and Colin Haselgrove for allowing me to empty his cupboard. Figures 1 and 2 were drawn by Yvonne Brown and the abstract was translated into German by Bettina Hartas.

REFERENCES

Bartley, D.D., Chambers, C. & Hart-Jones, B. 1976 The vegetational history of parts of south and east Durham. New Phytologist 77, 437 - 468.

Buurman, J. & Pals, J.P. 1974 Some remarks on prehistoric flax in the Netherlands. Berichten voor het Oudheidkundig Bodemonderzoek 24, 107 - 111.

Chambers, C. 1978 A radiocarbon dated pollen diagram from Valley Bog, on the Moor House National Nature Reserve. New Phytologist 80, 273 - 280.

Chapman, S.B. 1964 The ecology of Coom Rigg Moss, Northumberland. Journal of Ecology 52, 299 - 315.

Clack, P.A.G. 1982 'The northern frontier: farmers in the military zone.' In Miles, D. (ed.) The Romano- British countryside. Studies in rural settlement and economy. British Archaeological Reports, British Series 103, 377 - 402.

Clapham, A.R., Tutin, T.G. & Warburg, E.F. 1962 Flora of the British Isles. Cambridge, Cambridge University Press.

Colledge, S.M. 1982 Organic material from pits F 246 and 247. in Stanford, S.C. Bromfield, Shropshire - Neolithic, Beaker and Bronze Age sites 1966 - 1979. Proceedings of the Prehistoric Society 48, 286 - 287.

Davies, G. & Turner, J. 1979 Pollen diagrams from Northumberland. New Phytologist 82, 783 - 804.

Donaldson, A.M. & Turner, J. 1977 A pollen diagram from Hallowell Moss, near Durham City, U.K. Journal of Biogeography 4, 25 -33.

Evans, J.G. 1975 The environment of early man in the British Isles. London, Paul Elek.

Field, N.H., Matthews, C.L. & Smith, I.F. 1964 New Neolithic sites in Dorset and Bedfordshire. Proceedings of the Prehistoric Society 30, 352 - 381.

Gates, T. 1982a Excavations at Hallshill Farm, East Woodburn, Northumberland. Archaeological Reports for 1981, Universities of Newcastle and Durham, 7 - 9.

Gates, T. 1982b Farming on the frontier : Romano- British fields in Northumberland. In Clack, P. & Haselgrove, S. (eds.) Rural settlement in the Roman North. Durham, Council for British Archaeology, group 3, 21 - 42.

Gates, T. 1983 Unenclosed settlements in Northumberland. in Chapman, J.C. & Mytum, H.C. (eds.) Settlement in North Britain, 1000 BC - AD 1000. Papers presented to George Jobey, Newcastle upon Tyne, December 1982. British Archaeological Reports, British Series 118, 103 - 148.

Godwin, H. 1975 History of the British flora. Cambridge, Cambridge University Press.

Harding, D.W. 1979 Air survey in the Tyne-Tees region, 1969 - 79. In Higham, N.J. (ed.) The changing past. Manchester, 21 - 30.

Haselgrove, C.C. 1982 Indigenous settlement patterns in the Tyne-Tees lowlands. In Clack, P. & Haselgrove, S. (eds.) Rural settlement in the Roman North. Durham, Council for

British Archaeology, group 3, 57 - 107.
Haselgrove, C.C. & Allon, V.L. 1982 An Iron Age settlement at West House, Coxhoe, County Durham. Archaeologia Aeliana, fifth series, 10, 25 - 51.
Helbaek, H. 1952 Early crops in southern England. Proceedings of the Prehistoric Society 18, 194 - 233.
Heslop, D. 1983 The excavations of an Iron Age settlement at Thorpe Thewles. In Cleveland County Council Archaeology Section (eds.) Recent excavations in Cleveland. Middlesbrough, 17 - 26.
Hillman, G. 1982 Appendix 6 : Charred remains of plants. in Britnell, W. The excavation of two round barrows at Trelestan, Powys. Proceedings of the Prehistoric Society 48, 198 - 200.
Hinton, P. 1982 Carbonised seeds. In Drewett, P. Later Bronze Age Downland economy and excavations at Black Patch, East Sussex. Proceedings of the Prehistoric Society 48, 382 - 390.
Hubbard, C.E. 1968 Grasses. Harmondsworth Penguin Books.
Jobey, G. 1982 Between Tyne and Forth: some problems. In Clack, P. & Haselgrove, S. (eds.) Rural settlement in the Roman North. Durham, Council for British Archaeology, group 3, 7 - 20.
Jones, M. 1978 The plant remains. In Parrington, M. (ed.) The excavation of an Iron Age settlement, Bronze Age ring-ditches and Roman features at Asville Trading Estate, Abingdon (Oxon.) 1974-76. Council for British Archaeology Research Report 28, 93 - 110.
Jones, M. 1980 Carbonised cereals from Grooved Ware contexts. Proceedings of the Prehistoric Society 46, 61 - 63.
Jones, M. 1981 The development of crop husbandry. In Jones, M. & Dimbleby, G. (eds.) The environment of man: the Iron Age to the Anglo-Saxon period. British Archaeological Reports, British Series 87, 95 -127.
Jones, M. forthcoming Archaeobotany beyond subsistence reconstruction. In Barker, G.W. & Gamble, C. (eds.) Beyond Domestication. London. Academic Press.
Manning, W.H. 1975 Economic influences on land use in the military areas of the Highland Zone during the Roman period. In Evans, J.G., Limbrey, S. & Cleere, H. (eds.) The effect of man on the landscape: the Highland Zone. CBA Research Report 11, Council for British Archaeology, 112 - 116.
Megaw, J.V.S. & Simpson, D.D.A. 1979 Introduction to British Prehistory. Leicester. Leicester University Press.
Miket, R. 1982 The excavation at Whitton Hill, Northumber- land, 1982. Unpublished typescript.
Miket, R. & O'Brien, C. forthcoming The excavations at Thirlings.
O'Brien, C. 1982 Excavations at Thirlings. Archaeological Reports for 1981. Universities of Newcastle and Durham, 44 - 45.
Pearson, M.C. 1960 Muckle Moss, Northumberland. Journal of Ecology48, 647 - 666.
Piggott, S. 1958 Native economies and the Roman occupation of North Britain. In Richmond, I.A. (ed.) Roman and native in North Britain. Edinburgh, Nelson, 1 - 27.
Roberts, B.K., Turner, J. & Ward, P.F. 1973 Recent forest history and land use in Weardale, Northern England. In Birks,

H.J.B. & West, R.G. (eds.) Quaternary Plant Ecology. Oxford, Blackwells, 207 - 221.

Turner, J. 1979 The environment of northeast England during Roman times as known by pollen analysis. Journal of Archaeological Science 6, 285 - 290.

Turner, J. 1981 The vegetation. In Jones, M. & Dimbleby, G. (eds.) The environment of man: the Iron Age to the Anglo-Saxon period. British Archaeological Reports, British Series 87, 67 - 73.

Turner, J. 1983 Some pollen evidence for the environ- ment of northern Britain, 1000 bc. to AD. 1000. In Chapman, J.C. & Mytum, H.C. (eds.) Settlement in North Britain, 1000 BC - AD 1000. Papers presented to George Jobey, Newcastle upon Tyne, December 1982. British Archaeological Reports, British Series 118, 1 - 27.

Turner, J., Hewetson, V.P., Hibbert, F.A., Lowry, K.H. & Chambers, C. 1973 The history of the vegetation and flora of Widdybank Fell and the Cow Green reservoir basin, Upper Teesdale. Philosophical Transactions of the Royal Society of London, Series B 265, 327 - 408.

Turner, J. & Kershaw, A.P. 1973 A Late- and Post-Glacial pollen diagram from Cranberry Bog, near Beamish, County Durham. New Phytologist 72, 915 - 928.

Veen, M. van der 1984 Sampling for seeds. In Zeist, W. van & Casparie, W.A. (eds.) Plants and ancient man. Studies in palaeoethnobotany. Proceedings of the VI Symposium of the International Workgroup for Palaeoethnobotany, 30 May - 3 June 1983, A.A. Balkema, Rotterdam, 193-199.

Veen, M. van der forthcoming a. Carbonised seeds, sample size and on-site sampling. (these proceedings)

Veen, M. van der forthcoming b The plant remains. In Heslop, D. (ed.) The excavations at Thorpe Thewles, Cleveland.

Veen, M. van der & Haselgrove, C.C. 1983 Evidence for pre-Roman crops from Coxhoe, Co. Durham. Archaeologia Aeliana, fifth series 11, 23-25.

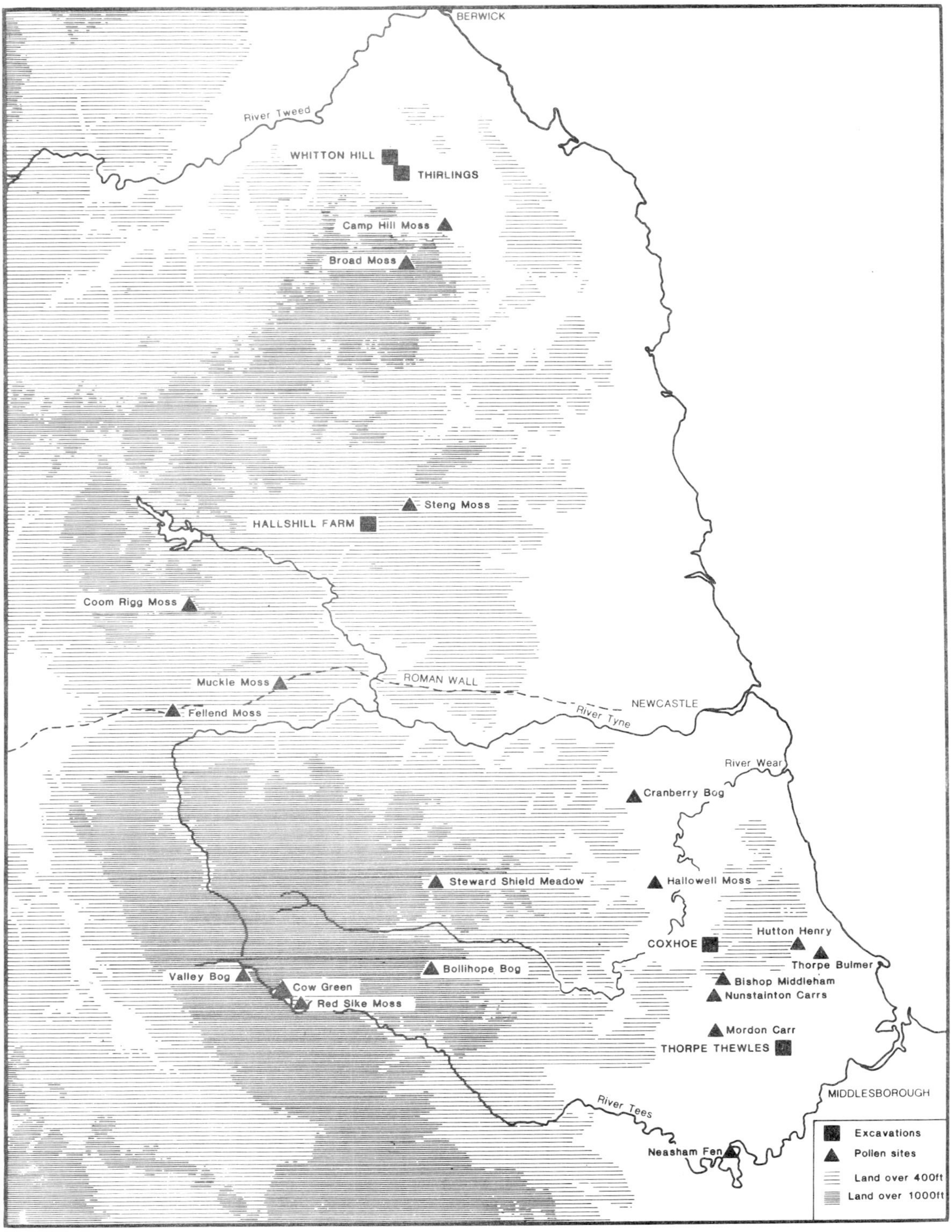

Figure 1. Distribution map showing the location of all pollen sites and excavations mentioned in the text.

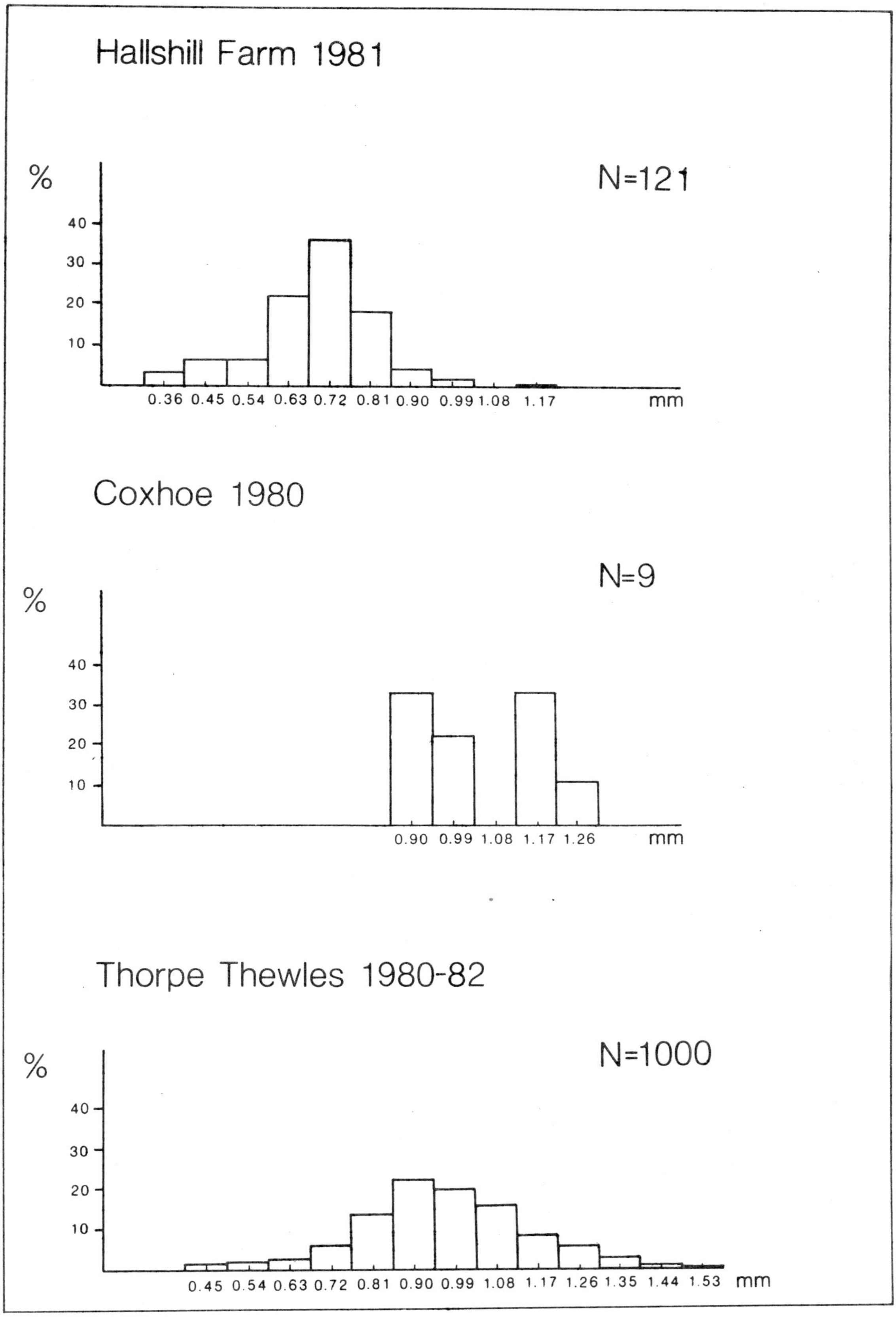

Figure 2. Histograms of glume-widths of *Triticum* sp., measured at the level of spikelet-articulation (N indicates the number of glumes measured for each histogram).

Whitton Hill 1982 Sample volume:	Central burial site 1. (14 litres)	Central burial site 2. (42 litres)	Ditch terminal site 1. (7 litres)	Misc.
Species:				
Hordeum sp. (barley)	3	-	-	1
Triticum dicoccum (emmer wheat)	1	-	-	-
Cerealia indet.	1	-	-	-
Corylus avellana (hazelnut)	-	-	5	3
Arrhenatherum elatius, var. bulbosum (couch tubers)	-	2	-	-
Plantago lanceolata (ribwort plantain)	2	2	-	-
Vicia sp. (vetch)	1	2	-	-
Caryophyllacaea indet.	-	2	1	-
Galium sp.	-	-	1	-
Indet.	-	1	1	1
Total	8	9	8	5

TABLE 1: Number of seeds for each sample from Whitton Hill.

Thirlings 1981 Sample volume:	1858 (42 litres)	1894 (28 litres)	1898 (42 litres)	1901 (56 litres)
Species:				
Hordeum sp. (barley)	-	-	-	1
Cerealia indet.	-	1	-	-
Corylus avellana (hazelnut)	341	410	826	10
Crataegus monogyna/ oxycanthoides (hawthorn)	-	-	1	0
Rubus cf. fruticosus (bramble)	1	-	-	-
Avena sp. (oat)	-	1	-	-
Bromus sp. (bromegrass)	-	-	-	1
Vicia sp. (vetch)	1	-	-	1
Chenopodium album (fat hen)	-	-	-	1
Stellaria media (chickweed)	-	-	-	1
Montia fontana, spp. chondrosperma (blinks)	-	-	-	1
Carex sp. (sedge)	-	-	-	1
Rumex sp.	-	-	-	8
Iridaceae/Liliaceae indet.	-	-	-	2
Indet.	1	-	1	2
Total	344	412	828	29

TABLE 2: Number of seeds for each sample from Thirlings

Hallshill Farm 1981	8	16	17	18	19	20
Sample volume:	(70 ltr)	(3 ltr)	(1 ltr)	(3 ltr)	(15 ltr)	(4 ltr)
Species:						
Triciticum dicoccum (emmer wheat)						
grains	33	-	-	1	-	-
glumebases	3	1	-	135	-	-
spikelet fork	1	-	-	-	-	-
rachis internodes	-	-	-	6	-	-
Triticum sp. (wheat)	22	-	-	2	1	-
Hordeum sp. (barley), grains	6	-	1	2	-	-
rachis internodes	-	-	-	7	1	-
Cerealia indet.	37	-	-	1	-	-
Avena sp. (oat)	1	1	-	-	-	-
Linum sp. (flax)	-	-	-	1	-	-
Chenopodiaceae indet.	-	-	-	3	-	-
Chenopodium sp.	-	-	-	1	-	-
Chenopodium album (fat hen)	-	1	-	-	-	-
Atriplex sp.	-	-	-	2	-	-
Plantago media (hoary plantain)	1	-	-	-	-	-
Rumex acetosella (sheep's sorrel)	2	-	-	-	-	-
Carex sp. (sedge)	-	-	-	-	-	2
Base of rhizome	-	-	-	-	-	1
Indet.	1	-	-	1	2	-
Total	107	3	1	162	4	3

TABLE 3: Number of seeds for each sample from Hallshill Farm

Coxhoe 1980	506	504	1013
Sample volume:	(28 litres)	(17.5 litres)	(21 litres)
Species:			
Triticum sp. (wheat) grains	1	-	-
glumebases	6	-	5
Hordeum sp. (barley)	-	-	1
Cerealia indet.	4	-	1
Sieglingia decumbens (heath grass)	-	-	2
Gramineae indet.	-	1	-
Indet.	1	1	-
Total	12	2	9

TABLE 4: Number of seeds for each sample from Coxhoe

GRAZING DAMAGE TO SCOTS PINE (PINUS SYLVESTRIS) BY RED DEER (CERVAS ELEPHAS) IN HIGHLAND SCOTLAND DURING THE MID-FLANDRIAN: A SHORT NOTE.

R.G.W. Ward
Sir John Cass Quaternary Research Unit
City of London Polytechnic
Calcutta House
Old Castle Street
London E1 7NT

ABSTRACT

Characteristic scars attributable to deer grazing damage have been identified on cross-sections of mid-Flandrian Scots Pine preserved in the peat bogs of Rannoch Moor, Scotland. Future maps of such scar tissue frequencies may have notable biogeographic and archaeological significance for our understanding of Mesolithic subsistence, herding and the ecology of the highland moors and forests.

INTRODUCTION

Dendrochonology is principally known to archaeologists as a powerful tool for precisely dating wood in standing structures or waterlogged sites (Bannister and Robinson 1975). However since its inception, study of living and sub-fossil tree stumps have been put to other uses. The effect of fire on tree cambium is well known and as long ago as 1910, F.E. Clements identified fire scars in water pine and used dendrochronological methods to establish the fire history of a coniferous forest. Fire histories identified in this manner have now been established in many regions (see Stokes and Dietrich 1980). Several other processes may leave characteristic scars or tissue types. These include soil erosion (e.g. La Marche 1963, 1968), mass-movement (Alestalo 1971, Schroder 1975, 1976), tectonic activity (La Marche and Wallace 1972), frost damage (La Marche 1970), flood frequencies and magnitudes (Sigafoos 1964, Helley and La Marche 1968, 1973), insect suppression (Morrow and La Marche 1978), air pollution (Thompson 1981) and avalanches (Potter 1969, Ward 1984).

SCAR TISSUE FROM GRAZING DEER ON SCOTS PINE

In the course of this latter work, Ward (1984) recognised scar tissues attributable to grazing by red deer (*Cervas elephas*) on Scots pine (*Pinus sylvestris*) in Glen Feshie in the Cairngorms of Scotland.

When exposed in cross-section, red deer damage shows up as a 'V' shaped notch in which a large quantity of bark has been trapped by subsequent growth. Occasionally 'streamers' of bark may be found within cross-sections. In both cases, an important characteristic of the scarring is that it is close to the pith, i.e. the damage occurred when the tree was young, and not more than 15 or 20 years old. These features are shown in Fig. 4 of the paper discussing avalanche frequency (Ward 1984).

SCAR TISSUE IN SUB-FOSSIL PINUS SYLVESTRIS

Recent work by the Dendrochronology Section of the Sir John Cass Quaternary Research Unit has concentrated attention on sub-fossil remains of *Pinus sylvestris*. Tree stumps and occasional trunks have been found preserved in mid-Flandrian peat deposits on the Black Mount Estate on the southern edge of Rannoch Moor. Two sites have been examined. Clashgour (NGR NN 259425) is at approximately 185m a.s.l., and the peat averages 1.5m deep, reaching a maximum of 3.0m. Coire Seilich (NGR NN 335464) is at approximately 305m a.s.l. and the peat averages 2.0m deep, with a maximum of 4.5m.

Clear evidence of grazing has been found in sub-fossil wood samples from Clashgour and Coire Seilich, showing that the presence of deer in an area can be demonstrated from sub-fossil trees (Plate 1).

DISCUSSION

It may be possible to suggest the relative abundance and distribution of deer in suitable sites from the frequency of grazing scars where tree stumps are preserved. Obviously as much deer grazing occurs on the moors above the treeline, observations on stumps will provide information on the edge of the grazing range. Such trace fossils have significance for improving our understanding of the ecology and subsistence strategies of both mesolithic hunter-gatherers and large vertebrates; especially in the highland zone where hunting, firing and grazing have long been recognised as important (e.g. Simmons 1975, Mellars 1976).

The first requirement is to establish the relationship between the size and range of modern deer populations and the frequency of damage to young scot's pine. This relationship may then be applied, with suitable caution, to the interpretation of sub-fossil forests in mesolithic landscapes. Second, it would be most useful to explore the precise character of scar tissues generated by other grazing animals such as horses.

At present, radiocarbon dates have not been obtained from the samples showing deer presence at Clashgour and Coire Seilich. However, a date not younger than 4000 years b.p. is anticipated.

REFERENCES

Alestalo, J. (1971) Dendrochronological interpretation of geomorphic processes, Societa Geographica Fennica 105, 1-140.

Bannister, B. and Robinson, W.J. (1975) Tree-ring dating in archaeology, World Archaeology 7 (2), 210-225.

Clements, F.E. (1910) The life history of water burn forests. United States Department of Agriculture Forest Service, Bulletin No. 79, 56pp.

Helley, E.J. and La Marche, V.C. Nr. (1968) A 400-year flood in northern California, United States Geological Survey Professional Paper 600-D, 34-37.

Helley, E.J. and La Marche, V.C. Jr. (1973) Historic flood information from northern California streams from geological and botanical evidence, United States Geological Survey Professional Paper 485-E, 1-16.

La Marche, V.C. Jr. (1963) Origin and significance of buttress roots of bristlecone pine, White Mountains, California, United States Geological Survey Professional Paper 474-C, 149-150.

La Marche, V.C. Jr. (1968) Rates of slope degradation as determined from botanical evidence, White Mountains, California, United States Geological Survey Professional Paper 352-I, 1341-1377.

La Marche, V.C. Jr. (1970) Frost-damage rings in sub-alpine conifers and their application to tree-ring dating problems. In Tree-Ring Analysis with Special Reference to Northwest America J.H.G. Smith and J. Worrall (eds) University of British Columbia Faculty of Forestry Bulletin No. 7, 99-100.

La Marche, V.C. Jr. and Wallace, R.E. (1972) Evaluation of effects on trees of past movements on the San Andreas Fault, Northern California. Geological Society of America Bulletin 83 (9), 2665-2676.

Mellars, P.A. (1976) Fire ecology, animal populations and man: a study of some ecological relationships in prehistory, Proceedings of the Prehistoric Society 42, 15-45.

Morrow, P.A. and La Marche, V.C. Jr. (1978) Tree-ring evidence for chronic insect suppression of productivity in sub-alpine Eucalyptus, Science 201 (4362), 1244-1246.

Potter, N. Jr. (1969) Tree-ring dating of snow avalanche tracks and geomorphic activity of avalanches, North Absaroka Mountains, Wyoming, Geological Society of America Paper No. 123, 141-165.

Schroder, J.F. Jr. (1975) Dendrogeomorphological analysis of mass movement Annals of the Association of American Geographers 7, 222-226.

Schroder, J.F. Jr. (1976) Dendrochronologic analysis of rock glaciers and snow avalanches, American Quaternary Association Abstracts 4, 115.

Sigafoos, R.S. (1964) Botanical evidence of floods and floodplain deposition, United States Geological Survey Professional Paper 485-A.

Simmons, I.G. (1965) Towards an ecology of mesolithic man in the uplands of Great Britain, Journal of Archaeological Science 2, 1-15.

Stokes, M.A. and Dietrich, J.H. (1980) Proceedings of the fire history workshop. October 20-24 1980, Tucson, Arizona. United States Department of Agriculture Forest Service General Technical Report, RM-81, 142 pp. Rocky Mountain Forest and Range Experiment Station, Fort Collins, Colorado, USA.

Thompson, M.A. (1981) Tree rings and air pollution: a case of Pinus monophylla growing in east-central Nevada, Environmental Pollution Series A 26, 251-266.

Ward, R.G.W. (1984) An estimate of avalanche frequency in Glen Feshie, Scotland, using tree rings. In Palaeoenvironmental Investigations, N.R.J. Fieller, D.D. Gilbertson and N.G.A. Ralph (eds). Oxford: British Archaeological Reports.

Plate 1. Scars characteristic of deer grazing and ripping on the trunk of a stump of Scots Pine in excess of 4,000 years old, Rannoch Moor, Scotland.

INDEX 1

abiotic factors : 74
abrasion : 67
Abscess : 45
absent teeth : 45
absolute dating : 132
absolute frequencies : 56, 143
absolute population numbers : 131
abundance : 135
abyssal depths : 121
acclimatization : 136
accumulation rates : 193
acetolysis : 182
acid brown earths : 190
acid soils : 198
acid solutions and bone preservation : 68
acidity : 68, 190
Actinobacillosis : 50
activity patterning : 27
adaptation : 136
adaptive responses : 136
adomantinoma : 50
aerial photography : 198, 212
aetiology : 44, 50
age data : 86
age estimation : 43
age structure : 86
age studies : 9
agriculture : 204, 215
air flow : 97, 101
alien species : 141
aliens : 101, 158
alkalinity : 89
allochthonous : 133, 141, 144
allochthonous component : 98
alluvium : 133
alveolus : 45, 49
amelobastoma : 50
ameloblasts : 46
amplitude in biome changes : 74
amplitude of change : 73, 74
anaerobic clays and silts : 89
ancient buildings : 98
Anglo-Saxon settlement : 207
animal bones : 65
animal husbandry : 198
anomalously wide tree growth rings : 222

ante-mortem : 45
ante-mortem loss : 44
anthropogenic biozones : 158
anthropogenically induced change : 192
anthropogenically induced environmental change : 158, 160
Antonine fort : 189
arable : 140, 143
arable agriculture : 197, 199, 209, 214
arable farming : 162
archaeological animal bone : 65
archaeozoology : 20, 55
articulation : 15, 22
assemblage formation : 98, 170
association analyses : 134
Asturian : 125
attrition : 42, 52
autecology : 131, 135, 137, 140
autochthones : 97
autochthonous : 143, 144
autochthonous assemblage : 133, 146
autotrophs : 75
Avalanche frequency : 221
Avalanche magnitude : 226
avalanche runout distances : 227
avalanche talus : 222
avalanche trimline : 222
avalanche velocities : 226
Avebury henge monument : 138, 139

background fauna : 97, 99
barriers : 145
base-line data : 99
base-line fauna : 97, 168
bathhouse : 70
beetles : 97, 134
bias : 16, 22, 31, 32, 38, 55, 56, 57, 59, 62, 65, 67, 72, 81, 91, 208
bio-sedimentary facies : 169
biogeography : 119, 121, 125, 131, 136, 140, 148, 191
biological capture-recapture methods : 57
biological clocks : 120
biomes : 141
biometrical studies : 134
biometrics : 20, 55
biostratigraphy : 132, 157, 190
biozones : 132, 157
bird droppings : 98, 99, 102
birds : 134
birds nests : 97
blanket peat : 202
blown sand : 133
blue-green algae : 125
boiling : 21
bone : 157
bone assemblages : 22
bone densities : 56

bone destruction : 31, 35
bone disposal : 65
bone preservation : 65, 68
bone remains : 119
bone shape : 16
bone shapes and sizes : 60
bone splitting : 14
bone structure : 5, 16, 66
bone survival : 35, 37
bone weights : 59
bones : 19, 31, 55, 81, 91
boneworking : 24
bony alveolus : 44
bovine pillars : 45
bracken : 55
bread wheat : 211
breakage : 42
breeding : 99
bronze age : 141, 161, 193
bulk density : 135, 174, 56
burial : 6
buried soils : 133, 138, 144, 146, 158
burnt bones : 84
butchery : 5, 8, 10, 16, 27, 33, 42, 48, 82, 85, 91

calcification : 46
calcite : 121
calcium : 43,136
calculus : 51, 52
Calluna cycle : 192
camouflage : 137
Capra : 34
Capture-Recapture Methods : 60
carbon fractionation : 123
carbonised grain : 197
carbonised plant remains : 197, 198, 205
carcase utilisation : 26, 27
Caries : 50
carnivore den bone samples : 32
carnivore dens : 32
carnivore kill-sites : 32
carnivores : 136
catchment : 148
cattle : 6, 19, 58, 81, 82
cattle grazing : 62
caves : 79
cellulose : 184
celtic bean : 211
Celtic-fields : 197
centipedes : 134
cereal grains : 198
cereal pollen : 198, 199
chalk : 139, 157, 159
chance : 145
chance finds : 194

changeover rates : 73, 75, 77
charcoal : 157
chewing : 37
chi-squared tests : 57, 135
chronology : 120, 122
circular double ditched enclosure : 81
clays : 183
clearance : 158, 160, 199
cleavage : 16
climatic : 157
climatic change : 73, 76, 132, 142, 157, 169
clustering : 6
coding systems : 27, 73
cold stage faunas : 77
Coleoptera : 97
colloids : 190
colluvial : 157
colluvium : 133, 157, 159, 160, 161, 162
colonisation : 141, 144, 162
colour : 82
colour and banding patterns : 137
comminution : 31, 72
community dynamics : 131
compaction : 172
comparative collection : 58
comparative methods : 59
comparative studies : 131
competition : 134
compound tooth : 50
compression : 147
compression wood : 222, 228
concentric sample pattern : 81
congenital porphyria - : 47
consumption : 21
contamination : 161
contemporary rockfall : 222
context : 133
context types : 65
contract work : 194
cookery : 37
coprolites : 33
correlation coefficients : 143
correlations : 157
corrosion : 158
corrosion : 31, 35, 36, 37
cottontails : 36
crop consumption : 206
crop plants : 197, 198, 205
crop production : 206
crushing : 31
crystal structure : 121
cultivated ground : 162
cultivation : 147, 158, 162, 214
cultural typicality : 82
cusps or pillars : 45
cut marks : 19

cuts : 7
cycles : 120

data logging systems : 24
data presentation : 134
data recording systems : 58
dating : 157
de-watering : 172
death assemblages : 102, 133, 141
decay : 190
decaying : 38, 100, 224, 225
definitions : 58
deflocculation : 181, 182
deforestation : 138, 139, 144, 148, 158, 161, 162, 191, 192, 193, 203, 214
degradation : 68, 81, 82, 83, 84, 86, 88, 89, 91
degraded bones : 81
deltoid and triceps muscles : 14
demography : 57
dendrochronology : 122, 221
Dental attrition : 42, 52, 47
dental calculus : 51
Dentigerous cysts : 50
deposition : 89, 119, 141, 190
depth : 91
derivation : 27, 67, 69, 97, 98, 119, 133, 135, 139, 141, 170, 190
destruction : 65
detritivores : 136
Developmental dystrophies : 43
Devensian : 73, 76, 78, 79, 142, 161
diagenetic change : 125
dielectric separation : 183
diet : 27, 43, 119
dietary preferences : 205
differential butchery : 11
differential degradation : 91, 92
differential destruction : 11, 198
differential preservation : 65, 68, 88
differential recovery : 85
differential survival : 68
digestion : 36
dilution : 135
disarticulation : 15, 20, 34
discard behaviour : 120
discolouration : 47
Dispersal: : 145
disposal practices : 11
dissolution : 125
distance effects : 97
distribution : 135
distribution and abundance : 136
disturbance of ventral margin : 49
Ditch infill : 189
ditch-fills : 144
ditches : 61, 66, 68, 83, 85, 89
diversity indices : 8, 143, 144

dog-pen bone samples : 32
dogs : 31
domestic mammals : 55
domestic ungulates : 55
domesticated animals : 198
double-counting : 35
down-washing : 190
dry valleys : 133, 157, 158, 160, 162
Durrington Walls henge monument : 138, 140

earthworm : 146
ecocatastrophes : 75, 78
ecological models : 162
ecological resolution : 131
ecological tolerance range : 74, 97
ecological tolerances : 168
ecology : 131, 136
economy : 120
ecophysiological stress : 131
ecosystem : 147, 162
ecosystem diversity : 77
emersion : 123
emmer wheat : 211
empirical approaches to molluscan studies : 148
empirical objectivity : 131
enamel production : 46
encrustations : 66
endogenous rhythms : 120
environmental archaeologists : 181
environmental change : 137, 144, 146, 167, 189
environmental history : 163
environmental reconstruction : 56, 132, 134, 135, 137, 141, 157, 190
environmentalism : 132
epiphyseal fusion : 16
epiphyses : 6, 81
epiphysial fusion data : 82
equilibrium communities : 141
erosion : 27, 61, 65, 66, 82, 158
estuaries : 124, 126
estuaries-salinity variation : 125
ethnoarchaeology : 20
ethnography : 119
evaluating recent human impacts : 97, 198
excavation : 83, 92, 162
excavation strategies : 81
exotics : 158
experimental design : 33
experimental studies : 32, 68
extinction : 75, 144

facies : 169
facies variation : 169
facultative omnivores : 136
faecal specimens : 35

faeces : 32
fallow : 143
fauna : 65
faunal assemblages : 131
faunal debris : 55
faunal mixing : 190
faunal survey : 172
faunal turnover rates : 75, 77
feature type : 81, 83, 85, 91
femur : 7
fibula : 33
field procedures : 69
field survey : 167
field systems : 197, 198
filleting : 23
filtration : 181
first millennium AD : 41
Fishers alpha index : 98
Flandrian : 157, 168, 174, 221
flotation : 206, 208
flow depth : 226
fluorine poisoning : 50
food : 136
food preferences : 136
food supply : 121
forest clearance : 158, 191, 198, 200, 204, 205, 215
forest regeneration : 198, 199, 200
fossil soils : 189
fragment frequencies : 88
fragment sizes : 65
fragmentation : 15, 19, 21, 22, 25, 36, 42, 58, 60, 61, 62, 81, 82
fragmentation patterns : 5
frequencies : 135
frequency data : 171
fresh breaks : 8
frontier studies : 56
frost : 159
froth filtration : 181
froth flotation : 183
fundamental niche : 136

genetic and congenital defects : 45
genetic changes : 136
genetic variability : 137
genetic variation : 149
genotypic variation : 137
gingivitis : 44
gleyed soils : 66
gnawing : 6, 22, 23, 27, 34, 61
gnawing/chewing : 48
goats : 34, 65
graduated stratigraphy : 193
grain : 197
grain impressions : 208
grassland : 137, 139, 146, 162

grassland plagioclimax : 162
gravity : 160
grazing : 57, 62, 143, 192, 224
Grooved Ware cultures : 214
ground water : 125
grouping : 6
growing season : 126
growth increment studies : 119, 121
growth-line analysis : 119, 120, 126
gullies : 66

habitat preferences : 73, 78, 168
habitat variation : 123
habitat-type : 73
habitatualism : 132
health : 41
heavy liquids : 182
herbivores : 136
herbivores : 47
heridity : 49
heterotrophs : 75
histograms : 134
hoe-cultivation : 198
Holocene (Post-glacial) : 221
Holocene : 159
horse : 81, 224
Hottentot villages : 32
house deposits : 98, 99, 102
human activity : 90, 126
human importation : 101
Humerus : 6
hunter-gatherer : 119, 124
husbandry practices : 62
huts : 60
Hydrofluoric acid maceration : 182
hydroxides : 84
hygrophilous species : 161
hypoplasia : 46

icterus : 47
identifiability Index : 58, 61
identification : 57, 84
immersion : 123
improvements to field and laboratory procedures : 173
increment cores : 223
index of bone degradation : 81, 84, 85, 88, 91
index of degradation : 83
index of identifiability : 55, 61
indicator species : 131, 135, 138
individual variation : 43
industrialisation : 167
infection : 43, 51
infill : 159
ingested soil : 43

innovation : 211
insect assemblages : 97
insect death assemblages : 97
insectivores : 102
insects : 97
inter species competition : 74
inter-site comparisons : 57
inter-specific competition : 134
inter-tidal deposits : 167, 168, 169
inter-tidal habitats : 167
inter-tidal molluscs : 167
interpretation : 132
intersite comparisons : 65
intersite variation : 57
interstadials : 78
intertidal : 126
intertidal zone : 119, 121
intra- and inter-site comparisons : 57
intra-dental attrition : 43
intra-site variability : 56, 60
intra-specific variation : 131
intrasite variation : 92
Ipswichian : 76
Ipswichian interglacial : 73, 168, 169, 174
Iron Age : 22, 81, 143, 161, 189, 193
Iron Age economy : 197
Iron Age Fort : 162
iron compounds : 66
iron oxides : 84
iron-panning : 89
isotope analysis : 126
isotopic composition of the water : 123
Isotopic exchange : 125
isotopic ratio : 123

King Index : 59, 61, 62

laboratory controls : 68
laboratory methods : 167
lake deposits : 189
land snail : 131
landscape change : 132
late Iron Age/Romano-British levels : 200
late-glacial : 157, 158, 159, 161
late-Victorian terraced two-storied houses : 99
lateral variation : 65, 68
latrine : 66, 67, 69, 70, 71
leaching : 67, 68, 81, 82, 83, 86, 89, 91
life assemblages : 133
life span : 57, 62
life tables : 146
lignin : 184
limb-bones : 33
limiting factors : 135, 136, 147

limpet : 122
linear : 69, 70
Linears : 67
Lloyds Bank site York : 98, 103
log transformation : 227
low-shore : 122
lumpy jaw : 50
lynchet : 135, 143

magnitude/frequency relationship : 221
Makapansgat : 35
mammal bones : 81
mammalian predators : 119
mammals : 73, 134
man : 159
mandibular fragments : 59
mandibular pathology : 42
marginal land : 214
marine molluscs : 119
marine molluscs : 62
marine species : 126
marrow extraction : 21
mastication : 41
matter : 100
maxillary teeth : 46
mean monthly sea temperatures : 124
meat estimate : 56
meat processing : 26
meat stripping : 8, 15
meat value : 62
medieaval agriculture : 204
medieval : 50
melody makers : 61
mesolithic : 123, 125
methodological safeguards : 55, 62
methods : 133
metrical variation : 149
micro-habitat diversity - : 144
microbial fauna : 190
micropore filtration : 181, 182
mid-Flandrian : 169
middens : 24, 57
migration : 141, 144, 157
migration routes : 145
millipedes : 134
mineralization : 83, 89
minerals : 65, 66
minimum number estimates : 31
minimum number of individuals : 35, 56, 59, 76, 77, 86, 99
mites : 134
MNI : 31, 36, 99
models : 132
modern buildings : 97
modern distributions : 73,75,97,99,119,121,136,137,140,147,167,169,191
modern faunal surveys : 44, 49, 52, 75, 97, 99, 103, 167

modern indoor deposits : 98
modern soils : 131
modern species distribution : 131
modern veterinary studies : 41
moles : 147
mollusca : 131, 132
molluscan biostratigraphy : 157
molluscan biozones : 132
molluscan frequencies : 56
molluscan species associations : 142
molluscs : 62, 157, 167
morphological variations : 122
multi-dimensional space : 136
multivariate analyses : 135

naked barley : 211
narrow rings : 223
natural hazards : 221, 227
naturally occurring environmental change : 158
necrosis : 45
negative evidence : 65
neolithic : 124, 125, 131, 138, 139, 140, 141, 144, 148
neolithic enclosures : 145, 148
neolithic henge monument : 135
neolithic land-use : 138
neolithic period : 204
neoplastic lesions : 50
neutral soil : 89
niche : 136
nomadism : 197
nominal scale of measurement : 171
nonsieving : 72
null hypothesis : 66
numbers of shells per unit weight of soil : 135
nutrient foramena : 48
nutrition : 41, 43, 119
nylon boulting-cloth : 182

oat : 211
observer bias : 72
occupation sites : 97
odontoma : 50
old rockfalls : 222
Oligodentia : 43
open-country : 158
oral bacteria : 44
oral palaeopathology : 41, 52
oral pathology : 41
originating species : 75
osteodontokeratic hypothesis : 32
out-breeding species : 137
overcrowding : 43
oxidation : 157
oxygen isotope analysis : 119, 120, 123

oxygen-isotope determinations : 132

palaeo-biomes : 74
palaeo-zoogeography : 74
palaeoecology : 52, 131, 132, 136, 137, 143, 157
palaeoeconomy : 5, 132
palaeogeography : 56
Palaeolithic : 73, 79
palaeopathology : 41
palaeosols : 131, 133, 138, 144, 146, 147, 158, 159, 161, 189, 193
palaeotemperature analysis : 132
paradontal disease : 44
parasites : 41
past distributions : 131
pastoral economy : 27
pastoral indicators : 199, 200
pastoralism : 197
pasture : 140
pasture juices : 43
pathology : 41
patterning : 37
peat : 189
penecontemporaneous re-working : 170
percentage frequencies : 131, 134, 135, 143
percentage frequency histograms : 140
periodontal disease : 43, 44, 51
periodontitis : 44
Petersen estimates : 57
pH : 68, 123, 136, 190
phenotypic variation : 137
phosphorus : 43
phytoliths : 53
pie charts : 223
pig : 55, 58, 81
pig-keeping : 55
pigmentation : 47
pigs per person : 57
pith : 223
pits : 61, 66, 69, 70, 85
plagioclimax : 161, 162
plant macrofossils : 157
podzols : 190, 193
political and economic stability : 201
pollen : 147, 157, 189
pollen analysis : 132, 181, 189, 193, 198
pollen analysis for the detection of arable agriculture : 211
pollen biozones : 157
pollen degradation : 190
pollen diagrams : 197, 198, 214
pollen entrainment : 190
Pollen extraction : 181
pollen spectrum : 190
pollen studies : 78
pollution : 167, 174
polymorphisms : 131, 137, 149

population : 86
population age structure : 9,16, 32, 68
population dynamics : 131, 134, 137, 138, 206
population fluctuations : 57, 73, 74, 75, 77, 79, 97, 168, 173
population increase and pressure : 205
population pressure : 214
population structure : 137
population structures : 65
post excavation methods : 52
post-depositional change : 125
post-excavation methods : 47
post-mortem changes : 48
post-mortem destruction : 5
post-mortem loss : 44
postglacial : 157-163
postholes : 85
pre-Quaternary : 181
pre-Quaternary palynology : 181
precision : 134
predation : 137
predators : 97
preferential destruction : 32, 170
presence/absence data : 65
presence/absence measurements : 171
preservation : 22, 25, 60, 61, 65, 66, 81, 84, 136
preservation categories : 72
preservation; : 133
preservational bias : 65
prey-predator relations : 134
primary products : 198
principal Components Analysis : 205
prolificity : 62
proxy data : 221
pulp cavity : 45
pulp cavity exposure : 48
pulpal haemorrhages : 47
pulpitis : 45
putrid pulpitis : 47
pyorrhoea : 44

quantification : 22, 27, 56, 131, 135, 149
quantitative analysis : 134
quantitative studies : 19
quaternary : 157, 167
Quaternary chronology : 73, 168
Quaternary distributions : 74, 97,119, 136, 137, 141, 147, 167, 191
Quaternary geologists : 181
Quaternary stratigraphy : 73, 76

rabbit : 36
radiocarbon : 132
radiocarbon dates : 161, 192
radiocarbon dating : 157
radius : 7, 33

rampart : 67, 69, 70, 71
random sampling : 212
rank order : 84
ranking procedures : 72
rates and scales of responses : 131
rates of response : 144
re-cycling : 171
re-deposition : 133
re-working : 67, 69, 190
reaction wood : 222
reaction wood growth : 223
realised niche : 136
recording : 27, 65
recording pathology : 41, 42
recording sheets : 42
recording systems : 24
recording tooth positions : 42
recovery techniques : 208
recrystallization : 125
recurrence intervals : 221, 226
red deer : 57
redistribution : 26
refuse disposal : 57
regeneration : 56
relative dating : 132
relative frequencies : 56
relative meat values : 57
relict shells : 139
rendzina soils : 146
reprecipitation : 125
representation : 56
rescue (salvage) archaeology : 133
research palynology : 181
residual material : 69
resolution : 126, 134
resorption of the alveolar cavity : 44
responses : 131
return periods : 226
reworking : 119, 135, 139, 171
rhythmic growth patterns : 126
river gravels : 81
road : 67, 68, 70
road casualties : 36
road surface : 66, 71
rockfall : 222
Roman : 5, 189
Roman army : 201, 204, 215
Roman forts : 189
Roman Lincoln : 5
Roman sites : 158, 161
Roman studies : 6
Roman withdrawal : 216
Romano-British : 65, 158
Romano-British deposits : 22
Romano-British sites : 65
roof-gutter assemblages : 98

rye : 211

salinity : 124
salt marshes : 169
salting : 21
saltmarsh : 62
sample reliability : 68
sample representativeness : 38, 86
sample size : 143
sampling : 16, 92, 132, 133, 206
sampling strategies : 16, 31, 38, 57, 58, 65, 69, 76, 97, 133, 143, 147, 167, 171, 191, 194, 206, 208
sands : 183
Saxon period : 216
scar tissue : 221, 223
scavengers : 86
scavenging : 22, 27, 85, 90, 91
sciurus : 33
scoring procedures : 42, 44, 45, 47, 48, 49, 142
Scots Pine : 225, 228
sea temperatures : 124
seasonal : 168
seasonal dating : 119, 121
seasonal effects : 62
seasonal indicators : 119
seasonal markers : 121
seasonal population fluctuations : 73,, 75, 97, 168
seasonality : 119
secondary butchery : 8
secondary carbonate : 125
sedimentation : 6
sedimentation rate : 135, 143, 192
sedimentology : 131
selection : 137
settlement : 120
settlement usage : 55
sewage pollution : 167
shape : 6
shear mouth : 43
sheep : 58, 81, 82
sheep mandibles : 42
shell destruction : 158
shell discard behaviour : 120
shell growth : 120
shell shape studies : 122
shellfish consumption : 120
shells : 119, 157
short-term population fluctuations : 168
sieving : 69, 72, 84
silts : 183
Sintered glass funnels : 182
site formation : 62
site occupation : 119
skeletal element representation : 82
skeletal elements : 81

skinning : 20
slaughtering : 5, 16, 82, 86
sodium pyrophosphate : 181, 182
soil : 158
soil corrosion : 38
soil deterioration : 202, 214
soil erosion : 158, 160
soil pollen : 189, 193
soil studies : 131
soils : 70, 204
solifluxion : 157, 158, 159, 160, 161
solution : 125, 158
South Street long barrow : 138, 139
spatial distribution : 81
spatial patterning : 81
species abundance : 65, 86, 131, 139
species composition : 65, 68
species diversity : 74
species diversity indices : 98, 134
species fluctuations : 57
species proportions : 65
species representation : 56, 62
species-area relationships : 148
species-associations : 131, 142
species-diversity indices : 148
species-fluctuations : 143
species-interactions : 137
spelt wheat : 211, 214
spiral fractures : 5
splintering patterns : 5
spring growth : 122
springtails : 134
squirrels : 33, 33
stability : 132
statistical differentiation : 73
statistical indices : 55
statistical techniques : 27
statistics : 19, 55, 56, 57, 62, 65, 66
stenoecious : 149
stenotopic : 147, 148
step mouth : 43
storage : 21
storage pits : 197
stratification : 133
stratified molluscan remains : 157
subjective classifications : 72
subsidence : 159
subsistence strategies : 28
subsoil hollows : 159
substantive uniformitarianism : 135, 136
subtidal specimens : 126
subtidal zone : 121
succession : 141
supernumary teeth : 45
survival rate : 35, 36, 37
swirling : 181, 183

systematic sampling : 222

taphonomy : 5, 8, 19,, 27, 31, 32, 55, 60, 61, 62, 65, 71, 73, 81, 91, 97, 98 119, 139, 141, 143, 146, 158, 169, 170, 189, 192, 223, 224
tartar : 47, 52
taxonomy : 58, 134, 171
teeth : 41, 72
temperature : 121, 123, 136
texture : 65
thatched barn : 98
theory : 131
tibia : 7, 33
tidal immersion : 120
tidal periodicity : 122
tillage : 139, 141, 143
tilled land : 141
tills (boulder clays) : 182
time resolution : 146
time-depth functions : 131
time-depth relations : 147
time-sequence : 190
time-span : 189
tooth eruption : 42
tooth presence/absence : 42
tooth rotation : 45
tooth wear : 42, 43
toothwear methodologies : 43
top soil stripping : 90
toxins : 41
traction epiphyses : 50
trade : 26
trampling : 13, 35, 61, 67, 86, 89, 91
transportation : 65, 98, 102, 119, 133, 135, 141, 170, 190
trauma : 50
travertine : 133
tree ring dating : 228
tree rings : 221
tropical shells : 124
tropical waters : 124
tropical zones : 121
tufa : 133, 162
tufaceous deposits : 162
turbulent friction coefficient : 227
turf rampart : 66
turves : 67, 189
twentieth century : 158

ulna : 7, 33
ultrasonic deflocculation : 181
ultrasound : 183
ungulate mandibles : 41
Upper Palaeolithic : 125
upper shore : 122

urban growth : 167
urban growth : 97
urban sites : 48, 50, 97, 98, 102
user bias : 72

variability : 27
vegetation dynamics : 192
vegetational changes : 189
vegetational history : 191
ventilated buildings : 103
vertebrate communities : 74
vertebrates : 73
village refuse bone samples : 32
voids : 147
volume measures : 57
vomited : 35

water pollution : 167, 174
water sieving : 38
water temperature : 123
water-table : 89
waterlogged contexts : 48
waterlogged deposits : 205
waterlogged sites : 102
waterlogging : 67, 157
waterlogging : 67
wattle : 97
wear resistance : 43
weathering : 22, 27, 38, 146
Weave mouth : 43
weed species : 162
weeds : 158
wild sheep : 45
wind : 225
winter grooves on bivalves : 121
woodland : 137, 158
woodland biozones : 158
woodland clearance : 138, 139, 144, 145, 146, 148, 158, 162, 191, 192
woodland distribution : 56
woodland fauna : 142
woodland management : 56
woodlice : 134
worm-sorting : 139

X-rays : 45

zonation : 132
zooarchaeology : 132

INDEX 2

A. nitidula : 137, 162
A. secale : 158, 159
Acari : 134
Acicula fusca : 159, 163
Actinobacillus ligniersi : 50
Actinomyces bovis : 50
Aegopinella nitidula : 137, 159
Aegopinella pura : 159
Alnus : 191

C. edule : 122
C. gigaxii : 160, 163
C. glaucum : 122
C. intersecta : 160, 161, 163
C. nemoralis : 137
C. tridentatum : 161, 162
C. virgata : 160, 163
Calluna : 191
Candidula gigaxii : 158
Candidula intersecta : 158
Canis lupus : 76
Carabidae : 142
Carychium tridentatum : 141, 159
Cepaea nemoralis (L.) : 136
Cepaea nemoralis : 137
Cerastoderma edule : 120, 123
Cerastoderma glaucum : 121, 125
Cernuella virgata : 158
Chilopoda : 134
Chrysosplenium : 183
Clausilia bidentata : 161
Clausiliidae : 158
Clinocardium nuttalli : 122
Cochicella acuta : 158
Cochlodina laminata : 161
Coelodonta antiquitatus : 73
Coelondonta antiquatis : 76
Coleoptera : 134, 142, 157
Collembolla : 134
Corbicula fluminalis : 169
Corylus : 191
Crocuta crocuta : 76

D. rotundatus : 159, 161
Diplopoda : 134

Discus rotundatus : 159
Discus ruderatus : 158

H. aspersa : 158, 163
H. cinctella : 158
H. itala : 143, 160, 161, 162, 163
Helicella itala (L.) : 135, 143, 160
Helicellids : 160, 163
Helicigona lapicida : 161
Helix aspersa : 158
Hydrobia ulvae : 167, 168, 169, 170, 173, 174
Hydrobia ventrosa : 169, 170, 171, 173
Hygromia limbata : 158

Isopoda : 134

Leuconia bidentata : 170
Leucophytia : 173
Leucophytia bidentata : 170
Littorina littorea : 125

M. cantiana : 161
M. cartusiana : 160, 161, 162, 163
Macoma balthica : 167, 168, 169, 170, 172, 173, 174
Mammuthus primigenius : 73, 76
Megaloceros giganteus : 76
Mollusca : 119
Monacha cantiana : 158, 161
Monacha cartusiana : 158

Nesovitrea hammonis : 137
Nesovitrea hammonis : 159

Oxychilus cellarius : 159

P. elegans : 138, 139, 140, 143, 144, 160, 162
P. intermedia : 125
Paludestrina stagnalise : 170
Patella tabularis : 126
Patella vulgata : 122, 125
Pectinidae : 121
Penitella penita : 122
Pinus sylvestris : 222, 223
Pisidium : 171
Plantago lanceolata : 160
Pomatias elegans : 131, 135, 138, 140, 141, 143, 144, 158
Punctum pygmaeum : 159
Pupilla muscorum (L.) : 135, 143, 146, 159

Quercus : 191

Rangifer tarandus : 76

Scrobicularia : 173
Solen vagina : 124

T. hispida : 162
Trichia hispida : 159

Ursus arctos : 76

V. costata : 161, 162
V. excentrica : 162
V. pulchella : 162
V. pygmaea : 162
Vallonia : 160
Vallonia costata : 146
Vallonia excentrica : 161
Vallonia pulchella : 161
Veronica hederifolia : 160
Vertigo pusilla : 161
Vertigo pygmaea : 146, 161

Zonitidae : 158, 161

INDEX 3

Aartswoud : 210
Ascott-under-Wychwood : 159
Asham Quarry : 157, 158, 159, 161, 162
Australopithecines : 32
Avebury : 138, 159
Avonmouth : 167, 168

Bailey : 127
Baltic : 125
Banjo : 60
Bar Hill : 189
Beachy Head : 162
Biggins : 72
Bishopstone : 143
Blue Circle Cement Company : 163
Boyd : 189
Brighton : 157, 158
Brighton Corporation : 163
Bristol : 41
Bristol and Gloucester Archaeological Society : 53
British Museum : 53, 163
Bronze Age : 198, 199, 203, 204
Bullock Down : 157
Burtle Beds : 171
Burtle Sand Beds : 171, 173

Caesar's : 197
Cairngorm Club Journal : 222
Cairngorm Mountains : 222
Candelaro river : 125
Cantabria : 124, 125
Castleford : 65, 68
Chilterns : 157
City of London Polytechnic : 221
Clark : 17
Cleveland : 199, 212
Coppergate : 98
Cow Gap : 162
Cow Green Reservoir : 201
Coxhoe : 211
Creag Mhigheachaidh : 222
Creswell Crags : 73
Cumbria : 201

Deith : 119
Denmark : 32, 125
Department of Education and Science : 127
Department of Human Environment : 131
Department of the Environment : 53
Devil's Dyke : 157, 158, 160, 161, 162
Dr D.T. Holyoak : 160, 163
Dr J.G. Evans : 149
Dr M.P. Kerney : 163
Dr R.B.G. Williams : 163
Dumball Island : 168, 170, 171, 172, 173
Durham : 199, 201

Exe Bridge : 50
Exeter : 50

Fourth International Flint Symposium : 163

Gilbertson : 127
Glen Feshie : 221, 222
Godwin Laboratory : 119
Gravel Banks : 171

Hadrian's Wall : 215
Hallshill Farm, East Woodburn : 209
Hardwick with Yelford : 81
Humber : 197
Huntingdonshire : 137

Indiana University : 31
Institute of Archaeology : 131
Institute of Terrestrial Ecology : 227
Iron Age : 56, 57, 204, 214
Italy : 125
Itford Hill : 157

Kenn, Somerset : 168

La Riera : 125
Lewes : 157
Lewes : 158
Limfjord : 125
Lincoln : 5, 6
Lincoln Archaeological Trust : 17
Lullingstone Villa, Kent : 161

Maltby : 19
Manfredonia : 124

Mann : 17
Masseria Valente : 124, 125
Mediterranean : 121, 126, 158
Mesolithic : 31
Mingie's Ditch : 81
Mollins : 189
Morton : 123
Munson : 31

Natural Environment Research Council : 163, 175, 227
Nature Conservancy Council : 227
Nelson Bay Cave, South Africa : 126
Neolithic : 55, 199, 206, 207, 210, 214
Netherlands : 210
Normans : 204
Norsminde : 125
North Downs : 157, 162, 163
North-East of England : 197
Northumberland : 199

O'Connor : 17
Owlesbury : 24
Oxford : 81
Oxfordshire : 81

Payne : 31
Pegsdon : 160
Peru : 32
Petersen Index : 60
Pin Hole Cave : 73, 76, 77, 78
Poynings : 160

Rake Bottom : 160
River Avon : 167
River Windrush : 81
Robin Hood's Bay, Yorkshire : 122
Roman : 97
Roman Wall : 215

Salisbury Plain : 157
Science and Engineering Research Council : 127
Scotland : 123
Scotland : 221
Severn Estuary : 167, 169
Sheffield : 65
Society of Antiquaries of London : 53
South Downs : 157, 159, 162, 163
Southeastern Italy : 124
Southern European : 158
Spain : 125
Stallibrass : 65

Stanwick : 197
Stonehenge : 138
Sussex : 143, 162
Swallow Cliff : 171

Tame Valley : 142
Tavoliere : 124
Tees : 199
Teesdale : 201
Thirlings, Northumberland : 207
Thorpe Thewles : 212
Torquay : 158
Tree Ring Laboratory : 227
Tyne and Wear : 199

University of Aberdeen : 227
University of Arizona : 227
University of Bristol : 167
University of Cambridge : 119
University of Durham : 197
University of Glasgow : 189
University of London : 131
University of Sheffield : 167, 175, 181
University of Southampton : 19, 55
Upper Thames Valley : 81

Vale of Brooks : 160
Van der Veen : 197
Viking-age : 97

Wachipaeri : 32
Warwickshire : 142
Weardale : 201
Wessex : 56, 61, 138, 148
West Hill : 49, 50
West Virginia : 38
Weston-in-Gordano : 170
Whitton Hill : 206
Wicken Bonhunt, Essex : 42
Wilson : 81
Wiltshire : 138
Woodbury : 197

York : 98